Cyclodextrin Chemistry and Toxicology

Cyclodextrin Chemistry and Toxicology

Editors

Marina Isidori
Margherita Lavorgna
Rosa Iacovino

MDPI • Basel • Beijing • Wuhan • Barcelona • Belgrade • Manchester • Tokyo • Cluj • Tianjin

Editors

Marina Isidori	Margherita Lavorgna	Rosa Iacovino
Department of	Department of	Department of
Environmental, Biological	Environmental, Biological	Environmental, Biological
and Pharmaceutical Sciences	and Pharmaceutical Sciences	and Pharmaceutical Sciences
and Technologies	and Technologies	and Technologies
University of Campania "L.	University of Campania "L.	University of Campania "L.
Vanvitelli"	Vanvitelli"	Vanvitelli"
Caserta	Caserta	Caserta
Italy	Italy	Italy

Editorial Office
MDPI
St. Alban-Anlage 66
4052 Basel, Switzerland

This is a reprint of articles from the Special Issue published online in the open access journal *Molecules* (ISSN 1420-3049) (available at: www.mdpi.com/journal/molecules/special_issues/CDs_Chemistry).

For citation purposes, cite each article independently as indicated on the article page online and as indicated below:

LastName, A.A.; LastName, B.B.; LastName, C.C. Article Title. *Journal Name* **Year**, *Volume Number*, Page Range.

ISBN 978-3-0365-4716-9 (Hbk)
ISBN 978-3-0365-4715-2 (PDF)

Contents

About the Editors

Marina Isidori

Marina Isidori is full Professor of Hygiene at University of Campania "L. Vanvitelli"–Italy. She is a biologist specialized in Human Nutrition and Food Science. The research field of Marina Isidori is focused on toxic, genotoxic effects and endocrine disruption activity of environmental and food pollutants in in vivo and in vitro experimental systems as well as in exposed humans trying to understand the connections between the environment and human health. Her expertise covers also the innovative packaging to extend the shelf-life of food by using antimicrobial and antioxidant natural compounds.

Margherita Lavorgna

Margherita Lavorgna is Associate Professor in Hygiene, Department of Environmental, Biological and Pharmaceutical Sciences and Technologies of the University of Campania "L. Vanvitelli". Her research field regards environmental toxicology through studies of acute and chronic toxicity on organisms of different trophic levels, the evaluation of mutagenic, genotoxic and endocrine interference activity of drugs and their derivatives, organic compounds products for secondary metabolism or isolated from plants, phenolic compounds from olive oil mill wastewaters, pesticides, aromatic hydrocarbons, heavy metals and alkylphenols. The studies are focused on environmental compartments as urban air, wastewater and surface water also addressing issues related to environmental quality and the risk to the environment and humans due to the presence of xenobiotics. The field of interest also extends to food mainly chemically contaminated by pesticides with potential genotoxic and estrogen-mimetic activities. Furthermore, her research is also focused on innovative technologies to preserve and improve the shelf-life of food by using active packaging materials.

Rosa Iacovino

Rosa Iacovino graduated in Chemistry at University of Naples "Federico II"discussing a thesis on chemical crystallography. Doctor of research in "Design and use of molecules with biotechnological interest"organized by the Second University of Naples. From 2004, she has been a researcher at the Department of Environmental, Biological and Pharmaceutical Sciences and Technologies of the University of Campania "L. Vanvitelli", Caserta (Italy). From 2008 to 2014, she obtained the position for teaching General and Inorganic Chemistry for the Degree Course in Pharmacy, and from 2015 for teaching General and Inorganic Chemistry for the Degree Course in Environmental Sciences. From 2006 to 2013, she was a component of the research doctorate board in "Design and Use of Molecules of Biotechnological Interest". Her research interests lie in peptide chemistry: the relationships between structure and activity of numerous peptide systems have been the object of her investigations, carried out with a variety of experimental and theoretical techniques, for the understanding of the mechanism of action of biologically relevant systems. Subsequently, she was interested in the following topics: design and synthesis of analogues of biologically active peptides; solid state structural characterization by X-ray diffraction of calixarenes and carexans; study of the conformational properties of functionalized β-cyclodextrin systems. Her research interests currently focus on the study of β-cyclodextrin inclusion complexes with molecules of biotechnological interest to understand their inclusion mechanism, also to evaluate their potential applications in the pharmaceutical and environmental fields. These studies were conducted

using spectroscopic techniques, X-ray diffraction and computational methods. She is the author
or co-author of more than 50 peer-reviewed publications in international scientific journals. She is
referee for several scientific journals.

Preface to "Cyclodextrin Chemistry and Toxicology"

Cyclodextrins (CDs) are cyclic oligosaccharides composed of six (α-CD), seven (β-CD), or eight (γ-CD) D (+) - glucose units linked by α-1,4 bonds. They have found wide application in numerous fields which, thanks to their typical toroidal structure consisting of an apolar internal cavity and a polar external surface, allows the physical inclusion of a wide range of molecules. Cyclodextrins have been the subject of several books, reviews and articles. This testifies to the great interest in these molecules and their derivatives due to the extensive use in various fields, such as biotechnology, chemistry, green chemistry, analytical chemistry, pharmacy, administration of drugs and pharmaceutical excipients, biology, medicine and biomedicine, food and beverage industry, cosmetic formulations/fragrance stabilization, chromatography, catalysis, biotechnology, agrochemistry and the remediation of polluted sites.

This collection of articles, through eleven contributions and three reviews, provides an overview of the applications of cyclodextrins, implements the information regarding the use of cyclodextrins and their inclusion complexes, considering both experimental and theorists approaches and using various scientific and technological tools. Furthermore, the importance of understanding the mechanisms involved in the formation of the complexes and of the resulting advantages, together with their possible integration within different application areas, including those related to data communication, are underlined.

In particular, the published articles have clearly highlighted the dynamism and breadth of the sectors involved, as well as the variety of approaches to the use of CDs: as food supplements and nutraceuticals by studying the impact on the grain dough and on the properties of bread, the effect on the antiplatelet capacity of proteins, development of new formulations with greater solubility, bioavailability and stability, the administration of drugs. The drug concentration plays an important role in the interaction with drug carriers that influence the kinetics of the release process and the toxicological effects. The Molecular Dynamics simulations are useful for evaluating the bioeffects of β-cyclodextrin-based nanosponges on two-dimensional cell cultures, also discussing the study of the cyclodextrin-based nanosystems in targeted cancer therapy. One of the reviews introduces the advances in cyclodextrin (pCD) polymer research, including their synthesis and applications in analytical separation science, materials science and biomedicine. Another review summarizes and highlights the results obtained in the field of cyclodextrin-based contrast agents for medical imaging. Finally, new perspectives are presented in the use of cyclodextrins in textiles, as smart materials for microbial control.

As guest editor of the Special Issue "Cyclodextrin Chemistry and Toxicology", we would like to express our deep appreciation to all the authors whose valuable work has been published in this issue, thus contributing to the success of the Special Issue. Despite the presence on the market of numerous food, cosmetic and pharmaceutical formulations, cyclodextrins continue to attract the interest of both the academic community and the industry and we believe that the rapid evolution of the different sectors considered is moving further in interdisciplinary way in the near future, this will give birth to new discoveries and applications based on innovative scientific insights carrying further technological advances, demonstrating that cyclodextrins, for research, still have a lot of interest.

Marina Isidori, Margherita Lavorgna, and Rosa Iacovino

Editors

Article

The Effect of α-, β- and γ-Cyclodextrin on Wheat Dough and Bread Properties

Anne-Sophie Schou Jødal [1,2], Tomasz Pawel Czaja [3,4], Frans W. J. van den Berg [3], Birthe Møller Jespersen [3] and Kim Lambertsen Larsen [1,*]

1   Section of Chemistry, Department of Chemistry and Bioscience, Aalborg University, DK-9220 Aalborg, Denmark; asj@bio.aau.dk
2   Lantmännen Unibake Denmark, DK-8700 Horsens, Denmark
3   Department of Food Science, Faculty of Science, University of Copenhagen, DK-1958 Frederiksberg, Denmark; tomasz.czaja@food.ku.dk (T.P.C.); fb@food.ku.dk (F.W.J.v.d.B.); bm@food.ku.dk (B.M.J.)
4   Department of Chemistry, University of Wrocław, 50-383 Wrocław, Poland
*   Correspondence: kll@bio.aau.dk; Tel.: +45-9940-8521

Abstract: Cyclodextrins (CDs) are cyclic oligosaccharides that have found widespread application in numerous fields. CDs have revealed a number of various health benefits, making them potentially useful food supplements and nutraceuticals. In this study, the impact of α-, β-, and γ-CD at different concentrations (up to 8% of the flour weight) on the wheat dough and bread properties were investigated. The impact on dough properties was assessed by alveograph analysis, and it was found that especially β-CD affected the viscoelastic properties. This behavior correlates well with a direct interaction of the CDs with the proteins of the gluten network. The impact on bread volume and bread staling was also assessed. The bread volume was in general not significantly affected by the addition of up to 4% CD, except for 4% α-CD, which slightly increased the bread volume. Larger concentrations of CDs lead to decreasing bread volumes. Bread staling was investigated by texture analysis and low field nuclear magnetic resonance spectroscopy (LF-NMR) measurements, and no effect of the addition of CDs on the staling was observed. Up to 4% CD can, therefore, be added to wheat bread with only minor effects on the dough and bread properties.

Keywords: cyclodextrins; alveograph; wheat dough; bread staling

Citation: Jødal, A.-S.S.; Czaja, T.P.; van den Berg, F.W.J.; Jespersen, B.M.; Larsen, K.L. The Effect of α-, β- and γ-Cyclodextrin on Wheat Dough and Bread Properties. *Molecules* **2021**, 26, 2242. https://doi.org/10.3390/molecules26082242

Academic Editors: Marina Isidori, Margherita Lavorgna and Rosa Iacovino

Received: 25 February 2021
Accepted: 10 April 2021
Published: 13 April 2021

Publisher's Note: MDPI stays neutral with regard to jurisdictional claims in published maps and institutional affiliations.

1. Introduction

Cyclodextrins (CDs) are cyclic non-reducing starch derivatives made by enzymes. The most commonly applied CDs are α-, β-, and γ-CD, which are cyclic oligosaccharides consisting of 6, 7, and 8 glucopyranose units, respectively. These three CDs are widely applied, as they have several beneficial attributes in, e.g., pharmaceuticals [1–3], foods [4–8], and cosmetics [9]; moreover, various health benefits have been observed when they are consumed [5,10]. Many of these effects originate from the ability of the CDs with their relatively hydrophobic cavity to form inclusion complexes with primarily lipophilic compounds or compounds with lipophilic moieties and thereby change the apparent properties of these [1,7]. When the CDs are applied in breadmaking, previous studies have found positive effects on bread quality (assessed by loaf volume) [11–13] and bread staling behavior [14,15]. In other food products, CDs are used as carriers and stabilizers of functional compounds, and they have, therefore, found multiple applications related to the extension of shelf life, food processing, and sensory improvement of food products [4–8].

Native α-, β-, and γ-CD are all considered non-toxic and safe for human consumption and have, therefore, received GRAS status (generally recognized as safe) [5,7]. The ADI (allowed daily intake) of α- and γ-CD is unspecified, while β-CD has been allocated with an ADI of 0–0.5 mg/kg body weight [16–23]. The approved use levels for bread, rolls, cakes, baking mixes, and refrigerated doughs have been set at 5, 2, and 1% (w/w) for α-, β-, and γ-CD, respectively [21–23].

CDs have multiple health effects, which makes them useful bioactive food supplements and nutraceuticals. There are various potentials in the application of CDs in bread, as bread is one of the most frequently consumed cereal products and a large source of available carbohydrates in the diet [24]. α- and β-CD can be considered as dietary fibers for controlling body weight and blood lipid profile, as the digestibility of α- and β-CD by the (human) amylolytic enzymes in the human gastrointestinal tract is negligible, while γ-CD is readily degraded [5,25–27]. α- and β-CD are instead partly fermented by gut microflora, and they have shown to be prebiotics that are able to improve the intestinal microflora [28]. Supplementation of CDs to starchy food has been shown to reduce their glycaemic index [29–33], which is considered favorable to health [24]. Health claims related to α-CD as dietary fiber and its ability to reduce post-prandial glycaemic responses have been permitted by The European Food Safety Authority [10].

A number of studies have investigated the effects of the addition of up to 3% pure CD of the flour weight on the wheat dough and bread performance [11–15]. The addition of CDs has been shown to change the mixing properties of the dough by increasing the water absorption and affecting the dough development time [11,13,15], increase the bread volume as well as texture and crumb structure with the addition of CD up to a certain concentration [11–14], and decrease the staling of the bread [14,15]. The effects of the addition of CD producing amylolytic enzymes (specifically cyclodextrin glycosyltransferases, CGTases, of various origin) on the properties of wheat bread have been investigated in various studies [34–37]. Here the CGTases were found to improve selected properties, including specific volume, texture, and staling rate of the resulting bread on par or superior to other amylolytic enzymes (e.g., commercial anti-staling enzymes). Some studies have attempted to quantify the amount of α-, β-, and γ-CDs produced by a CGTase from gluten-free baked bread [38,39] and found concentrations of up to around 59 mg CD/g crumb (sum of α-, β-, and γ-CD) [39].

As judged from the literature, immediate positive effects of supplementing industrial bread with CDs are improved bread volume and a pronounced anti-staling effect, which are key parameters for bread quality. However, the bigger potential for supplementation of wheat bread with CD may lie in their nutraceutical properties, including glycemic index reduction, as well as their prebiotic, anti-obesity, and anti-diabetic effects. Nevertheless, although previous studies have revealed potential positive effects relative to bread quality and shelf life, it is also evident that there is a limit to the amounts of a particular CD that can be supplemented to a wheat bread without compromising the key quality parameters of the products, processing suitability of the dough, and final product quality. In order to elucidate the effects of the CDs on processability and product quality of simple wheat bread, we have conducted a comparative study of the effects of the addition of α-, β-, and γ-CD in the range of 1 to 8% relative to wheat flour on both dough properties and bread quality. Dough properties were determined using alveograph and consistograph analysis, while bread properties were assessed by specific bread volume and staling measurements by texture analysis and low field nuclear magnetic resonance (LF-NMR).

2. Results and Discussion

2.1. Effect on Dough Properties

The effect of the α-, β-, and γ-CDs on flour water absorption was investigated by consistograph analysis, and the results can be seen in Figure 1. The water absorption increased when up to 4% CD was added after which water absorption was approximately constant or decreased again. The water absorption of the control dough (52.7%) increased the most by addition of 4% α-CD (55.4%), while it increased less for 4% γ-CD (54.1%) and 4% β-CD (53.4%). The decrease in water absorption at 8% CD compared to 4% CD was only significant for β-CD. In general, increased water absorption was expected with increasing CD concentration, as the CDs were added to a constant amount of flour, and the total mass (dry matter) did, therefore, increase. However, the water absorption did not increase proportionally to the amount of CD added, and it stagnated or decreased when going from 4% to 8%, dependent on the type of CD.

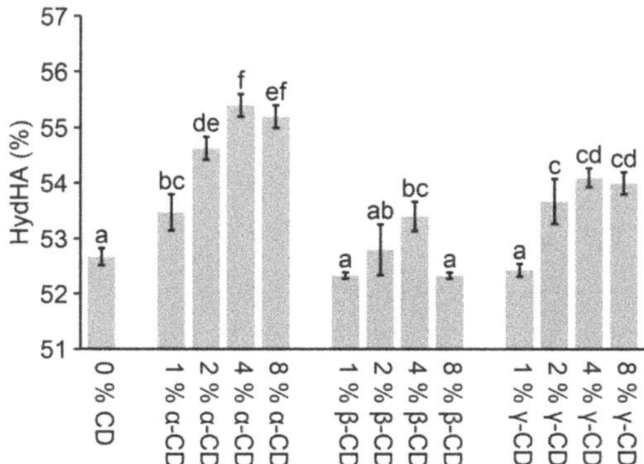

Figure 1. Water absorption for different types and concentrations of CD determined by the consistograph method. The water absorption is the hydration needed to obtain a dough with a maximum pressure of 2200 mbar. The error bars indicate the standard deviation. Different letters indicate the significant difference between the treatments ($p < 0.05$).

Similar tendencies have been found by other authors. Up to 3.4% increase in water absorption dependent on the CD concentration was observed for wheat doughs supplemented with up to 1.6% β-CD by Kim and Hill [11]. Likewise, Zhou et al. [13] observed up to 6.4% increase in water absorption with increasing CD concentration of up to 3.0% α- or γ-CD for a durum wheat flour dough. Of particular interest, Duedahl-Olesen et al. [15] found that an increase in water absorption by 7.3% and 8.0% for the addition of 3% α- and γ-CD, respectively, whereas a much lower increase in water absorption, was recorded when supplying glucose and maltooligosaccharides at the same level (wt%).

The effects of the different concentrations of α-, β-, and γ-CD on the biaxial extensional properties of the dough were tested by alveograph analysis. The results can be seen in Figure 2. The *P* value, which represents the tenacity of the dough, was approximately constant up to 2% CD (no significant difference from the control), after which it increased with increasing CD concentration for all three types of CD. A concentration of 8% α-, β-, and γ-CD caused an increase in *P* of 63%, 51%, and 43%, respectively. The biaxial extensibility of the dough measured by the *L* values decreased with increasing concentration of the three types of CDs. 8% addition of α-, β-, and γ-CD caused a decrease in *L* of 49%, 52%, and 37%, respectively. The deformation energy measured by the parameter *W* seemed to decrease for low concentrations of CDs, after which it increased again for higher CD concentrations. However, the decrease in *W* was only significant for β-CD, while the *W* values for α- and γ-CD supplemented doughs were not significantly different from the control for any concentration. The *W* value for all concentrations of β-CD was significantly below the value of the control. The *Ie* value, which is called the elasticity index, changed differently dependent on whether α- and γ-CD or β-CD were applied. The *Ie* value seemed to increase for a concentration of up to 4% α- and γ-CD. However, only the *Ie* value for 4% α-CD was significantly different from the control. At 8% α- and γ-CD, the *Ie* value decreased significantly. The *Ie* value decreased with increasing concentration of β-CD. The alveograph results reveal that addition of CDs entails a stiffer and less extensible dough, as *P* increases, while *L* decreases. Addition of α- and γ-CD did not change the strength of the dough significantly, as indicated by *W*, while addition of β-CD resulted in a weaker dough. This was further substantiated by the *Ie* values obtained for the β-CD series of doughs, as according to Kitissou [40], *Ie* is related to the gluten network quality of the dough. However, the addition of 8% α- and γ-CD also resulted in a significant decrease of the *Ie* value.

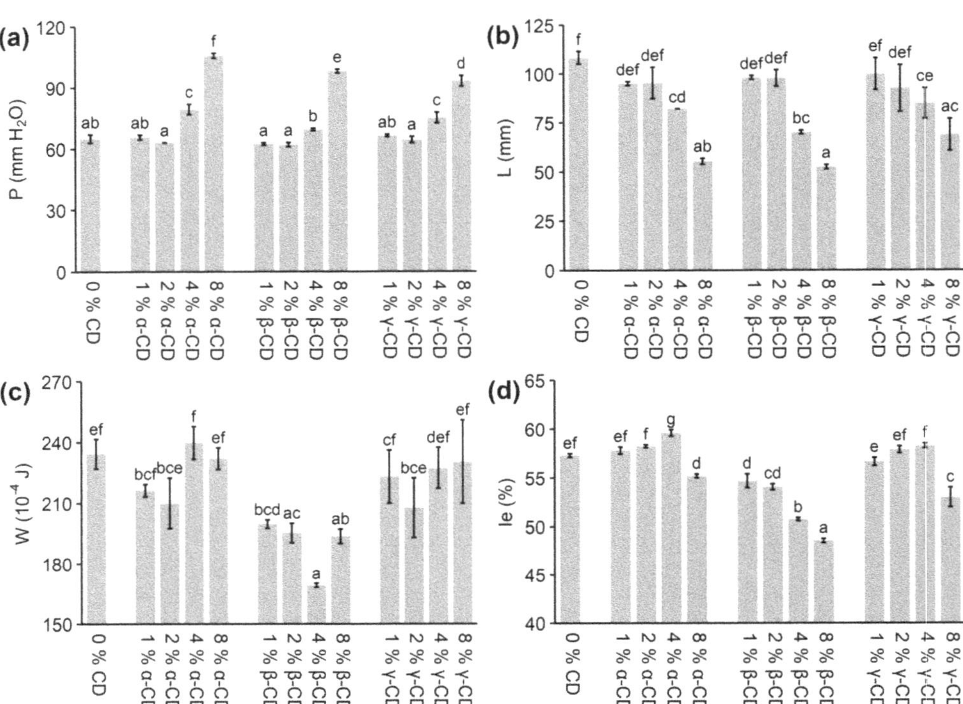

Figure 2. Effect of the different types and concentrations of CD on the dough extensional properties as determined by the alveograph method. The results for the alveograph parameters *P* (**a**), *L* (**b**), *W* (**c**), and *Ie* (**d**) are shown. The error bars indicate the standard deviation. Different letters indicate the significant difference between the treatments ($p < 0.05$).

While a few studies have investigated the effect of CDs on the mixing properties of the dough, the effect on the extensional properties of the dough has only been studied to a limited extent. Zhou et al. [12] investigated the effect of β-CD on the dough using the extensograph. They found that 0.5–1.5% β-CD increased the maximum resistance to deformation compared to the control, while the maximum resistance to deformation decreased for 2.0–3.0% β-CD. The extensibility increased slightly up to 1.0% β-CD, after which it decreased slightly up to 3.0% β-CD. The results from the alveograph method and the extensograph method cannot be directly compared due to differences in the sample and analysis conditions. However, the study by Zhou et al. [12] supports that at least elevated amounts of β-CD resulted in a weaker dough, probably through a weakening of the gluten network.

The addition of CD to wheat dough has multiple effects, as the CDs can affect both water distribution and the other flour constituents. The water in wheat dough interacts with the different constituents of dough, but the water availability is in general limited [41]. CDs contain multiple hydroxyl groups, which are able to form hydrogen bonds with the water. The addition of CDs might, therefore, limit the availability of water and thereby affect the gluten network development, which would be observed as changes in the mixing and extensional properties of the dough, including the HydHA value and the alveograph parameters. However, our results and other studies indicate that the effects of CDs are also caused by their direct influence on other components of the dough matrix and not just a shift in the distribution of water. Duedahl-Olesen et al. [15] found that α- and γ-

CD resulted in higher water absorption during mixing compared to an equal amount of glucose or non-cyclic maltooligosaccharides. If the higher water absorption should only be attributed to the water-binding capacity of the hydrophilic CDs, similar effects should be expected using their non-linear counterparts, as the water binding capacity is considered to be comparable in the dough matrix with its limited water availability. Furthermore, in the alveograph analysis, doughs with similar consistencies according to the HydHA values were analyzed, which was considered to reduce the effect of the varying water absorption on the results. This suggests that the large changes that were observed in the resultant parameters cannot solely be explained by differences in varying water absorption.

One characteristic that distinguishes CDs from smaller carbohydrates and, to some extent starches, is their general ability to form inclusion complexes by exchanging water in the cavity with a hydrophobic molecule or part of a molecule. In this process, complex formation is mainly driven by the release of "enthalpy rich" cavity-bound water and hydrophobic interaction (removal of ordered low entropy, high enthalpy water around the hydrophobic guest) [42]. Both driving forces would be expected to be favorable in an environment with low water activity. The CDs are (relatively rigid) cyclic oligosaccharides, and they are, therefore, capable of forming rather stable inclusion complexes with a range of primarily lipophilic molecules [1,2,5]. The CDs might interact with lipophilic molecules (e.g., lipids) and lipophilic parts of molecules, e.g., lipophilic parts of gluten proteins, but the strength and selectivity will be dependent on the cavity size of the specific CD. In essence, α-CD is most suitable for complex formation with linear aliphatic molecules (such as lipids), β-CD is suitable for complex formation with aromatic molecules, and γ-CD is suitable for larger aromatic molecules [1,2,5]. Although this leads to a considerable degree of selectivity, the complex-forming ability of the CDs is somewhat general, as typically all three CDs will be able to form a complex with a given (preferably lipophilic) molecule, but with different association constants.

The starch in the dough might also be affected by the addition of CDs. It has previously been suggested that β-CD might disrupt the amylose-lipid complex formation as well as it might form amylose-β-CD and amylose-lipid-β-CD complexes [14,43–45]. This might change the crystallinity of the starches and thereby cause an indirect change in the distribution of water, which has been suggested to affect the mixing properties [15]. However, the disruption of amylose-lipid complex formation and formation of complexes with CDs have primarily been observed for starch, which was at least partly gelatinized. In the dough, most of the starch is organized in starch granules, and the accessibility of the starch is, therefore, limited [46]. The effects of interactions between CDs and starch are, therefore, assumed to be smaller for the dough compared to the bread where the starches have been subjected to extensive gelatinization.

The CDs might also interact with the gluten proteins in the dough due to their ability to form weak inclusion complexes with proteins, which might affect the development and the properties of the gluten network. This was identified by Zhou et al. [12], who found that the addition of β-CD to wheat dough changed the secondary structure of the gluten proteins by increasing the proportion of α-helixes and decreasing the proportion of β-sheets. α-, β-, and γ-CD have been shown to be able to influence the behavior of proteins [47–50], but β-CD causes by far the largest effects, which have been explained by a relatively large affinity towards solvent exposed aromatic amino acids [47,51]. The interaction between β-CD and aromatic amino acids reduces the formation of protein-protein interactions by hydrophobic interaction in aqueous solutions [47,48]. All CDs had a significant influence on the viscoelastic properties of the dough, but especially β-CD was revealed to have large effect on the value of the alveograph parameters W and Ie, which are among other things dependent on the gluten network quality. These effects might be caused by interaction between the gluten proteins and β-CD, leading to changes in the strength of potential protein-protein interactions by non-covalent interactions, including hydrophobic interactions. The results indicate that the addition of a large amount of β-CD leads to a lower gluten network quality as assessed from the rheological properties. However, Zhou et al. [12] suggested that the addition of up to 1.5% β-CD positively affected the gluten network, as the maximum dough tensile resistance in extension increased. In addition to the effects of β-CD, the addition of CDs dilutes the protein content in the dough, which might also decrease the strength of the gluten network. This might be a contributing cause to why the addition of 8% of any of the CDs results in a lower than expected increase in the water absorption as well as the low *Ie* values in the alveograph analysis.

2.2. Effect on Bread Properties

Baking experiments with up to 8% addition of α-, β-, and γ-CD were made on a domestic bread maker (Breadmaking I) to investigate the effect on the bread volume. The results can be seen in Figure 3. Although the addition of CDs affected the water absorption, we decided to apply constant water addition in the bread doughs to minimize the number of variables. For the different types of CD, the largest specific bread volumes were observed for 4% α-CD, 2% β-CD, and 2% γ-CD, which resulted in an increase of 14%, 9%, and 7% in specific bread volume, respectively. However, only the bread with 4% α-CD were significantly larger than the control sample. When higher concentrations of the three types of CDs were added, the specific bread volume decreased, especially when β-CD was applied. All bread with the exception of the 8% α-CD, and 4% and 8% β-CD supplemented version, displayed acceptable crust and crumb structure as perceived by visual and manual inspection (see Supplementary Materials). In contrast, for the exceptions, it was observed that the crumb of the bread had partially collapsed, had an irregular crumb, and was very dense at the bottom. An irregular and uneven crust was observed at the top of the bread as if air had escaped. This indicates that the gluten network had been adversely effected by the addition of CD, preventing the development of a suitable gluten network with sufficient stability from supporting the dough foam.

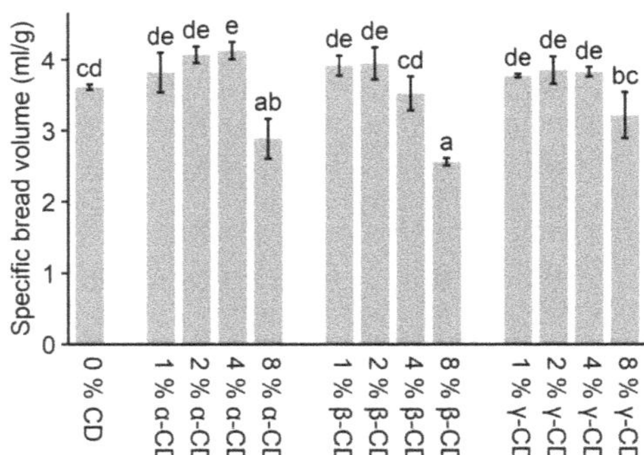

Figure 3. Specific bread volume for wheat bread with the addition of different types and concentrations of CD (Breadmaking I). The error bars indicate the standard deviation. Different letters indicate the significant difference between the treatments ($p < 0.05$).

Similar bread was produced using a kitchen mixer (Breadmaking II). 8% addition of CD was omitted, as this concentration resulted in a significant decrease in specific bread volume. During the preparation of the dough pieces for the analysis, it was noted that the dough stickiness increased with increasing CD concentration. In contrast to the bread produced in the domestic bread maker (Breadmaking I), no significant difference in specific bread volume between the bread with and without CD could be observed. The specific bread volumes in this trial ranged from 3.6–3.9 mL/g.

These results only partly confirm the results presented in other studies, where, in general, significant increases in bread volume could be observed in the range of 1–3% added CD. Kim and Hill [11] showed that an increase in bread loaf volume of 12% could be obtained in the range of 0.8 to 1.4% β-CD added for wheat bread. Mutsaers and Eijk [34] reported a 14–20% increase in loaf volume for two types of wheat bread supplemented with 1.5–2% β-CD. The addition of β-CD was found to be on par with the addition of shortening (3%), CGTase, and amylase in an American straight dough process judged from bread loaf volume [34]. Zhou et al. [12] found a slight increase in specific loaf volume until 1.5% β-CD after which the specific volume decreased below the specific volume of the control without β-CD. Zhou et al. [13] found a maximum increase in a specific volume at 2% α-CD and 3% γ-CD in a study in which the range of added CD was 0.5 to 3%. Both Zhou et al. [12] and Zhou et al. [13] observed a change in bread crumb pore distribution towards smaller and more uniform pores for the bread supplemented with either 2% α-CD, 1.5% β-CD, or 3% γ-CD. Furthermore, multiple studies have used CGTases in the production of wheat bread and found bread volume increments, which is assigned to the production of CDs [34,35,37]. Although our results, at least for the Breadmaking II data set, did not fully corroborate the data obtained on a domestic bread maker (Breadmaking I), we can partly confirm the tendency that the addition of small amounts (1–2%) CD may lead to an increase in bread volume. On the other hand, our results clearly demonstrate that the addition of larger amounts of CD, e.g., >4%, leads to a loss of bread volume compared to the control. However, the changes in bread volume are, as shown, somewhat dependent on production conditions and procedure. Although significant increases in bread volume based on the addition of CDs could not be unequivocally verified, our results underline that acceptable bread with respect to bread volume and quality may be achieved for additions of all three native CDs up to at least 4%.

The results obtained for the effects of CDs on bread volume, in essence, corroborates the tendencies obtained from the alveograph analysis of the doughs, including minor increases in parameters correlated to bread quality (volume; e.g., *Ie*) at low CD concentrations, followed by large decreases at high CD concentration. This is to some extent expected since both alveograph analysis and the foam producing step in breadmaking involves bubble inflation causing biaxial extension of the dough matrix [52]. This substantiates that the tendencies found for the effects of CD on bread volume are caused by the interaction of the CD with the proteins in the gluten network, facilitating minor improvements of the network quality (as judged by bread volume) at low concentration and larger adverse effects at high CD concentration.

To evaluate the effects of the CDs on the staling of the bread, the bread crumb from bread stored at room temperature were analyzed by texture analysis and LF-NMR to detect changes in the firmness and in the water distribution, respectively.

The result of the texture analysis can be seen in Figure 4. The firmness of the bread crumb gradually increased with longer storage time for all the bread. Increasing firmness of bread crumb is often used as a measure of bread staling [53]. No significant difference in the firmness measurements between the bread with and without CD during the storage could be found, indicating that the three types of CD did not retard the staling of the bread as judged by firmness.

To further elucidate a potential effect on staling, LF-NMR analysis on breadcrumb was conducted. The use of LF-NMR in food science is well established [54]. The LF-NMR data were analyzed to label discrete exponential decays, representing distinct water populations. Three populations of protons were identified in all bread samples with relaxation time T_{21} varying between 0.5–2.1 ms, T_{22} 2.8–7.7 ms, and T_{23} 17.4–36.6 ms (see Supplementary Materials). The ranges of T_{2n} values are similar to those presented in the literature [55,56]. The relaxation time T_{21} represents the least-mobile proton population and, therefore, the most tightly bound, and vice versa, T_{23} represents the most mobile proton population. No apparent systematic development of the T_{2n} values during storage was found. The corresponding M_n-values, which are the abundances of the three proton populations, presented in Figure 5 indicate the relative concentration of the different proton populations. The figures show a stable distribution up until 7 days of storage for all treatments with a reproducible signal. After 7 days, the samples display considerable variation among the triplicate measurements within each treatment, suggesting variations within the bread crumb. A weak tendency of proton exchange between the two faster relaxation times, T_{21} and T_{22}, is also observed, in which it should be noted that M_n is a relative indicator. Selective loss of water during the staling process will thus give the same impression.

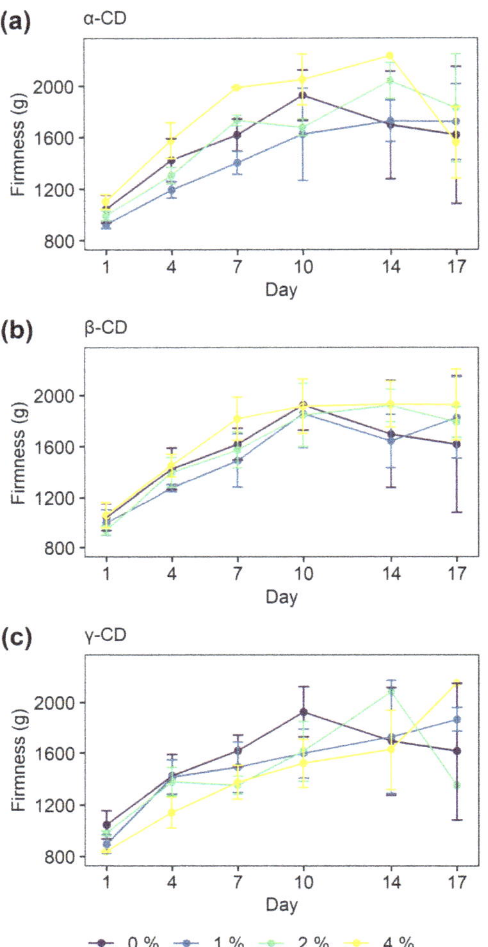

Figure 4. Firmness of bread without and with the addition of α-CD (**a**), β-CD (**b**), and γ-CD (**c**) for different storage time. The error bars indicate the standard deviation.

No systematic change in the distribution of water populations could be observed between the control and the CD supplemented bread. This corroborates the firmness studies carried out on the same series of bread (Figure 4). However, the literature suggests that an anti-staling effects may be achieved by the addition of CDs to wheat bread. β- and γ-CD have been found to have a small (but significant) retarding effect on the staling rate of wheat bread stored at room temperature, while no significant decrease in staling has been observed for the addition of α-CD [14,15]. Tian et al. [14] suggested that retarding effect of β-CD on staling was caused by the formation of an amylose-lipid-β-CD complex, which retard the transformation of the crystalline starch types in the crumb. Furthermore, the retarding effect of β-CD on the retrogradation of various starches has been presented in several studies [43,44,57–59]. The addition of CGTases has also in multiple studies been shown to inhibit the staling in bread, but it is also presumed to be linked to the amylolytic activity of the enzymes and not solely the effects of the CD produced [35–38].

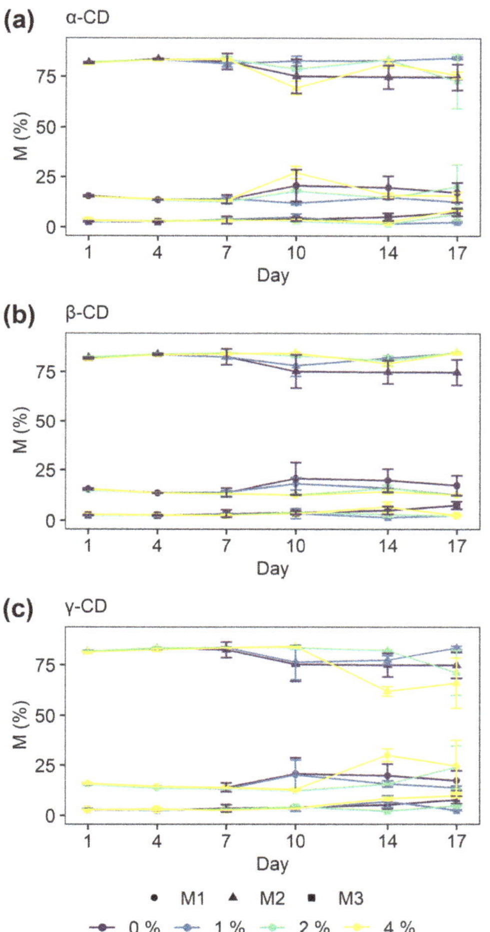

Figure 5. The relative abundances (M_n) of the three different proton populations in the breads with different types and concentrations of α-CD (**a**), β-CD (**b**) and γ-CD (**c**) for different storage time. The error bars indicate the standard deviation.

Despite that several studies have found an anti-staling effect by the addition of CDs in bread, no such effect could be identified in this study. This discrepancy may originate in differences in ingredients, water addition, processing methods, and method for assessing a potential anti-staling effect, as several of the abovementioned studies apply DSC measurements together with texture analysis. It may also underline that an anti-staling effect of CDs, if any, may be small and lower than the random variation of the experiments.

3. Materials and Methods

3.1. Materials

Two batches of commercial wheat flour (Lantmännen Cerealia) were obtained from Lantmännen Unibake, Hatting, Denmark. The first batch, which was used for consistography, alveography, and bread baking for volume measurement, had a moisture content of

14.2% and a protein content of 12.6% (dry matter basis). The second batch, which was used for bread baking for staling measurements, had a moisture content of 14.1% and a protein content of 14.5% (dry matter basis). The moisture content of the flour was determined according to AACC method 44–15.02 [60], and the protein content was determined by the Kjeldahl method as described in AACC method 46–11.02 [61]. α-cyclodextrin (food grade), β-cyclodextrin (pharma grade), and γ-cyclodextrin (food grade) were provided by Wacker Chemie. Sodium chloride was from VWR, and dry yeast (Lesaffre) was obtained from the local supermarket.

3.2. Consistographic and Alveographic Analysis

The flour was tested with and without 1, 2, 4, or 8% (of flour weight) of α-, β-, and γ-CD. The consistograph and alveograph measurements was made on AlveoLab (Chopin Technologies, Villeneuve-La-Garenne, France). All measurements were made in triplicate. The different concentrations and types of CDs were tested using the constant hydration consistograph test AACC method 54–50.01 [62]. In short, CDs were added on top of the required amount of flour, and a dough with a fixed water content was made. During mixing, the consistency of the dough was measured by monitoring the pressure on one side of the mixer. The maximum dough consistency was used to find the water absorption value HydHA, which was the hydration equivalent to a maximum pressure of 2200 mbar on the basis of 15% H_2O (flour basis).

The flour with different CD type and concentrations was analyzed by alveography as described in the AACC method 54–30.02 [63], except the amount of flour and the water addition were based on the HydHA value determined in the consistograph analyses [64]. The parameters P (tenacity, related to the maximum height of the curve), L (biaxial extensibility, length of the curve at bubble rupture), W (deformation energy, related to the area under the curve), and Ie (elasticity index, ratio between the height of the curve at 40 mm and the maximum height) was found from the bubble inflation air pressure curves.

3.3. Breadmaking I

Bread with and without 1, 2, 4, or 8% (based on flour weight) of α-, β-, and γ-CD were made in triplicates using the recipe in Table 1. The bread was made on a domestic breadmaker (model 48319, Morphy Richards, Swinton, United Kingdom) using the settings program 5 (French bread), 450 g loaf, and 5 in crust darkness. In this program, the dough was kneaded for 13 min, rested for 40 min, kneaded again for 17 min, proofed for 30 min, kneaded shortly, proofed for a further 50 min, and baked for 60 min. The bread volume was measured using the mustard seed displacement method. The volume was calculated by subtracting the volume of seeds held by a container with a baked product from that of the volume of seeds without the baked product. The volume of the mustard seeds was determined by their weight using a density of 0.759 g/mL. All weights were determined on an analytical scale (Kern PFB 1200-2A, Balingen, Germany). The specific bread volume was found by dividing the bread volume with the weight of the bread.

Table 1. Recipe for bread used in Breadmaking I and Breadmaking II.

Ingredient	Ratio in Grams
Flour	100
Tap water	60
Sodium chloride	1
Dry yeast	0.8
α-, β-, or γ-CD	0, 1, 2, 4, or 8

3.4. Breadmaking II

Bread for assessment of staling rate was made using the recipe in Table 1, except doughs with 8% CD were omitted. Doughs were made for each type and concentration of CD and mixed for 5 min with a kitchen mixer (Kenwood Chef XL, Havant, UK) at mixing speed 3 with a final dough temperature of 26 °C. The dough was divided into 300 g pieces, rounded, and molded by hand, proofed at ambient temperature (26 °C) under a linen cover for 90 min and baked at 200 °C for 12 min in an oven (Rational SCC WE 101, Heerbrugg, Switzerland). The loaves were cooled at room temperature for 1 h, after which the breads were weighed, and bread volume was measured using the mustard seed displacement method allowing the specific bread volume to be calculated. Immediately hereafter, the breads were brushed with sodium benzoate solution, sealed in plastic bags, and stored at 19 °C for aging studies.

3.5. Crumb Firmness Measurements

A texture analyzer TA.XTplus (Stable Micro Systems, Surrey, UK) was used to measure force-time curves according to the AACC standard 74–09.01 [65] with modifications. At day 1, 4, 7, 10, 14, and 17, bread slices (2.5 cm thick) were compressed to a deformation level of 40% of the original sample height by a 25 mm cylindrical probe (P25) at a test speed of 1.7 mm/s. The peak force of compression was reported as firmness (g). Measurements in triplicates were used for the evaluation of the bread staling.

3.6. Low Field Nuclear Magnetic Resonance Spectroscopy Measurements

To identify possible differences in the water distribution (or rather proton populations) in the bread supplemented with different concentrations and types of CDs, bread samples from day 1,4,7,10,14, and 17 were analyzed by ^{1}H low field nuclear magnetic resonance (LF-NMR) spectroscopy (MQR Spectro-P spectrometer, Oxford Instruments, Oxfordshire, UK, operating at 20 MHz). Approximately 1.5 g of bread crumb was sampled from a slice of bread using a cork borer and placed into a glass tube for NMR measurements. Data were collected using a Carr–Purcell–Meiboon–Gill (CPMG) sequence at 25 °C with the parameters: Recycle delay of 5 s, τ-delay of 100 µs, and 16 scans averaged. Data from 6000 echoes were acquired with a receiver gain of 5.0. All measurements were prepared in triplicates on distinct samples. Transverse relaxation times (T_{2n}) of different relaxation components were obtained using an in-house MATLAB (version R2019a, The Math-Works) script designed for fitting the relaxation curves to a series of exponential decays according to Equation (1).

$$I(t) = \sum_{n=1}^{N} M_n \cdot e^{-t/T_{2n}} \tag{1}$$

In which $I(t)$ is the echo intensity as a function of relaxation time, N is the number of relaxation components, the transverse relaxation time for site n is T_{2n}, and the corresponding abundance is M_n. N is determined by visual inspection of the residuals after model fitting. Each LF-NMR recording was fitted individually. All M_n-values were presented as percentage of total intensity to eliminate sample size differences.

3.7. Statistical Analysis

Statistical analysis of the results was carried out in R (version 3.6.1., R Core Team) using analysis of variance (ANOVA) with Tukey's multiple comparison procedure with a significance level of 5%.

4. Conclusions

α-, β-, and γ-CD affects the mixing and extensional properties of wheat dough dependent on concentration and CD type. β-CD displayed the largest effects, which may be caused by its potentially stronger (compared to α- and γ-CD) direct interaction with the proteins in the gluten network. The addition of up to 4% CD did not significantly affect the bread volume, in general, expect 4% α-CD, which resulted in a minor, significant increase in bread volume in one of the breadmaking trials. No significant effect of the CDs on staling of the bread could be detected. The results suggest that up to 4% CD can be added to bread with only minor effects on dough properties and without a significant decrease in the bread quality. This opens up for the use of CD supplemented wheat bread for nutraceutical purposes.

Supplementary Materials: Pictures of CD-supplemented breads and a figure with relaxation time (T_{2n}) from the LF-NMR measurements are available in the supplementary materials.

Author Contributions: Conceptualization, A.-S.S.J. and K.L.L.; methodology, A.-S.S.J., K.L.L., and B.M.J.; formal analysis, A.-S.S.J., T.P.C., and F.W.J.v.d.B.; investigation, A.-S.S.J., T.P.C., B.M.J., and K.L.L.; writing—original draft preparation, A.-S.S.J. and K.L.L.; writing—review and editing, T.P.C., B.M.J., and F.W.J.v.d.B.; visualization, A.-S.S.J. and T.P.C.; supervision, B.M.J. and K.L.L.; project administration, K.L.L.; funding acquisition, K.L.L. All authors have read and agreed to the published version of the manuscript.

Funding: This research was in part funded by Innovation Fund Denmark, grant number 5189-00062B.

Data Availability Statement: Data available on request to the corresponding author.

Acknowledgments: Wacker Chemie (Burghausen, Germany) is acknowledged for generous and kind donation of cyclodextrins. The authors wish to thank Elsebeth Juhl Pedersen (Department of Chemistry and Bioscience, Aalborg University) for conducting the alveograph and consistograph analysis, and Lisbeth Dahl (Department of Food Science, Faculty of Science, University of Copenhagen) for performing the flour analysis and texture analysis.

Conflicts of Interest: The authors declare no conflict of interest.

Sample Availability: Samples of the compounds are not available from the authors.

References

1. Loftsson, T. Cyclodextrins in Parenteral Formulations. *J. Pharm. Sci.* **2021**, *110*, 654–664. [CrossRef] [PubMed]
2. Jansook, P.; Ogawa, N.; Loftsson, T. Cyclodextrins: Structure, physicochemical properties and pharmaceutical applications. *Int. J. Pharm.* **2018**, *535*, 272–284. [CrossRef] [PubMed]
3. Dhiman, P.; Bhatia, M. Pharmaceutical applications of cyclodextrins and their derivatives. *J. Incl. Phenom. Macrocycl. Chem.* **2020**, *98*, 171–186. [CrossRef]
4. Astray, G.; Gonzalez-Barreiro, C.; Mejuto, J.C.; Rial-Otero, R.; Simal-Gándara, J. A review on the use of cyclodextrins in foods. *Food Hydrocoll.* **2009**, *23*, 1631–1640. [CrossRef]
5. Fenyvesi, É.; Vikmon, M.; Szente, L. Cyclodextrins in Food Technology and Human Nutrition: Benefits and Limitations. *Crit. Rev. Food Sci. Nutr.* **2016**, *56*, 1981–2004. [CrossRef]
6. Dos Santos, C.; Buera, P.; Mazzobre, F. Novel trends in cyclodextrins encapsulation. Applications in food science. *Curr. Opin. Food Sci.* **2017**, *16*, 106–113. [CrossRef]
7. Matencio, A.; Navarro-Orcajada, S.; García-Carmona, F.; López-Nicolás, J.M. Applications of cyclodextrins in food science. A review. *Trends Food Sci. Technol.* **2020**, *104*, 132–143. [CrossRef]
8. Tian, B.; Xiao, D.; Hei, T.; Ping, R.; Hua, S.; Liu, J. The application and prospects of cyclodextrin inclusion complexes and polymers in the food industry: A review. *Polym. Int.* **2020**, *69*, 597–603. [CrossRef]
9. Feng, T.; Zhuang, H.; Yang, N. Cyclodextrins in Parenteral Formulations. In *Cyclodextrins: Preparation and Application in Industry*; Jin, Z., Ed.; World Scientific Publishing: Singapore, 2018; pp. 143–207.
10. EFSA. Panel on Dietetic Products Nutrition and Allergies. Scientific Opinion on the substantiation of health claims related to alpha cyclodextrin and reduction of post prandial glycaemic responses (ID 2926, further assessment) pursuant to Article 13(1) of Regulation (EC) No 1924/2006. *EFSA J.* **2012**, *10*, 2713. [CrossRef]
11. Kim, H.O.; Hill, R.D. Modification of Wheat Flour Dough Characteristics by Cycloheptaamylose. *Cereal Chem.* **1984**, *61*, 406–409.
12. Zhou, J.; Yang, H.; Qin, X.; Hu, X.; Liu, G.; Wang, X. Effect of β-Cyclodextrin on the Quality of Wheat Flour Dough and Prebaked Bread. *Food Biophys.* **2019**, *14*, 173–181. [CrossRef]
13. Zhou, J.; Ke, Y.; Barba, F.J.; Xiao, S.; Hu, X.; Qin, X.; Ding, W.; Lyu, Q.; Wang, X.; Liu, G. The addition of α-cyclodextrin and γ-cyclodextrin affect quality of dough and prebaked bread during frozen storage. *Foods* **2019**, *8*, 174. [CrossRef] [PubMed]
14. Tian, Y.Q.; Li, Y.; Jin, Z.Y.; Xu, X.M.; Wang, J.P.; Jiao, A.Q.; Yu, B.; Talba, T. Beta-Cyclodextrin (Beta-CD): A new approach in bread staling. *Thermochim. Acta* **2009**, *489*, 22–26. [CrossRef]
15. Duedahl-Olesen, L.; Zimmermann, W.; Delcour, J.A. Effects of low molecular weight carbohydrates on farinograph characteristics and staling endotherms of wheat flour-water doughs. *Cereal Chem.* **1999**, *76*, 227–230. [CrossRef]
16. JECFA. Summary of Evaluations Performed by the Joint FAO/WHO Expert Committee on Food Additives: Beta-Cyclodextrin. Available online: http://www.inchem.org/documents/jecfa/jeceval/jec_465.htm (accessed on 18 February 2021).
17. JECFA. Safety Evaluation of Certain Food Additives: Gamma-Cyclodextrin. Available online: http://www.inchem.org/documents/jecfa/jecmono/v042je11.htm (accessed on 18 February 2021).

18. JECFA. Summary of Evaluations Performed by the Joint FAO/WHO Expert Committee on Food Additives: Alpha-Cyclodextrin. Available online: http://www.inchem.org/documents/jecfa/jeceval/jec_464.htm (accessed on 18 February 2021).
19. JECFA. *Safety Evaluation of Certain Food Additives*; World Health Organization: Geneva, Switzerland, 2006; Volume 54.
20. JECFA. Evaluations of the Joint FAO/WHO Expert Committee on Food Additives: Gamma-Cyclodextrin. Available online: https://apps.who.int/food-additives-contaminants-jecfa-database/chemical.aspx?chemID=2067 (accessed on 18 February 2021).
21. FDA. GRAS Notice GRN No. 46 Gamma-Cyclodextrin. Available online: https://www.cfsanappsexternal.fda.gov/scripts/fdcc/index.cfm?set=GRASNotices&id=46 (accessed on 18 February 2021).
22. FDA. GRAS Notice GRN No. 74 Beta-Cyclodextrin. Available online: https://www.cfsanappsexternal.fda.gov/scripts/fdcc/index.cfm?set=GRASNotices&id=74 (accessed on 18 February 2021).
23. FDA. GRAS Notice GRN No. 155 Alpha-Cyclodextrin. Available online: https://www.cfsanappsexternal.fda.gov/scripts/fdcc/index.cfm?set=GRASNotices&id=155 (accessed on 18 February 2021).
24. Scazzina, F.; Siebenhandl-Ehn, S.; Pellegrini, N. The effect of dietary fibre on reducing the glycaemic index of bread. *Br. J. Nutr.* **2013**, *109*, 1163–1174. [CrossRef]
25. Spears, J.K.; Karr-Lilienthal, L.K.; Grieshop, C.M.; Flickinger, E.A.; Wolf, B.W.; Fahey, G.C. Pullulans and γ-cyclodextrin affect apparent digestibility and metabolism in healthy adult ileal cannulated dogs. *J. Nutr.* **2005**, *135*, 1946–1952. [CrossRef]
26. Koutsou, G.A.; Storey, D.M.; Bär, A. Gastrointestinal tolerance of γ-cyclodextrin in humans. *Food Addit. Contam.* **1999**, *16*, 313–317. [CrossRef]
27. Lumholdt, L.R.; Holm, R.; Jorgensen, E.B.; Larsen, K.L. In vitro investigations of α-amylase mediated hydrolysis of cyclodextrins in the presence of ibuprofen, flurbiprofen, or benzo[a]pyrene. *Carbohydr. Res.* **2012**, *362*, 56–61. [CrossRef]
28. Spears, J.K.; Karr-Lilienthal, L.K.; Fahey, G.C. Influence of supplemental high molecular weight pullulan or γ-cyclodextrin on ileal and total tract nutrient digestibility, fecal characteristics, and microbial populations in the dog. *Arch. Anim. Nutr.* **2005**, *59*, 257–270. [CrossRef]
29. Suzuki, M.; Sato, A. Nutritional Significance of Cyclodextrins: Indigestibility and Hypolipemic Effect of α-Cyclodextrin. *J. Nutr. Sci. Vitaminol.* **1985**, *31*, 209–223. [CrossRef]
30. Lai, C.-S.; Chow, J.; Wolf, B.W. Methods of Using Gamma Cyclodextrin to Control Blood Glucose and Insulin Secretion. U.S. Patent No. 7,423,027, 9 September 2008.
31. Raben, A.; Andersen, K.; Karberg, M.A.; Holst, J.J.; Astrup, A. Acetylation of or β-cyclodextrin addition to potato starch: Beneficial effect on glucose metabolism and appetite sensations. *Am. J. Clin. Nutr.* **1997**, *66*, 304–314. [CrossRef]
32. Zhan, J.; Tian, Y.; Tong, Q. Preparation and slowly digestible properties of β-cyclodextrins (β-CDs)-modified starches. *Carbohydr. Polym.* **2013**, *91*, 609–612. [CrossRef] [PubMed]
33. Schmid, G.; Reuscher, H.; Antlsperger, G. Method for reducing the glycemic index of food. European Patent Office EP 1 447 013 A1, 18 August 2004.
34. Mutsaers, J.H.G.M.; Eijk, J.H. Van Process for increasing the volume of a baked product. US Patent Number 5,916,607, 29 June 1999.
35. Shim, J.-H.; Kim, Y.-W.; Kim, T.-J.; Chae, H.-Y.; Park, J.-H.; Cha, H.; Kim, J.-W.; Kim, Y.-R.; Schaefer, T.; Spendler, T.; et al. Improvement of cyclodextrin glucanotransferase as an antistaling enzyme by error-prone PCR. *Protein Eng. Des. Sel.* **2004**, *17*, 205–211. [CrossRef]
36. Shim, J.-H.; Seo, N.-S.; Roh, S.-A.; Kim, J.-W.; Cha, H.; Park, K.-H. Improved Bread-Baking Process Using Saccharomyces cerevisiae Displayed with Engineered Cyclodextrin Glucanotransferase. *J. Agric. Food Chem.* **2007**, *55*, 4735–4740. [CrossRef] [PubMed]
37. Jemli, S.; Ben Messaoud, E.; Ayadi-zouari, D.; Naili, B.; Khemakhem, B.; Bejar, S. A beta-cyclodextrin glycosyltransferase from a newly isolated Paenibacillus pabuli US132 strain: Purification, properties and potential use in bread-making. *Biochem. Eng. J.* **2007**, *34*, 44–50. [CrossRef]
38. Gujral, H.S.; Haros, M.; Rosell, C.M. Starch Hydrolyzing Enzymes for Retarding the Staling of Rice Bread. *Cereal Chem.* **2003**, *80*, 750–754. [CrossRef]
39. Gujral, H.S.; Guardiola, I.; Carbonell, J.V.; Rosell, C.M. Effect of Cyclodextrin Glycosyl Transferase on Dough Rheology and Bread Quality from Rice Flour. *J. Agric. Food Chem.* **2003**, *51*, 3814–3818. [CrossRef] [PubMed]
40. Kitissou, P. Un nouveau paramètre alvéographique: l'indice d'élasticité (Ie). *Ind. Céréales* **1995**, *92*, 9–17.
41. Fessas, D.; Schiraldi, A. Water properties in wheat flour dough. I: Classical thermogravimetry approach. *Food Chem.* **2001**, *72*, 237–244. [CrossRef]
42. Rekharsky, M.V.; Inoue, Y. Complexation thermodynamics of cyclodextrins. *Chem. Rev.* **1998**, *98*, 1875–1917. [CrossRef]
43. Tian, Y.; Li, Y.; Manthey, F.A.; Xu, X.; Jin, Z.; Deng, L. Influence of β-cyclodextrin on the short-term retrogradation of rice starch. *Food Chem.* **2009**, *116*, 54–58. [CrossRef]
44. Tian, Y.; Xu, X.; Li, Y.; Jin, Z.; Chen, H.; Wang, H. Effect of β-cyclodextrin on the long-term retrogradation of rice starch. *Eur. Food Res. Technol.* **2009**, *228*, 743–748. [CrossRef]
45. Kim, H.O.; Hill, R.D. Physical characteristics of wheat starch granule gelatinization in the presence of cycloheptaamylose. *Cereal Chem.* **1984**, *61*, 432–435.
46. Goesaert, H.; Brijs, K.; Veraverbeke, W.S.; Courtin, C.M.; Gebruers, K.; Delcour, J.A. Wheat flour constituents: How they impact bread quality, and how to impact their functionality. *Trends Food Sci. Technol.* **2005**, *16*, 12–30. [CrossRef]

47. Otzen, D.E.; Knudsen, B.R.; Aachmann, F.; Larsen, K.L.; Wimmer, R. Structural basis for cyclodextrins' suppression of human growth hormone aggregation. *Protein Sci.* **2002**, *11*, 1779–1787. [CrossRef] [PubMed]
48. Aachmann, F.L.; Otzen, D.E.; Larsen, K.L.; Wimmer, R. Structural background of cyclodextrin-protein interactions. *Protein Eng.* **2003**, *16*, 905–912. [CrossRef]
49. Matilainen, L.; Larsen, K.L.; Wimmer, R.; Keski-Rahkonen, P.; Auriola, S.; Järvinen, T.; Jarho, P. The Effect of Cyclodextrins on Chemical and Physical Stability of Glucagon and Characterization of Glucagon/γ-CD Inclusion Complexes. *J. Pharm. Sci.* **2008**, *97*, 2720–2729. [CrossRef] [PubMed]
50. Bajorunaite, E.; Cirkovas, A.; Radzevicius, K.; Larsen, K.L.; Sereikaite, J.; Bumelis, V.A. Anti-aggregatory effect of cyclodextrins in the refolding process of recombinant growth hormones from Escherichia coli inclusion bodies. *Int. J. Biol. Macromol.* **2009**, *44*, 428–434. [CrossRef]
51. Aachmann, F.L.; Larsen, K.L.; Wimmer, R. Interactions of cyclodextrins with aromatic amino acids: A basis for protein interactions. *J. Incl. Phenom. Macrocycl. Chem.* **2012**, *73*, 349–357. [CrossRef]
52. Dobraszczyk, B.J.; Morgenstern, M.P. Rheology and the breadmaking process. *J. Cereal Sci.* **2003**, *38*, 229–245. [CrossRef]
53. Gray, J.A.; Bemiller, J.N. Bread Staling: Molecular Basis and Control. *Compr. Rev. Food Sci. Food Saf.* **2003**, *2*, 1–21. [CrossRef]
54. Webb, G.A. *Modern Magnetic Resonance*, 2nd ed.; Springer: Boston, MA, USA, 2018.
55. Engelsen, S.B.; Jensen, M.K.; Pedersen, H.T.; Nørgaard, L.; Munck, L. NMR-baking and multivariate prediction of instrumental texture parameters in bread. *J. Cereal Sci.* **2001**, *33*, 59–69. [CrossRef]
56. D'Avignon, D.A.; Hung, C.-C.; Pagel, M.T.L.; Hart, B.; Bretthorst, G.L.; Ackerman, J.J.H. 1H and 2H NMR Studies of Water in Work-Free Wheat Flour Doughs. In *NMR Applications in Biopolymers*; Springer: Boston, MA, USA, 1990; pp. 391–414.
57. Tian, Y.; Li, Y.; Jin, Z.; Xu, X. Comparison tests of hydroxylpropyl β-cyclodextrin (HPβ-CD) and β-cyclodextrin (β-CD) on retrogradation of rice amylose. *LWT Food Sci. Technol.* **2010**, *43*, 488–491. [CrossRef]
58. Tian, Y.; Yang, N.; Li, Y.; Xu, X.; Zhan, J.; Jin, Z. Potential interaction between β-cyclodextrin and amylose-lipid complex in retrograded rice starch. *Carbohydr. Polym.* **2010**, *80*, 581–584. [CrossRef]
59. Gunaratne, A.; Ranaweera, S.; Corke, H. Thermal, pasting, and gelling properties of wheat and potato starches in the presence of sucrose, glucose, glycerol, and hydroxypropyl beta-cyclodextrin. *Carbohydr. Polym.* **2007**, *70*, 112–122. [CrossRef]
60. AACC Method 44-15.02. Moisture—Air-Oven Methods. In *AACC Approved Methods of Analysis*; Cereals & Grains Association: St. Paul, MN, USA, 2009.
61. AACC Method 46-11.02. Crude Protein—Improved Kjeldahl Method, Copper Catalyst Modification. In *AACC Approved Methods of Analysis*; Cereals & Grains Association: St. Paul, MN, USA, 2009.
62. AACC Method 54-50.01. Determination of the Water Absorption Capacity of Flours and of Physical Properties of Wheat Flour Doughs, Using the Consistograph. In *AACC Approved Methods of Analysis*; Cereals & Grains Association: St. Paul, MN, USA, 2009.
63. AACC Method 54-30.02. Alveograph Method for Soft and Hard Wheat Flour. In *AACC Approved Methods of Analysis*; Cereals & Grains Association: St. Paul, MN, USA, 2009.
64. Dubois, M.; Dubat, A.; Launay, B. *The AlveoConsistograph Handbook*, 2nd ed.; AACC International: St. Paul, MN, USA, 2008.
65. AACC Method 74-09.01. Measurement of Bread Firmness by Universal Testing Machine. In *AACC Approved Methods of Analysis*; Cereals & Grains Association: St. Paul, MN, USA, 2009.

Article

Exploring Charged Polymeric Cyclodextrins for Biomedical Applications

Noemi Bognanni [1], Francesco Bellia [2], Maurizio Viale [3], Nadia Bertola [3] and Graziella Vecchio [1,4,*]

1 Dipartimento di Scienze Chimiche, Università degli Studi di Catania, Viale A. Doria 6, 95125 Catania, Italy; noemibognanni91@gmail.com
2 Istituto di Cristallografia, CNR, via P. Gaifami 18, 95126 Catania, Italy; francesco.bellia@cnr.it
3 IRCCS Ospedale Policlinico San Martino, U.O.C. Bioterapie, L.go R. Benzi 10, 16132 Genova, Italy; maurizio.viale@hsanmartino.it (M.V.); nadia.bertola@gmail.com (N.B.)
4 Consorzio Interuniversitario di Ricerca in Chimica dei Metalli nei Sistemi Biologici (CIRCMSB), via Celso Ulpiani, 27, 70126 Bari, Italy
* Correspondence: gr.vecchio@unict.it; Tel.: +39-9-5738-5064

Abstract: Over the years, cyclodextrin uses have been widely reviewed and their proprieties provide a very attractive approach in different biomedical applications. Cyclodextrins, due to their characteristics, are used to transport drugs and have also been studied as molecular chaperones with potential application in protein misfolding diseases. In this study, we designed cyclodextrin polymers containing different contents of β- or γ-cyclodextrin, and a different number of guanidinium positive charges. This allowed exploration of the influence of the charge in delivering a drug and the effect in the protein anti-aggregant ability. The polymers inhibit Amiloid β peptide aggregation; such an ability is modulated by both the type of CyD cavity and the number of charges. We also explored the effect of the new polymers as drug carriers. We tested the Doxorubicin toxicity in different cell lines, A2780, A549, MDA-MB-231 in the presence of the polymers. Data show that the polymers based on γ-cyclodextrin modified the cytotoxicity of doxorubicin in the A2780 cell line.

Keywords: aggregation; cancer; doxorubicin; nanoparticles

Citation: Bognanni, N.; Bellia, F.; Viale, M.; Bertola, N.; Vecchio, G. Exploring Charged Polymeric Cyclodextrins for Biomedical Applications. *Molecules* **2021**, *26*, 1724. https://doi.org/10.3390/molecules26061724

Academic Editors: Marina Isidori, Margherita Lavorgna and Rosa Iacovino

Received: 13 February 2021
Accepted: 16 March 2021
Published: 19 March 2021

Publisher's Note: MDPI stays neutral with regard to jurisdictional claims in published maps and institutional affiliations.

1. Introduction

Cyclodextrins (CyDs) are cyclic oligosaccharides of α-1,4-linked D(+)-glucopyranose with the unique property to act as molecular containers. They have been used in biomedical applications for their ability to include drugs or several biomolecules, such as cholesterol [1–4].

CyD properties can be modulated through their chemical modification. In recent years, CyDs represented an important nanocarrier family, thus developing into sophisticated drug delivery systems [5–7]. Nanoparticles based on CyD, have allowed encapsulation of drugs to protect them and improve their bioavailability. These systems have higher stability constants than those of the corresponding CyD monomers; they show excellent properties in drug release kinetics, mechanical properties and stimuli-responsiveness [8–11]. Furthermore, these polymeric systems can be used in a clinical setting, such as controlled drug and gene delivery systems [12]. Successful examples of linear CyD polymers specifically designed as drug carriers are Cyclosert and CALAA-01 [13]. CyD polymers modified with choline, amino or carboxylic groups have been investigated to increase their drug loading ability.

Due to the ability to include aromatic molecules of appropriate sizes, CyDs have also been investigated as protein chaperones [14–16]. β-CyD reduced the β-amyloid aggregation in vitro at millimolar concentration. The protective effects of β-CyD were also proven in vivo. The interaction between Amyloid beta peptide (Aβ) and α-, β- and γ-CyD was correlated to the ability to include Phe19 and Phe20 side chain of Aβ [16]. Other studies have suggested the higher antiaggregant activity of some functionalised CyDs with aromatic moieties [17]. CyD dimers were also found to act as inhibitors of Aβ_{40} aggregation at 1 mM concentration [18,19]. We investigated amino-cyclodextrin oligomers as promising CyD derivatives in inhibiting the aggregation of Aβ at micromolar concentration [20].

Inspired by the properties of multi-cavity systems and the importance of functionalisation to improve CyD properties, in this paper we report the synthesis of new linear polymers of β- and γ-CyD with different contents of guanidinium positive charge and number of CyD cavities. We also assayed the polymer systems as antiaggregant agents and as drug carriers to explore the effects of multi-cavity systems and their functionalisation for biological applications (Figure 1).

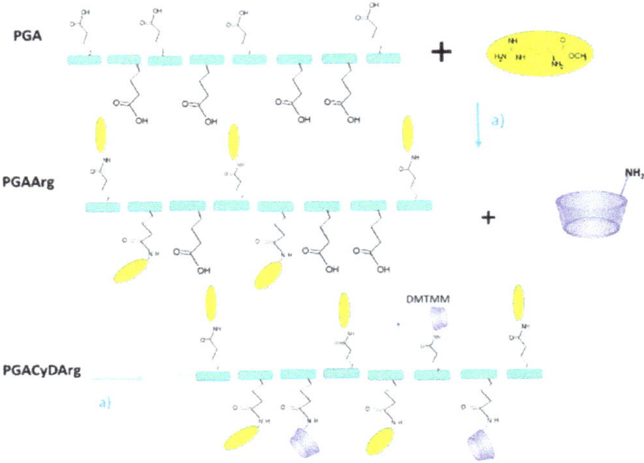

Figure 1. PGA-cyclodextrin (CyD)-Arg polymers, PGA is N-butyl-polyglutamate, a) 4-(4,6-Dimethoxy-1,3,5-triazin 2-yl)-4-methylmorpholinium chloride (DMTMM), H_2O pH 8, rt, 12 h.

It is well known that cancer cells have a deficiency of Arg amino acid and their requirement for arginine is higher than that for other amino acids. Certain tumours lose the ability to synthesise arginine dependently. Therefore, arginine depletion can be considered the weak point in cancer treatment for arginine auxotrophic tumours [21,22]. We used doxorubicin (DOX) as a drug model to test the activity of the new CyD derivatives as nanocarriers. Several studies have highlighted that CyDs can stabilise DOX in solution and enhance its dissolution rate [23,24]. Complexation of DOX with CyDs can increase permeability across the blood–brain barrier, due to the disruption of the membrane [25].

2. Results and Discussion

2.1. Synthesis and Characterisation

Charged CyD polymers were synthesised in water, starting from the polypeptide N-butyl-polyglutamate (PGA), ArgOCH₃ and CyD 3-amino derivatives, using 4-(4,6-Dimethoxy-1,3,5-triazin 2-yl)-4-methylmorpholinium chloride (DMTMM) as the condensing agent, as reported elsewhere [26]. We found that this method gave high conjugation yield using a green synthetic route. This procedure is appropriate for modulating the number of CyD cavities and charged groups of Arg in the PGA backbond [27].

We synthesised various polymers with different amounts of Arg and CyD units to explore the effect on polymer properties.

All the new polymers were characterised by NMR (Figures S1–S11). In Figure 2 the NMR spectra of PGAβCyDArg1 and PGAβCyDArg4 are reported. ^{1}H NMR spectra of all the derivatives show common patterns; the protons of CyD resonate at 5 ppm (H-1), and 4.0-3.4 ppm (H-3,-6,-5,-4,-2). Protons of arginine and the glutamic acid side chain of PGA resonate at 3.3 ppm and 2.5-1.8 ppm region. Butyl protons of PGA are also evident between 1.5 ppm and 1.0 ppm. We determined the number of CyD units linked to the PGA backbone for each polymer derivative by calculating the integral ratios of the signal of Hs-1 of CyD, the signal of the ethylene chain protons of PGA or the N-buthyl chain protons. Moreover, the integral ratio of signal due to the γ-CH$_2$ of Arg moieties at 3.3 ppm and the signals of PGA ethylenic protons or the N-buthyl protons was used to value the number of Arg moieties grafted to the polymer. The results obtained from NMR for each bioconjugate are reported in Table 1.

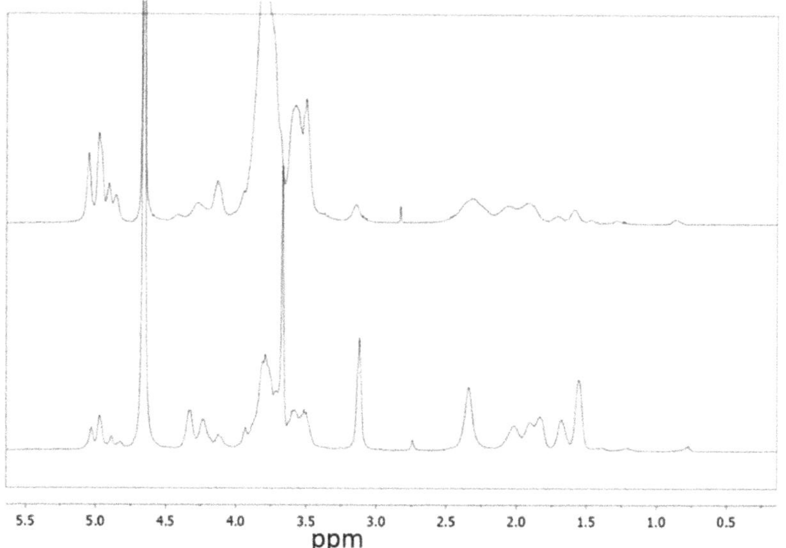

Figure 2. ^{1}H NMR spectra of PGAβ-CyDArg 1 (top), PGAβCyDArg4 (bottom) (D$_2$O, 500 MHz).

Table 1. Features of PGA-CyD-Arg derivatives.

Polymer	CyD Units	Arg Units	Z Potential (mV)	Mw (Da)
PGAβCyDArg1	19 ± 1	4 ± 1	8 ± 1	25,500
PGAβCyDArg2	15 ± 1	7 ± 1	7.7 ± 0.5	21,700
PGAγCyDArg3	12 ± 1	10 ± 1	2.3 ± 0.5	21,100
PGAβCyDArg4	6 ± 1	15 ± 1	45 ± 5	13,600
PGAγCyDArg5	5 ± 1	15 ± 1	37 ± 3	13,300

The ^{13}C NMR spectra of the derivatives show signals due to guanidium carbons at about 160 ppm and signals around 174 ppm due to the carboxyl group of PGA and arginine methyl ester, in addition to the signals of CyD units in the aliphatic region.

CyD polymers were also characterised by dynamic light scattering (DLS) and Zeta potential values were also measured (Table 1). The hydrodynamic diameters increase with the number of cavities linked to the PGA backbone. The Z potential values increased when the number of Arg units increased from negative values (-58 mV for PGA alone) to positive values, in keeping with the progressive increase in positive charges due to the guanidinium groups.

Spectrometric measurements were also carried out further to characterise the structural features of the new polymers. The MALDI spectra recorded in linear mode (Figure S12) mainly contain a wideband; the m/z values of the highest peaks match to those obtained by the NMR studies (Table 1), within the experimental errors, thus confirming the calculated molecular weight (Mw) of the new CyD polymers.

As for PGAβCyDArg1, the MALDI spectrum is resolved into several components (Figure 3). The average difference between two successive relative peaks is 1280 ± 20. This value suggests that the repeat unit contains both the CyD (MW 1135) and glutamic acid (MW 147) moieties, as expected.

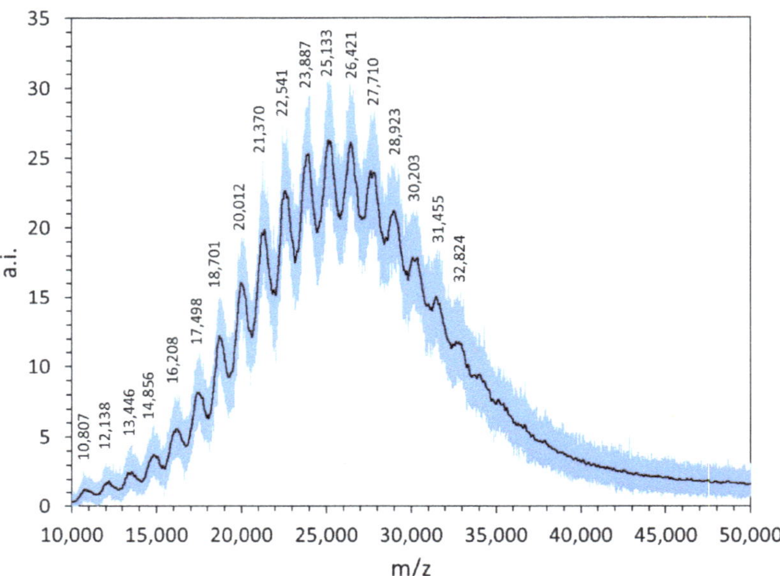

Figure 3. MALDI-TOF MS spectrum of PGAβCyDArg1. The raw spectrum (gray line) was properly smoothed (black line) in order to obtain the m/z values of all the relative peaks.

2.2. Antiaggregant Activity

The abnormal aggregation of Aβ is one of the main hallmarks of Alzheimer's disease (AD). In the pathological pathway, the amyloid peptide firstly forms soluble and highly toxic oligomers; then, the growing dimensions of the aggregated species lead to the formation of fibrillary and insoluble structures, mainly accumulated into brain plaques [28]. Finding new molecules able to inhibit the formation of amyloid-type aggregates represents an important strategy to prevent the onset of AD or attenuate the development of this devastating disorder. Therefore, we tested the effect of all the PGA polymers on the self-induced aggregation of Aβ by using a turn-on fluorescent dye Thioflavin T (ThT), sensitive to the formation of fibril species.

The fluorescence data recorded for the amyloid-type aggregation of Aβ fit to a sigmoid curve. The maximum fluorescence gain ($F_{max} - F_0$) is 33 ± 1 and the lag phase lasts 24 ± 2 h. When the compounds of interest are co-incubated with Aβ, the kinetic parameters of the aggregation process could be modified due to non-covalent interactions between Aβ and the PGA polymers. $F_{max} - F_0$ is proportional to the amount of Aβ fibrils, whereas during the lag phase (t_{lag}) only small aggregated species form. As a consequence, the lower $F_{max} - F_0$ is and/or the higher t_{lag} is, the better the antiaggregant activity.

Figure 4 shows the $F_{max} - F_0$ values obtained by the aggregation of Aβ alone (control (CTRL)) or in the presence of each PGA polymer. Several amounts of the compounds have been tested, the (Aβ)/(Polymer) molar ratio ranging from 1:1 to 1:8.

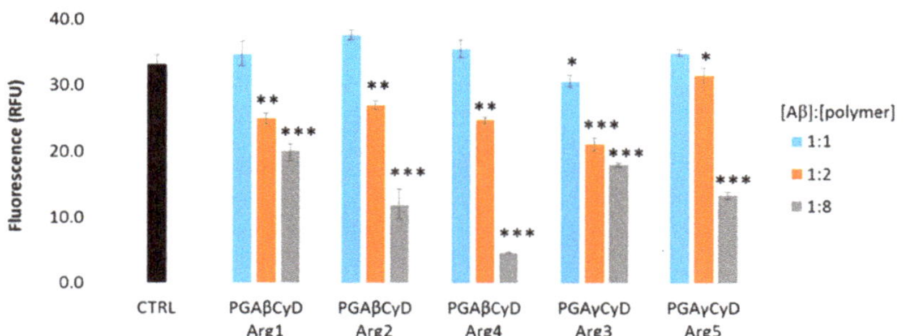

Figure 4. Maximum fluorescence gain values ($F_{max} - F_0$) of the samples containing $A\beta_{1-40}$ (20 µM) alone (control (CTRL)) or in the presence of the PGA polymers, the (Aβ)/(Polymer) molar ratio ranging from 1:1 to 1:8. (* $p < 0.05$, ** $p < 0.01$, *** $p < 0.001$ vs. CTRL, ANOVA test).

The co-incubation of any polymer with Aβ in a 1:1 molar ratio induced little or no effect on the final extent of the aggregation process. Higher amounts of the polymers significantly decrease the $F_{max} - F_0$ values and the antiaggregant activity is exerted in a dose-dependent manner, as reported in Figure 5 in the case of PGAβCyDArg2.

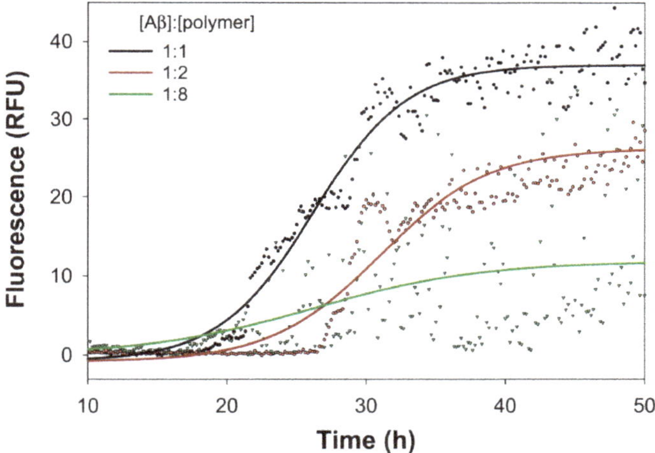

Figure 5. Representative kinetic profiles of the amyloid aggregation due to the co-incubation of $A\beta_{1-40}$ (20 µM) with PGAβCyDArg2, the (Aβ)/(Polymer) molar ratio ranging from 1:1 to 1:8. Single points represent the experimental data, whereas the straight lines are the fitted curves (adjusted R^2 is 0.9766, 0.9481 and 0.7227 for 1:1, 1:2, 1:8 (Aβ)/(Polymer) molar ratio, respectively).

As for the PGA polymers containing β-CyD (PGAβCyDArg1, PGAβCyDArg2 and PGAβCyDArg4), the antiaggregant activity is very comparable among the polymers, when the Aβ/polymer molar ratio was 1:2. Instead, these β-CyD-containing polymers differently affect the amyloid aggregation extent when the 1:8 Aβ/polymer molar ratio was tested. In particular, it seems that the greater the number of Arg units, the greater the inhibition effect of the amyloid aggregation. Such a trend is also observed when the γ-CyD-containing polymers were taken into account (PGAγCyDArg3 and PGAγCyDArg5). However, the inhibition activity of these γ-CyD polymers is slightly lower than that shown by the corresponding β-CyD polymers (PGAβCyDArg2 and PGAβCyDArg4). Such a difference could be reasonably ascribed to the structural differences between the β- and γ-CyD cavities that, in turn, could affect the non-covalent interaction between the amyloid peptide and the polymers. A similar trend has been reported for single CyDs [14].

The lag phase of the amyloid aggregation was not significantly modified by the polymers (data not shown), meaning that the interaction between the Aβ and these PGA polymers is not altered by the number of CyD and Arg units.

Above all, the antiaggregant activity towards the self-induced formation of amyloid aggregated species exerted by the PGACyD polymers could reasonably be due to the effects of both the CyD cavity and charged units (Arg and PGA). The effect of positive and negative charges of carbohydrate polymers on the on-pathway Aβ aggregation has been recently ascertained [29], thus corroborating these results.

2.3. Solubility Experiments

We explored the different affinity of the polymers for the guest DOX by solubility experiments. The effect of the polymers on the solubility can provide a comparison of the affinity for a guest. Data were reported in Figure S13. We found that water solubility of DOX (2.2×10^{-4} M) increased in the presence of all the polymers at concentration 25 mg/mL. Particularly the polymers based on γ-CyD showed an effect higher than that of β-CyD polymers at physiological (pH 7.4). PGAγCyDArg4 with more γ-CyD cavities was also more effective than that with a lower numerv of CyDs. This trend is in keeping with the highest affinity of γCyD cavities for DOX [26,30].

2.4. Antiproliferative Activity (MTT Assay)

New polymers were studied as drug delivery systems for the topoisomerase inhibitor DOX. We performed cell proliferation assays on A2780, A549 and MDA-MB-231 cancer cell lines. Complexes polymer/DOX (1:10 molar ratio) were prepared and assayed. Data obtained are reported in Table 2. Polymers alone did not show toxicity for cells (data not shown).

Table 2. Half maximal inhibitory concentration (IC_{50}) values (nM) of doxorubicin (DOX) in the presence of CyD polymers in human tumor cells.

Cell Line	PGAβCyDArg1	PGAβCyDArg2	PGAγCyDArg3	PGAβCyDArg4	PGAγCyDArg5	DOX
A2780 [a]	4.7 ± 1.7 [b]	5.9 ± 1.6	12.7 ± 2.4 [c]	10.0 ± 1.7 [d]	11.7 ± 0.4 [e]	7.7 ± 3.9
A549	55.2 ± 10.0	52.2 ± 10.1	70.0 ± 16.6	52.6 ± 4.4	56.2 ± 2.9	54.6 ± 19.2
MDA-MB-231	37.7 ± 11.8	40.9 ± 6.5	50.6 ± 22.5	60.2 ± 15.2	65.3 ± 16.7	40.9 ± 13.8

[a] $p = 0.0003$, as detected by ANOVA; [b] $p = 0.0657$ vs. DOX; [c] $p = 0.0001$ vs. PGAβCyDArg1; [d] $p = 0.0016$ vs. PGAβCyDArg1; [e] $p = 0.0007$ vs. PGAβCyDArg1, all calculated by Bonferroni/Dunn post-hoc analysis of data.

Overall, the data show that the polymers did not change the antiproliferative activity of DOX significantly. In fact, in A549 and MDA-MB-231 cell lines, the half maximal inhibitory concentration (IC_{50}) values do not change significantly depending on the type of functionalisation. Conversaly, in A2780 cells the complexes with PGAγCyDArg3 ($p = 0.0028$), and PGAβCyDArg5 (trend, $p = 0.0738$) showed higher IC_{50} values compared to free DOX.

In the case of the polymers based on γ-CyD, the higher affinity for DOX suggested by solubility data may explain the effect on the cytotoxicity, as reported for similar systems studied by us [26]. The reduction in the antiproliferative effect was observed for many systems and only in vivo studies may provide information on the potential of the drug carriers [31].

Only PGAβCyDArg1/DOX showed a slight trend towards a higher antiproliferative activity ($p = 0.0657$) than free DOX and a significant difference compared to PGAγCyDArg3, PGAγCyDArg4, and PGAβCyDArg5 (Table 2).

3. Materials and Methods

3.1. Materials

The water-soluble polymer butyl-polyglutamate (20) sodium salt (3 KDa, PGA) was acquired from IRIS Biotech gmbh. 3A-amino-3A-deoxy-2A(S),3A(R)-β-cyclodextrin (βCyD3NH2), 3A-amino-3A-deoxy-2A(S),3A(R)-γ-cyclodextrin (γCyD3NH2) and DOX were acquired from TCI (Tokyo Chemical Industry). L-Arginine methyl ester dihydrochloride was acquired by SIGMA Aldrich. Aβ_{1-40} (Bachem) was properly treated, as previously reported [32]. Inclusion complexes of DOX with CyD polymers were prepared by mixing the stock solution of DOX with solutions of CyD polymers for 2 h.

3.1.1. Synthesis of PGAβCyDArg1

βCyD3NH$_2$ (50 mg in 1 mL of H$_2$O), and DMTMM (18.26 mg in 350 μL) were added to PGA (6.61 mg in 350 μL) every 30 min in three aliquots. The pH of the reaction mixture was adjusted to 8. After 24 h, ArgOCH$_3$(2 mg) and DMTMM (18 mg) were added to the solution (during 30 min). The reaction mixture was stirred at r.t. for 24 h.

The polymer was isolated with Sephadex G-15 column chromatography. The various fractions collected were examined using TLC, (eluent PrOH/AcOEt/H$_2$O/NH$_3$ 5:2:3:1). The main product was characterised by NMR spectroscopy.

^{1}H NMR (500, in D$_2$O) δ (ppm): 5.20-4.80 (H-1 of CyD); 4.28 (s, CH Glu); 4.2 (m, CH Arg); 4.2–3.2 (m, H-3, -6, -5, -2,-4 of CyD, OCH$_3$); 3.2 (m, γ CH$_2$ Arg); 2.60–1.50 (m, β- and γ-CH$_2$ PGA); 1.35 (m, CH$_2$ butyl chain of PGA); 1.26 (m, CH$_2$ butyl chain of PGA), 0.88 (m, CH$_3$ butyl chain of PGA).

^{13}C NMR (125 MHz, in D$_2$O) δ (ppm): 24.5 (β-CH$_2$ of Arg), 26.8 (α-CH$_2$ of Arg), 31.8 (β-CH$_2$ of PGA), 40.5 (δ CH$_2$ butyl chain of PGA), 52.3 (CH Arg), 52.9 (C-2 of CyD and OCH$_3$ of Arg), 53.0 (CH of Glu), 60.0 (C-6 of CyD), 71.6 (C-3 of CyD), 73.0 (C-5 of CyD), 80.4 (C-4 of CyD), 101–105 (C-1 of CyD), 160 (C=N of Arg), 173–174 (CNH PGA-CyD, PGA-Arg), 174.72 (CO methyl ester of Arg).

Dimension (DLS, Z average): 49 ± 5; PDI (DLS): 0.4; Zeta potential: 8 ± 1 mV (pH = 7.4).

The other polymers were synthesised in the same manner as PGAβCyDArg1 with different amounts of the reagents.

3.1.2. Synthesis of PGAβCyDArg2

The synthesis was carried out as described above with PGA (7 mg), DMTMM (18 mg), ArgOCH$_3$ (6 mg), DMTMM (10 mg) and βCyD3NH$_2$ (50 mg).

^{1}H NMR (500 MHz, in D$_2$O) δ (ppm): 5.16-4.75 (H-1 of CyD); 4.28 (s, CH Glu); 4.20 (m, CH Arg); 4.10–3.20 (m, H-3, -6, -5, -2, -4 of CyD, OCH$_3$); 3.13 (m, γ-CH$_2$ Arg); 2.58–1.60 (m, γ-CH$_2$ PGA,); 1.35 (m, CH$_2$ butyl chain of PGA), 1.6 (m, CH$_2$ butyl chain of PGA); 1.28 (m, CH$_2$ butyl chain of PGA), 0.89 (m, CH$_3$ butyl chain of PGA).

Dimension (DLS, Z average): 35 ± 2 nm; PDI (DLS): 0.5; Zeta potential: 7.7 ± 0.5 mV (pH = 7.4).

3.1.3. Synthesis of PGAγCyDArg3

The synthesis was carried out as for PGAβCyDArg2 with PGA (10 mg), DMTMM (19 mg), ArgOCH$_3$ (8 mg), DMTMM (20 mg) and γCyDNH$_2$ (67 mg).

^{1}H NMR (500 MHz, in D$_2$O) δ (ppm): 5.20–4.77 (m, H-1 of CyD); 4.32 (m, CH Arg); 4.23 (s, CH Glu); 4.12 (m, H-3A of CyD); 3.93–3.49 (m, H-3, -6, -5, -2, -4 of CyD and OCH$_3$ of Arg); 3.12 (γ-CH$_2$ Arg); 2.57–1.45 (β- and δ- CH$_2$ PGA, CH$_2$ Arg); 1.38 (m, CH$_2$ butyl chain of PGA), 1.21 (m, CH$_2$ butyl chain of PGA); 0.77 (m, CH$_3$ butyl chain of PGA).

Dimension (DLS, Z average): 29 ± 3 nm; PDI (DLS): 0.4; Zeta potential: 2.3 ± 0.5 mV (pH = 7.4).

3.1.4. Synthesis of PGAβCyDArg4 and PGAγCDArg5

The synthesis was carried out as described above with PGA (25 mg), DMTMM (48 mg), ArgOCH$_3$ (33 mg), DMTMM (21 mg) and βCyD3NH$_2$ (59 mg) or γCyD3NH$_2$ (72 mg).

PGAβCyDArg4

^{1}H NMR (500 MHz, in D$_2$O) δ (ppm): 5.22–4.80 (m, H-1 of CyD); 4.32 (m, CH Arg); 4.23 (s, CH Glu); 4.12 (m, H-3A of CyD); 3.93–3.50 (m, H-3, -6, -5, -2, -4 of CyD and OCH$_3$

of Arg); 3.12 (γ-CH$_2$ Arg); 2.60–1.49 (β- and δ- CH$_2$ PGA, CH$_2$ Arg); 1.39 (m, CH$_2$ butyl chain of PGA), 1.20 (m, CH$_2$ butyl chain of PGA); 0.78 (m, CH$_3$ butyl chain of PGA).

^{13}C NMR (125 MHz, in D$_2$O) δ (ppm): 24.4 (β-CH$_2$ of Arg), 27.6 (α-CH$_2$ of Arg), 31.3 (β-CH$_2$ of PGA), 40.5 (δ CH$_2$ butyl chain of PGA), 52.34 (CH Arg), 52.8 (C-2 of CyD and OCH$_3$ of Arg), 52.9 (CH of Glu), 60.0 (C-6 of CyD), 71.4 (C-3 of CyD), 72.5 (C-5 of CyD), 80.3 (C-4 of CyD), 101. 7 (C-1 of CyD), 160 (C=N of Arg), 173–174 (C-NH PGA-CyD, PGA-Arg), 174.7 (CO methyl ester of Arg).

Dimension (DLS, Z average): 79 $\pm$ 8 nm; PDI (DLS): 0.6; Zeta potential: 45 $\pm$ 5 mV (pH = 7.4).

PGAγCyDArg5

^{1}H NMR (500 MHz, in D$_2$O) δ (ppm): 5.22–4.70 (m, H-1 of CyD); 4.32 (m, CH Arg); 4.23 (s, CH Glu); 4.12 (m, H-3-A of CyD); 4.05–3.32 (m, H-3, -6, -5, -2, -4 of CyD and OCH$_3$ of Arg); 3.12 (γ-CH$_2$ Arg); 2.59-1.46 (β- and δ- CH$_2$ PGA, CH$_2$ Arg); 1.37 (m, CH$_2$ butyl chain of PGA), 1.21 (m, CH$_2$ butyl chain of PGA); 0.78 (m, CH$_3$ butyl chain of PGA).

^{13}C NMR (125 MHz, in D$_2$O) δ (ppm): 24.4 (β-CH$_2$ of Arg), 27.7 (α-CH$_2$ of Arg), 31.3 (β-CH$_2$ of PGA), 40.4 (δ CH$_2$ butyl chain of PGA), 52.4 (CH Arg), 52.8 (C-2 of CyD and OCH$_3$ of Arg), 52.9 (CH of Glu), 60.0 (C-6 of CyD), 71.5 (C-3 of CyD), 72.0 (C-5 of CyD), 80.4 (C-4 of CyD), 101. 7 (C-1 of CyD), 160 (C=N of Arg), 173–174 (CNH PGA-CyD, PGA-Arg), 174.72 (CO methyl ester of Arg).

Dimension (DLS, Z average): 59 $\pm$ 6 nm; PDI (DLS): 0.6; Zeta potential: 37 $\pm$ 3 mV (pH = 7.4).

3.2. Instrumentation

^{1}H and ^{13}C NMR spectra were recorded at 25 °C with a VARIAN UNITY PLUS-500 spectrometer at 499.9 and 125.7 MHz, respectively, using Varian library standard pulse programs. All samples were prepared in deuterated solvents (D$_2$O); ^{1}H NMR spectra were referred to the HOD signal and ^{13}C NMR spectra to acetone (external reference). In all the experiments, the pulse at 90° lasted about 7 µs. 2D experiments (COSY, HSQC and HMBC) were acquired using 1K data points and 256 increments.

UV-Vis spectra were recorded with a VersaWave microvolume UV/Vis spectrophotometer (Expedeon, Ottawa, ON, Canada). The molar absorptivity of DOX 10,410 (mol^{-1} L cm^{-1}) at 482 nm was used.

3.2.1. Dynamic Light Scattering and Zeta Potential Measurements

Dynamic light scattering (DLS) and zeta potential (ZP) measurements were performed at 25 °C with a Zetasizer Nano ZS (Malvern Instruments, Oxford, UK) operating at 633 nm (He–Ne laser). The mean hydrodynamic diameter (d) of the NPs was calculated from intensity measurement after averaging the five measurements. The samples (1 mg/mL) were diluted in phosphate buffer (pH = 7.4) prepared in ultrapure water filtered (0.2 µm).

3.2.2. Mass Spectrometry

MALDI-TOF MS experiments were performed using an AB SCIEX MALDI-TOF/TOF 5800 Analyzer (AB SCIEX, Foster City, CA, USA) equipped with a nitrogen UV laser (λ = 337 nm) pulsed at a 20 Hz frequency by using a set up previously described [26]. Briefly, the mass spectrometer operated in the linear mode and the laser intensity set above the ionisation threshold (4500 in arbitrary units). Mass spectra were processed using Data Explorer 4.11 software (Applied Biosystems, Warrington, UK). 2,5-di-hydroxybenzoic acid (DHB) was used as the matrix, dissolved in water/acetonitrile 1:1 containing 0.03% of CF$_3$COOH. Molar-mass averages (M$_n$ and M$_w$) values were also calculated using Data Explorer software (Applied Biosystems, Warrington, UK).

3.3. Cell Culture and Antiproliferative Assay

A2780 (ovarian carcinoma), A549 (lung carcinoma) and MDA-MB-231 (breast carcinoma) cells (all obtained from ICLC, Genova, Italy) were grown as monolayers in Roswell Park Memorial Institute (RPMI 1640) or Dulbecco's Modified Eagle's Medium (DMEM) media (EuroClone, Pero, Italy) supplemented with 10% fetal bovinum serum (FBS) (Euroclone), antibiotics (EuroClone), and non-essential amino-acids (only DMEM, EuroClone). For the assay, cells plated into flat-bottomed 96-well microtiter plates were treated after 6–8 h with the complexes (five 1:5 scalar solutions, 20 µL, starting from 1 µM concentration). Seventy-two hours later, cells were analysed by the 3-(4,5-dimethylthiazol-2-yl)-2,5-diphenyltetrazolium Bromide (MTT) assay as described elsewhere [33].

IC_{50} values were calculated from the analysis of single concentration–response curves. Final values are the mean of 4–12 experiments.

3.4. Aβ Aggregation Assay

The antiaggregant effect of the newly synthesised compounds on the self-induced aggregation process of Aβ was assayed as previously reported [32]. Briefly, Aβ (20 µM), ThT (60 µM) and the compounds of interest were incubated in phosphate buffered saline (pH 7.4) at 37 °C in a multiplate reader (Varioskan Flash, Thermo Scientific, Leiden, The Netherlands). The fluorimetric readings (excitation and emission wavelengths were 450 nm and 480 nm, respectively) were collected every 10 min up to 60 h. Data of all the measurements, carried out in triplicate, were fitted to Equation (1).

$$F(t) = F_0 + \frac{F_{max} - F_0}{1 + e^{\frac{t-t}{k}}} \tag{1}$$

$F_{max} - F_0$ is the higher fluorescence increment recorded all over the aggregation process; the lag phase (t_{lag}) is the time interval preceding the formation of amyloid-type species sensitive to ThT. The t_{lag} values were calculated by using Equation (2)

$$t_{lag} = t - 2/k \tag{2}$$

The parameters of each set of measurements were expressed as the mean ± SD.

3.5. Solubility Experiments

DOX hydrochloride (0.017 M, water solution) was added to the solutions containing different concentrations (25 mg/mL, 12 mg/ML, 6 mg/mL) of all the polymers in phosphate buffer (50 mM, pH 7.4). The suspensions were sonicated for 3 min, incubated at 25 °C in the dark and centrifugated after 2 h. DOX was determined in the supernatant with UV/Vis spectroscopy, at 482 nm.

3.6. Statistical Analysis

For statistical analysis the one way ANOVA was used followed by the post-hoc Bonferroni/Dunn analysis of data.

Supplementary Materials: The following are available online, Figures S1–S11: NMR spectra of the polymers, Figure S12: MALDI-TOF MS spectra of the PGACyDArg polymers, Figure S13: Solubility of DOX in the presence of the PGACyDArg polymers.

Author Contributions: Conceptualisation, G.V., methodology, N.B. (Noemi Bognanni), F.B., M.V., N.B. (Nadia Bertola), G.V.; formal analysis, N.B. (Noemi Bognanni); investigation, G.V., N.B. (Noemi Bognanni), F.B., N.B. (Nadia Bertola) and M.V.; data curation, G.V., N.B. (Noemi Bognanni), F.B., N.B. (Nadia Bertola) and M.V.; writing—original draft preparation, G.V., N.B. (Noemi Bognanni), F.B., and M.V.; writing—review and editing, G.V., N.B. (Noemi Bognanni), F.B. and M.V.; project administration, G.V., funding acquisition, M.V. and G.V. All authors have read and agreed to the published version of the manuscript.

Funding: This research received no external funding.

Institutional Review Board Statement: Not applicable.

Informed Consent Statement: Not applicable.

Data Availability Statement: Please refer to suggested Data Availability Statements in the section "MDPI Research Data Policies" at https://www.mdpi.com/ethics.

Acknowledgments: The authors acknowledge support from Università degli Studi di Catania (Piano di incentivi per la ricerca di Ateneo 2020/2022 (Pia.ce.ri.) and from Italian Ministry of Health (Ricerca Corrente).

Conflicts of Interest: The authors declare no conflict of interest.

Abbreviations

A2780	Human ovarian carcinoma
A549	Human Caucasian lung carcinoma
Aβ	Amyloid beta
AD	Alzheimer's disease ArgOCH$_3$, arginine methyl ester
CyD	Cyclodextrin
CTRL	control
DHB	2,5-di-hydroxybenzoic acid
DLS	Dynamic light scattering
DMEM	Dulbecco's Modified Eagle's Medium
DMTMM	4-(4,6-Dimethoxy-1,3,5-triazin 2-yl)-4-methylmorpholinium chloride
DOX	Doxorubicin F
FBS	Fetal Bovinum Serum
IC$_{50}$	half maximal inhibitory concentration
MTT	3-(4,5-dimethylthiazol-2-yl)-2,5-diphenyltetrazolium Bromide
MDA-MB-231	Human Caucasian breast adenocarcinoma
PGA	N-butyl-polyglutamate
PGACyDArg	N-butyl-polyglutamate Cyclodextrin Arginine
RPMI	Roswell Park Memorial Institute
ThT	Thioflavin T

References

1. Oliveri, V.; Vecchio, G. Metallocyclodextrins in Medicinal Chemistry. *Future Med. Chem.* **2018**, *10*, 663–677. [CrossRef]
2. Jansook, P.; Ogawa, N.; Loftsson, T. Cyclodextrins: Structure, Physicochemical Properties and Pharmaceutical Applications. *Int. J. Pharm.* **2018**, *535*, 272–284. [CrossRef] [PubMed]
3. Muankaew, C.; Loftsson, T. Cyclodextrin-Based Formulations: A Non-Invasive Platform for Targeted Drug Delivery. *Basic Clin. Pharmacol. Toxicol.* **2018**, *122*, 46–55. [CrossRef]
4. Garrido, P.F.; Calvelo, M.; Blanco-González, A.; Veleiro, U.; Suárez, F.; Conde, D.; Cabezón, A.; Piñeiro, Á.; Garcia-Fandino, R. The Lord of the NanoRings: Cyclodextrins and the Battle against SARS-CoV-2. *Int. J. Pharm.* **2020**, *588*. [CrossRef]
5. Jacob, S.; Nair, A.B. Cyclodextrin Complexes: Perspective from Drug Delivery and Formulation. *Drug Dev. Res.* **2018**, *79*, 201–217. [CrossRef] [PubMed]
6. Santos, A.C.; Costa, D.; Ferreira, L.; Guerra, C.; Pereira-Silva, M.; Pereira, I.; Peixoto, D.; Ferreira, N.R.; Veiga, F. Cyclodextrin-Based Delivery Systems for in Vivo-Tested Anticancer Therapies. *Drug Deliv. Transl. Res.* **2021**, *11*, 49–71. [CrossRef]
7. Zhang, D.; Lv, P.; Zhou, C.; Zhao, Y.; Liao, X.; Yang, B. Cyclodextrin-Based Delivery Systems for Cancer Treatment. *Mater. Sci. Eng. C* **2019**, *96*, 872–886. [CrossRef]
8. Hoang Thi, T.T.; Du Cao, V.; Nguyen, T.N.Q.; Hoang, D.T.; Ngo, V.C.; Nguyen, D.H. Functionalized Mesoporous Silica Nanoparticles and Biomedical Applications. *Mater. Sci. Eng. C* **2019**, *99*, 631–656. [CrossRef]
9. Xu, S.; Wang, P.; Sun, Z.; Liu, C.; Lu, D.; Qi, J.; Ma, J. Dual-Functionalization of Polymeric Membranes via Cyclodextrin-Based Host-Guest Assembly for Biofouling Control. *J. Memb. Sci.* **2019**, *569*, 124–136. [CrossRef]
10. Kulkarni, A.; Caporali, P.; Dolas, A.; Johny, S.; Goyal, S.; Dragotto, J.; Macone, A.; Jayaraman, R.; Fiorenza, M.T. Linear Cyclodextrin Polymer Prodrugs as Novel Therapeutics for Niemann-Pick Type C1 Disorder. *Sci. Rep.* **2018**, *8*, 9547. [CrossRef]
11. Liao, R.; Lv, P.; Wang, Q.; Zheng, J.; Feng, B.; Yang, B. Cyclodextrin-Based Biological Stimuli-Responsive Carriers for Smart and Precision Medicine. *Biomater. Sci.* **2017**, *5*, 1736–1745. [CrossRef]
12. van de Manakker, F.; Vermonden, T.; van Nostrum, C.F.; Hennink, W.E. Cyclodextrin-Based Polymeric Materials: Synthesis, Properties, and Pharmaceutical/Biomedical Applications. *Biomacromolecules* **2009**, *10*, 3157–3175. [CrossRef]

13. Heidel, J.D.; Schluep, T. Cyclodextrin-Containing Polymers: Versatile Platforms of Drug Delivery Materials. *J. Drug Deliv.* **2012**, *2012*, 1–17. [CrossRef] [PubMed]
14. Serno, T.; Geidobler, R.; Winter, G. Protein Stabilization by Cyclodextrins in the Liquid and Dried State. *Adv. Drug Deliv. Rev.* **2011**, *63*, 1086–1106. [CrossRef]
15. Otzen, D.E.; Knudsen, B.R.; Aachmann, F.; Larsen, K.L.; Wimmer, R. Structural Basis for Cyclodextrins' Suppression of Human Growth Hormone Aggregation. *Protein Sci.* **2009**, *11*, 1779–1787. [CrossRef] [PubMed]
16. Oliveri, V.; Vecchio, G. Cyclodextrins as Protective Agents of Protein Aggregation: An Overview. *Chem. Asian J.* **2016**, *11*, 1648–1657. [CrossRef] [PubMed]
17. Oliveri, V.; Bellia, F.; Pietropaolo, A.; Vecchio, G. Unusual Cyclodextrin Derivatives as a New Avenue to Modulate Self- and Metal-Induced Aβ Aggregation. *Chem. Eur. J.* **2015**, *21*, 14047–14059. [CrossRef]
18. Wahlström, A.; Cukalevski, R.; Danielsson, J.; Jarvet, J.; Onagi, H.; Rebek, J.; Linse, S.; Gräslund, A. Specific Binding of a β-Cyclodextrin Dimer to the Amyloid β Peptide Modulates the Peptide Aggregation Process. *Biochemistry* **2012**, *51*, 4280–4289. [CrossRef] [PubMed]
19. Wang, Z.; Chang, L.; Klein, W.L.; Thatcher, G.R.J.; Venton, D.L. Per-6-Substituted-per-6-Deoxy β-Cyclodextrins Inhibit the Formation of β-Amyloid Peptide Derived Soluble Oligomers. *J. Med. Chem.* **2004**, *47*, 3329–3333. [CrossRef]
20. Giglio, V.; Bellia, F.; Oliveri, V.; Vecchio, G. Aminocyclodextrin Oligomers as Protective Agents of Protein Aggregation. *Chempluschem* **2016**, *81*, 660–665. [CrossRef] [PubMed]
21. Zou, S.; Wang, X.; Liu, P.; Ke, C.; Xu, S. Arginine Metabolism and Deprivation in Cancer Therapy. *Biomed. Pharmacother.* **2019**, *118*, 109210. [CrossRef]
22. Fuchs, S.M.; Rutkoski, T.J.; Kung, V.M.; Groeschl, R.T.; Raines, R.T. Increasing the Potency of a Cytotoxin with an Arginine Graft. *Protein Eng. Des. Sel.* **2007**, *20*, 505–509. [CrossRef]
23. Mizusako, H.; Tagami, T.; Hattori, K.; Ozeki, T. Active Drug Targeting of a Folate-Based Cyclodextrin-Doxorubicin Conjugate and the Cytotoxic Effect on Drug-Resistant Mammary Tumor Cells In Vitro. *J Pharm Sci* **2015**, *104*, 2934–2940. [CrossRef]
24. Viale, M.; Vecchio, G.; Monticone, M.; Bertone, V.; Giglio, V.; Maric, I.; Cilli, M.; Bocchini, V.; Profumo, A.; Ponzoni, M.; et al. Fibrin Gels Entrapment of a Poly-Cyclodextrin Nanocarrier as a Doxorubicin Delivery System in an Orthotopic Model of Neuroblastoma: Evaluation of In Vitro Activity and In Vivo Toxicity. *Pharm. Res.* **2019**, *36*, 1–12. [CrossRef]
25. Tian, B.; Hua, S.; Liu, J. Cyclodextrin-Based Delivery Systems for Chemotherapeutic Anticancer Drugs: A Review. *Carbohydr. Polym.* **2020**, *232*, 115805. [CrossRef]
26. Oliveri, V.; Bellia, F.; Viale, M.; Maric, I.; Vecchio, G. Linear Polymers of β and γ Cyclodextrins with a Polyglutamic Acid Backbone as Carriers for Doxorubicin. *Carbohydr. Polym.* **2017**, *177*, 355–360. [CrossRef] [PubMed]
27. Giachino, C.; Viale, M.; Vecchio, G. Exploring the Functionalization of Polymeric Nanoparticles Based on Cyclodextrins for Tumor Cell Targeting. *ChemistrySelect* **2019**, *4*, 13025–13028. [CrossRef]
28. Haass, C.; Selkoe, D.J. Soluble Protein Oligomers in Neurodegeneration: Lessons from the Alzheimer's Amyloid β-Peptide. *Nat. Rev. Mol. Cell Biol.* **2007**, *8*, 101–112. [CrossRef]
29. Greco, V.; Naletova, I.; Ahmed, I.M.M.; Vaccaro, S.; Messina, L.; La Mendola, D.; Bellia, F.; Sciuto, S.; Satriano, C.; Rizzarelli, E. Hyaluronan-Carnosine Conjugates Inhibit Aβ Aggregation and Toxicity. *Sci. Rep.* **2020**, *10*, 1–14. [CrossRef] [PubMed]
30. Anand, R.; Malanga, M.; Manet, I.; Manoli, F.; Tuza, K.; Aykaç, A.; Ladavière, C.; Fenyvesi, E.; Vargas-Berenguel, A.; Gref, R.; et al. Citric Acid-γ-Cyclodextrin Crosslinked Oligomers as Carriers for Doxorubicin Delivery. *Photochem. Photobiol. Sci.* **2013**, *12*, 1841–1854. [CrossRef] [PubMed]
31. Giglio, V.; Viale, M.; Bertone, V.; Maric, I.; Vaccarone, R.; Vecchio, G. Cyclodextrin Polymers as Nanocarriers for Sorafenib. *Investig. New Drugs* **2018**, *36*, 370–379. [CrossRef] [PubMed]
32. Grasso, G.I.; Bellia, F.; Arena, G.; Satriano, C.; Vecchio, G.; Rizzarelli, E. Multitarget Trehalose-Carnosine Conjugates Inhibit Aβ Aggregation, Tune Copper(II) Activity and Decrease Acrolein Toxicity. *Eur. J. Med. Chem.* **2017**, *135*, 447–457. [CrossRef] [PubMed]
33. Viale, M.; Cordazzo, C.; de Totero, D.; Budriesi, R.; Rosano, C.; Leoni, A.; Ioan, P.; Aiello, C.; Croce, M.; Andreani, A.; et al. Inhibition of MDR1 Activity and Induction of Apoptosis by Analogues of Nifedipine and Diltiazem: An in Vitro Analysis. *Investig. New Drugs* **2011**, *29*, 98–109. [CrossRef] [PubMed]

Article

Risperidone/Randomly Methylated β-Cyclodextrin Inclusion Complex—Compatibility Study with Pharmaceutical Excipients

Laura Sbârcea [1,2], Ionuț-Mihai Tănase [3], Adriana Ledeți [1,2,*], Denisa Cîrcioban [1,2], Gabriela Vlase [4], Paul Barvinschi [5], Marinela Miclău [6], Renata-Maria Văruţ [7], Oana Suciu [8,*] and Ionuţ Ledeți [1,2,3]

[1] Department Pharmacy I, Faculty of Pharmacy, "Victor Babeş" University of Medicine and Pharmacy, 2 Eftimie Murgu Square, 300041 Timisoara, Romania; sbarcea.laura@umft.ro (L.S.); circioban.denisa@umft.ro (D.C.); ionut.ledeti@umft.ro (I.L.)

[2] Advanced Instrumental Screening Center, Faculty of Pharmacy, "Victor Babes" University of Medicine and Pharmacy, Romania, 2 Eftimie Murgu Square, 300041 Timisoara, Romania

[3] Faculty of Industrial Chemistry and Environmental Engineering, Politehnica University of Timisoara, Vasile Parvan Street 6, 300223 Timisoara, Romania; ionut.tanase@student.upt.ro

[4] Research Centre for Thermal Analysis in Environmental Problems, West University of Timisoara, Pestalozzi Street 16, 300115 Timisoara, Romania; gabriela.vlase@e-uvt.ro

[5] Faculty of Physics, West University of Timisoara, 4 Vasile Parvan Blvd, 300223 Timisoara, Romania; pc_barvi@yahoo.fr

[6] National Institute for Research and Development in Electrochemistry and Condensed Matter, Dr. A. Păunescu-Podeanu 144, 300587 Timisoara, Romania; marinela.miclau@gmail.com

[7] Department of Physical-Chemistry, University of Medicine and Pharmacy Craiova, 2–4 Petru Rares Str., 200349 Craiova, Romania; rennata_maria@yahoo.com

[8] Department of Medicine XIV, Faculty of Medicine, "Victor Babeş" University of Medicine and Pharmacy, 2 Eftimie Murgu Square, 300041 Timisoara, Romania

* Correspondence: afulias@umft.ro (A.L.); suciu.oana@umft.ro (O.S.)

Citation: Sbârcea, L.; Tănase, I.-M.; Ledeți, A.; Cîrcioban, D.; Vlase, G.; Barvinschi, P.; Miclău, M.; Văruţ, R.-M.; Suciu, O.; Ledeţi. I. Risperidone/Randomly Methylated β-Cyclodextrin Inclusion Complex—Compatibility Study with Pharmaceutical Excipients. *Molecules* **2021**, *26*, 1690. https://doi.org/10.3390/molecules26061690

Academic Editors: Marina Isidori, Margherita Lavorgna and Rosa Iacovino

Received: 15 February 2021
Accepted: 15 March 2021
Published: 17 March 2021

Publisher's Note: MDPI stays neutral with regard to jurisdictional claims in published maps and institutional affiliations.

Abstract: Risperidone (RSP) is an atypical antipsychotic drug used in treating schizophrenia, behavioral, and psychological symptoms of dementia and irritability associated with autism. The drug substance is practically insoluble in water and exhibits high lipophilicity. It also presents incompatibilities with pharmaceutical excipients such as magnesium stearate, lactose, and cellulose microcrystalline. RSP encapsulation by randomly methylated β-cyclodextrin (RM-β-CD) was performed in order to enhance drug solubility and stability and improve its biopharmaceutical profile. The inclusion complex formation was evaluated using thermal methods, powder X-ray diffractometry (PXRD), universal-attenuated total reflectance Fourier transform infrared (UATR-FTIR), UV spectroscopy, and saturation solubility studies. The 1:1 stoichiometry ratio and the apparent stability constant of the inclusion complex were determined by means of the phase solubility method. The compatibility between the supramolecular adduct and pharmaceutical excipients starch, anhydrous lactose, magnesium stearate, and cellulose microcrystalline was studied employing thermoanalytical tools (TG-thermogravimetry/DTG-derivative thermogravimetry/HF-heat flow) and spectroscopic techniques (UATR-FTIR, PXRD). The compatibility study reveals that there are no interactions between the supramolecular adduct with starch, magnesium stearate, and cellulose microcrystalline, while incompatibility with anhydrous lactose is observed even under ambient conditions. The supramolecular adduct of RSP with RM-β-CD represents a valuable candidate for further research in developing new formulations with enhanced bioavailability and stability, and the results of this study allow a pertinent selection of three excipients that can be incorporated in solid dosage forms.

Keywords: risperidone; inclusion complex; randomly methylated β-cyclodextrin; compatibility studies; excipients; solubility; stability

1. Introduction

Risperidone (RSP), 3-[2-[4-(6-fluoro-1,2-benzoxazol-3-yl)piperidin-1-yl]ethyl]-2-methyl-6,7,8,9-tetrahydropyrido[1,2-a]pyrimidin-4-one (Figure 1), is a benzoxazole derivative used in treating schizophrenia, behavioral, and psychological symptoms of dementia and irritability associated with autism [1,2]. This atypical antipsychotic drug blocks the serotonin-2 (5TH2) and dopamine-2 (D2) receptors in the brain. RSP is a member of the class of pyridopyrimdines that is practically insoluble in water and presents high lipophilicity (log *P* of 3.49), which makes it a class II candidate of the Biopharmaceutics Classification System (BCS) [1,3,4]. It exhibits the tendency of forming polymorphs [1], which could present different solubility, dissolution rates, and stability, leading to variations in the biopharmaceutical profile of its drug substance [5].

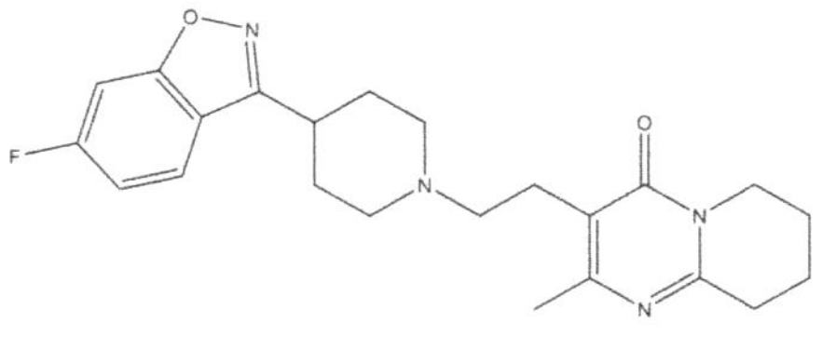

Figure 1. Chemical structure of risperidone.

Solubility is an essential feature of drugs, being one of the most critical and important characteristics that influence their bioavailability. Among the various approaches used to enhance the solubility of BCS II class drugs, encapsulation of the active pharmaceutical substances in cyclodextrins is a valuable strategy [5].

Cyclodextrins (CDs) are cyclic glucose oligomers consisting of six, seven, or eight glucose units (α, β, and γ-cyclodextrin), linked by 1,4-α-glycosidic bonds. CDs possess a hydrophobic internal cavity that provides a microenvironment for appropriate sized molecules and a hydrophilic outer surface responsible for their water-solubility. This particular structure of CDs confers them multiple applications in the pharmaceutical field, food, cosmetics, textile, and chemistry industry based on their property of forming guest–host inclusion complexes [6–10]. The inclusion complexation leads to an increase in the solubility of insoluble drug substances, including the antiviral drug remdesivir [11] to improve the chemical stability, the biological activity, and the bioavailability of guest molecules, to prevent drug–excipient or drug–drug interactions, to reduce/eliminate the unpleasant taste or odors and also ocular and gastrointestinal irritation [10,12–22]. Therefore, the encapsulation of the drug in the CD cavity results in a remarkable improvement of physicochemical, biopharmaceutical properties, and therapeutic potential of the guest [23–25]. Despite its low aqueous solubility, β-cyclodextrin (β-CD) has achieved pharmaceutical relevance due to its availability, lack of toxicity, appropriate internal cavity size for a wide variety of drug substances, and economic advantages [6,21,26]. The random substitution of any β-CD hydroxyl group creates a disruption of stable hydrogen bond system around the CD rim generating an intensive enhancement of its aqueous solubility. Thus, several CD derivatives of pharmaceutical interest have been developed, among them methylated β-CD [27,28].

Several papers have reported the interaction between RSP and CDs such as β-CD, hydroxypropyl-β-CD (HP-β-CD), and methyl-β-CD, in solid state and in solution [4,29,30]. In addition, the solubility of RSP in aqueous solution of α-, β-, γ- and HP-β-CDs has been evaluated [31]. In our recent paper, we have investigated in both solution and solid state the encapsulation of RSP in two methylated CDs, heptakis(2,6-di-O-methyl)-β-cyclodextrin (DM-β-CD) and heptakis(2,3,6-tri-O-methyl)-β-cyclodextrin (TM-β-CD) [22].

The efficiency, safety, quality, and stability of pharmaceutical dosage forms, which are the result of the active pharmaceutical ingredients (API) combination with excipients, are of major importance in the drug development process. A proper formulation design involves the selection of suitable excipients; although these pharmacologically-inactive substances are considered inert molecules, during the formulation stage and/or under storage of final product, interactions may occur even in solid state, leading to a diminution of concentration of API [32–34]. Potential interactions between an API and excipients have to be evaluated because the incompatibilities between the components of a dosage form can affect the bioavailability, stability, potency, and safety of drug products [35,36]. According to the International Conference on Harmonisation (ICH) Q8 recommendations, a drug substance/excipient compatibility study should be evaluated as a part of pharmaceutical development [37].

Within this framework, the aim of this study was to investigate the encapsulation of RSP in randomly methylated β-CD (RM-β-CD) and to evaluate the compatibility of supramolecular adduct with selected pharmaceutical excipients. According to our knowledge, there is no study focused on RSP inclusion complex compatibility with excipients. In the present paper, the RSP/RM-β-CD inclusion complex has been characterized using solubility studies, thermal methods, spectroscopic techniques, and molecular modeling studies. Later, the interaction between the binary system and excipients, namely starch (STR), microcrystalline cellulose (CEL), magnesium stearate (MgSTE), and anhydrous lactose (LCT) has been studied by means of thermoanalytical tools (TG—thermogravimetry/DTG—derivative thermogravimetry/HF—heat flow), powder X-ray diffractometry (PXRD), and universal attenuated total reflectance Fourier transform IR spectroscopy (UATR-FTIR).

2. Results and Discussion

2.1. Inclusion Complex Characterization

2.1.1. Phase Solubility Studies

The stoichiometry of the RSP/RM-β-CD inclusion complex and its stability constant were investigated by means of phase solubility studies. The phase solubility diagram of RSP with RM-β-CD was obtained at 25 °C by plotting the apparent equilibrium concentration of the drug against RM-β-CD concentration. The apparent solubility of RSP increased linearly ($R^2 = 0.9980$) as a function of RM-β-CD concentration in the 0–50 mM concentration range, as shown in Figure 2. The RSP solubility enhancement in phosphate buffer 0.1 M (pH 7.4) confirms the interaction between the drug substance and the CD. The linear relation between RSP concentration and RM-β-CD concentration indicates an A_L type phase solubility diagram defined by Higuchi and Connors [38]; also, the slope value (0.1965) is less than unity, revealing that a soluble inclusion complex in 1:1 molar ratio was formed between the guest and host molecule in phosphate buffer 0.1 M (pH 7.4). The apparent stability constant ($K_{1:1}$) of the inclusion complex calculated from the slope of the phase solubility diagram, using Equation (1) is 173.38 ± 5.54 M^{-1}; this value is within the range of 100 and 5000, which is considered ideal for the formation of an inclusion complex that may improve the bioavailability profile [39,40].

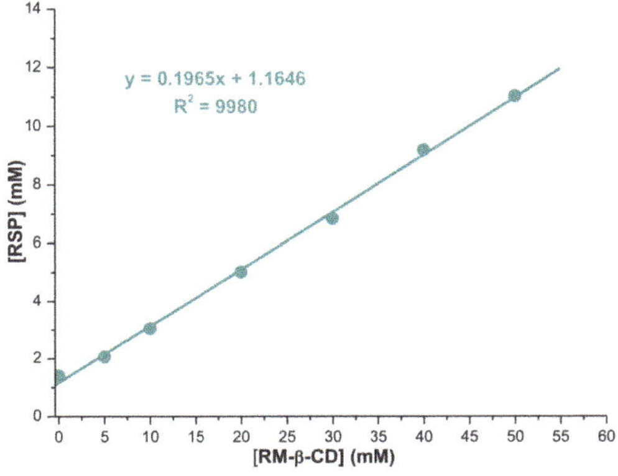

Figure 2. Phase solubility diagram of risperidone (RSP) in the presence of randomly methylated β-cyclodextrin (RM-β-CD) in phosphate buffer 0.1 M, pH 7.4, at 25 °C.

2.1.2. Molecular Modeling

Molecular modeling is a powerful tool employed by theoretical chemistry for quantitative predictions on guest–host interaction. The molecular docking analysis was performed using the Autodock 4.2.6 software together with the AutoDockTools [41]. The software applies a semi-empirical free energy force field and grid-based docking to assess conformations during docking process. The force field includes six pair-wise assessments (*V*) and an estimate of the conformational entropy lost upon binding ($\Delta Sconf$):

$$\Delta G = \left(V^{L-L}_{bound} - V^{L-L}_{unbound} \right) + (V^{T-T}_{bound} - V^{T-T}_{unbound}) + \left(V^{T-L}_{bound} - V^{T-L}_{unbound} + \Delta S_{conf} \right) \quad (1)$$

where *L* makes mention of the "ligand" and *T* refers to the "target" in a ligand–target docking calculation. Each of the pair-wise energetic terms includes evaluations for dispersion/repulsion, hydrogen bonding, electrostatics and desolvation [42]. Following the redocking analysis, the root-mean-square deviation (RMSD) values lower than 0.4 Å have been calculated, suggesting the robustness and repeatability of the docking analysis.

The binding free energy value calculated for RSP/RM-β-CD inclusion complex (1:1) was −3.26 kcal/mol. Figure 3 presents the theoretical RSP/RM-β-CD inclusion complex, as rendered in the PyMOL [43] and Discovery Studio molecular visualization systems, simulated in a 1:1 molar ratio.

Analyzing the 3D images of the RSP/RM-β-CD (1:1) interaction, we noticed the presence of two Pi–sigma interactions, between pyrimidin-4-one cycle and the hydrogen (hydrogen methyl group) of a glucopyranose (2.28 and 2.73 Å). Two non-classical hydrogen bonds occur between the nitrogen group from the 4,5-dihydro-isoxazole heterocycle and the hydrogen from position 4 of a carbohydrate moiety. Four non-classical hydrogen bonds occur between the 1,2-oxazole heterocycle and the carbohydrate moiety hydrogen, with lengths of 2.3 Å approximately.

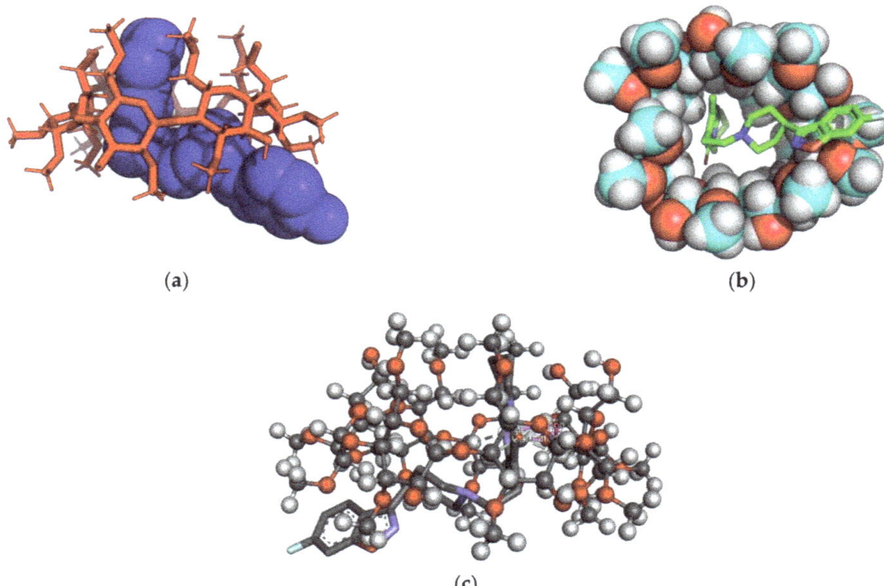

(a)

(b)

(c)

Figure 3. Molecular docking for 1:1 RSP/RM-β-CD inclusion complex. Figures (**a**,**b**) show the inclusion complex from the secondary face of RM-β-CD's cavity. The RSP molecule is represented in green and blue spheres while RM-β-CD is represented in red sticks (**a**); RSP is presented in sticks colored by element, RM-β-CD is presented in spheres colored by element (**b**); Image (**c**) show contacts between RSP and RM-β-CD, RSP is colored by element, while CD is presented in balls and sticks.

2.1.3. Thermal Analysis

In order to evaluate the interaction between RSP and RM-β-CD in solid state, the thermal behavior of parent substances, their physical mixture (PM), and kneaded product (KP) have been investigated using TG, DTG, and HF. The thermoanalytical curves of RSP, CD, RSP/RM-β-CD binary systems obtained by physical mixture (PM) and by kneading (KP) are shown in Figure 4a–d.

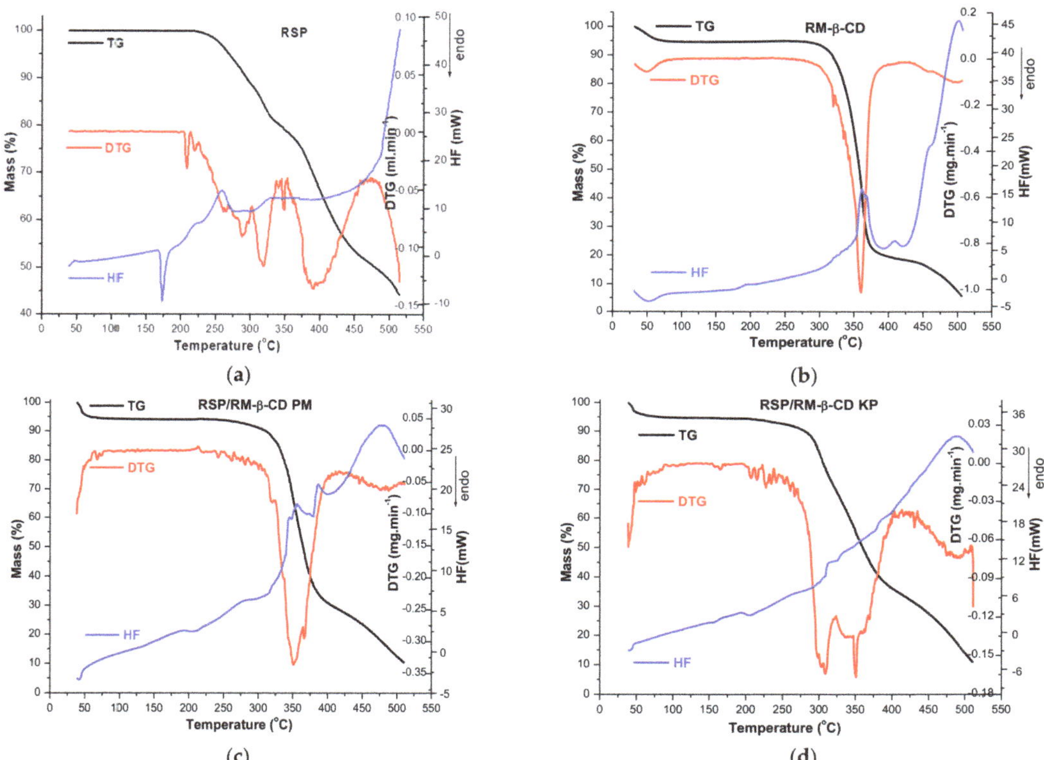

Figure 4. The thermal profile (TG-thermogravimetry/DTG-derivative thermogravimetry/HF-heat flow) of: RSP (**a**); RM-β-CD (**b**); RSP/RM-β-CD physical mixture (**c**); and kneaded product (**d**) in air atmosphere (100 mL/min), temperature range of 40–500 °C and heating rate of 10 °C/min.

Detailed interpretation of the RSP thermal profile is presented in our previous paper [22]. The drug substance exhibits thermal stability up to 206 °C, which is a temperature that marks the onset of its decomposition; then, a continuous mass loss process is observed up to 510 °C (Δm = 55.6%). As the DTG curve shows, RSP thermal decomposition takes place in three stages: the first one is noticed between 200 and 226 °C (peak at 209 °C); the second is in the temperature range of 226–350 °C (main peak at 319 °C); and the last one is between 350 and 474 °C (peak at 391 °C) [22]. RSP melting is revealed as an endothermic peak at 173 °C in the HF curve (Figure 4a) [30,44]. In addition, a small exothermic peak is noticed at 259 °C that corresponds to the first process of RSP thermal degradation; above 474 °C, the degradation occurs rapidly confirmed by the exothermic effect on HF and the rapid mass loss on the TG curve [22].

The thermoanalytical curves of RM-β-CD reveal a small mass loss (Δm = 5.3%) between 40 and 85 °C and an endothermic effect (peak at 52.0 °C) which relates to the crystallization water loss (Figure 4b). A stability stage of CD is observed in the temperature range of 85–270 °C, but above 270 °C, the mass loss continues, and the degradation process takes place as the exothermic event (peak at 361.0 °C) of the HF curve indicates [19].

The thermal curves of RSP/RM-β-CD binary systems present significant differences as compared to those of the pure substances. Both the endothermic melting peak of RSP and the RSP exothermic effect are no more present neither in the HF curve of PM nor in that of KP. In addition, the endothermic events at 361.0 °C and 500 °C from the HF curve of RM-β-CD are displaced at lower temperature in HF curves of PM (357 °C and 477 °C, Figure 4c) and KP (the first event disappeared, second at 487 °C, Figure 4d). On the other hand, the TG curves of binary systems reveal a decrease in thermal stability of RSP (Δm = 55.6%) in temperature range of 38–510 °C in both PM (Δm = 89.7%) and KP (Δm = 89.2%); this phenomenon may be a consequence of the crystallinity reduction of drug substance as a result of interaction with RM-β-CD [45]. Furthermore, the decomposition pathway of RSP in the PM and KP differs from that of pure drug substances as the DTG curves show; two distinct regions can be noticed in the DTG of KP, while the DTG of PM shows only one stage.

Thermal methods are valuable tools frequently used to investigate the interaction between CD and drug substances and to prove the encapsulation of guest molecule in the host cavity [46–48]. The guest–host interaction is characterized by changes in the thermal profiles of guest substances in the inclusion complex. The melting point of the guest molecule, which is embedded in the CD cavity, generally shifts to a different temperature and decreases its intensity or disappears [14,21,49]. The above-mentioned changes in the RSP thermal profile in the RSP/RM-β-CD binary products provide evidence for an interaction between the drug substance and CD as a result of inclusion complex formation.

2.1.4. Powder X-ray Diffractometry

The diffraction profiles of RSP, RM-β-CD, and their RSP/RM-β-CD PM and KP are depicted in Figure 5.

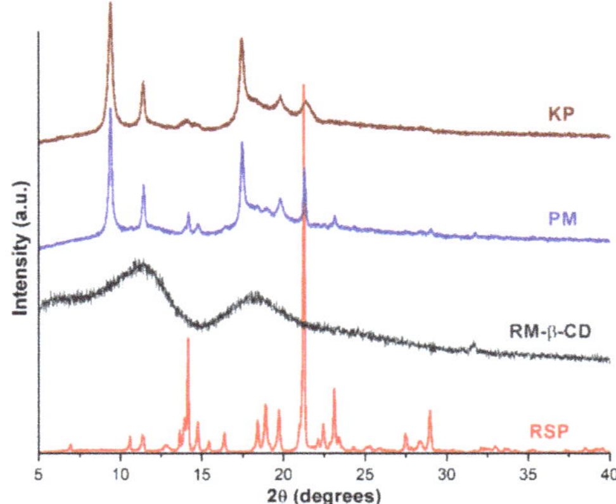

Figure 5. Powder X-ray diffractometry (PXRD) pattern of RSP, RM-β-CD, and RSP/RM-β-CD binary products physical mixture (PM) and kneaded product (KP).

The RSP crystalline nature is emphasized by the two crystalline reflections of high intensity at 14.19 and 21.27 2θ in addition to other characteristic reflections at 14.79; 16.42; 18.44; 18.93; 19.74; 23.15; and 29.00 2θ [22]. The diffraction pattern of RM-β-CD reveals two broad peaks and many undefined, diffused peaks of low intensities, reflecting its amorphous nature [19,50]. The PXRD pattern of both RSP/RM-β-CD PM and KP present a marked diminution of RSP characteristic diffraction peaks along with the disappearance of several RSP characteristic reflections in the diffractogram of both PM (at 16.42; 18.44;

18.93 2θ) and KP (at 14.79; 16.42; 18.44; 18.93; 23.15 and 29.00 2θ), which indicate a reduction in drug crystallinity in the binary products. Furthermore, new peaks are observed in the diffraction patterns of the RSP/RM-β-CD binary systems both PM (at 9.42; 11.43; 17.40 2θ) and KP (at 9.42; 11.42; 17.44 2θ). These data suggest that interaction occurs between the drug substance and the CD and demonstrate the inclusion complex formation in solid state, confirming the results obtained using thermal methods.

2.1.5. UATR-FTIR Spectroscopy

Universal-attenuated total reflectance Fourier transform infrared (UATR-FTIR) spectra of RSP, RM-β-CD, and their corresponding PM and KP are presented in Figure 6.

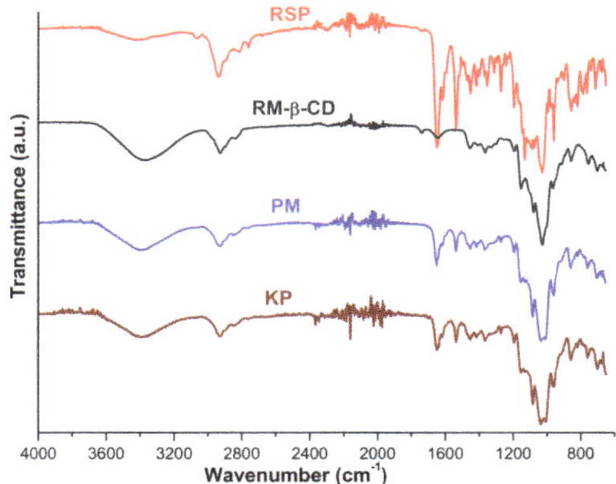

Figure 6. Universal-attenuated total reflectance Fourier transform infrared (UATR-FTIR) spectra of RSP, RM-β-CD, and RSP/RM-β-CD binary systems PM and KP.

The UATR-FTIR spectrum of RSP presents characteristic bands at 3063, 2936, 2812, 2759, 1648, 1534, 1449, 1414, 1352, 1130, 1027, 959, 854, and 816 cm^{-1} that have been assigned to the functional groups from the drug structure in our previous study [22]. RM-β-CD shows a broad absorption band in the 3600–3100 cm^{-1} region corresponding to the O-H stretching vibration from the non-methylated hydroxyl moieties and a large region below 1500 cm^{-1} which exhibits distinct peaks, which is most probably characteristic to the cyclodextrin ring [19,51].

In the spectral patterns of binary products, several differences are noticed as compared with those of the parent compounds. Thus, the band assigned to the C-N stretching vibration shifted from 1352 cm^{-1} in the RSP spectrum to 1364 cm^{-1} in both PM and KP spectra. In addition, the C=O stretching vibration (from tetrahydropyrido-pyrimidinone ring) characteristic band from 1648 cm^{-1} in the drug substance spectrum is displaced to 1649 cm^{-1} and 1642 cm^{-1} in the PM and KP spectra and is markedly reduced in intensity in both binary compounds. In addition, the spectral band attributed to aliphatic C-H stretching vibration from 2936 cm^{-1} in RSP shifted to 2929 and 2924 cm^{-1} in the spectral pattern of PM and KP, respectively. In the spectral region of 1050–1000 cm^{-1} the parent substances, RSP and RM-β-CD, exhibit bands at 1028 and 1027 cm^{-1}, while in the spectral pattern of binary products, two bands can be observed, at 1009 and 1035 cm^{-1} in PM and at 1006 and 1037 cm^{-1} in KP, respectively. A marked reduction in intensity and a shift to different wavenumbers (1535 cm^{-1}) in the PM and KP spectra is also noticed for the band assigned to the C=C stretching vibration of the aromatic ring from 1534 cm^{-1} in the RSP spectrum. Furthermore, the bands from 2812, 2759, and 1130 cm^{-1} in the RSP spectrum disappeared in binary KP and PM spectra.

The UATR-FTIR spectroscopy pointed out a decreasing in the intensity of RSP characteristic bands along with the shifting to different wavenumbers and the disappearance of several peaks in the spectral pattern of PM and KP. These data give evidence about the interaction between the antipsychotic drug substance and RM-β-CD.

2.1.6. Solubility Profile of RSP/RM-β-CD Inclusion Complexes

The solubility of the drug substance in the inclusion complex has been evaluated using the shake-flask method [18,19,52]. The drug concentration in saturated solution was assessed using UV spectrophotometry. RM-β-CD in phosphate buffer 0.1 M (pH 7.4) does not present absorption in the spectral range of 210–310 nm (Figure 7a); an RSP calibration curve accomplished using the absorbance values from 277 nm at 25 °C (Figure 7b) was employed in order to quantify the drug in the inclusion complex.

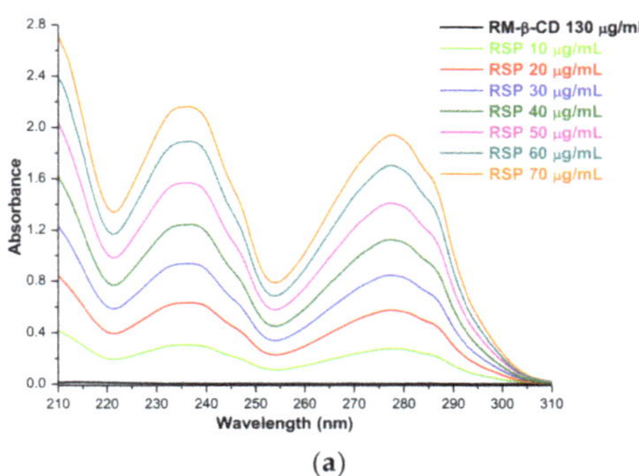

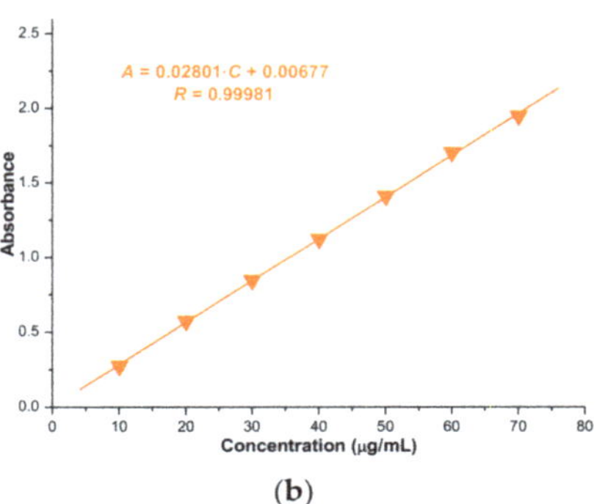

(a) (b)

Figure 7. The absorption spectra of RM-β-CD and RSP in phosphate buffer 0.1 M (pH 7.4), in the spectral range of 210–310 nm, at 25 °C (**a**); RSP calibration curve at 277 nm (**b**).

The solubility of the included RSP as KP obtained as an average of five experimental determinations is 1392.949 ± 0.016 µg/mL. In standard controlled experiments, clear solution was obtained when 29.31 mg of RSP/RM-β-CD KP were dissolved in 5 mL 0.1 M phosphate buffer (pH 7.4) at room temperature.

The results of the saturation solubility studies reveal RM-β-CD's ability to increase RSP solubility in phosphate buffer 0.1 M (pH 7.4) by 2.58-fold as compared with free RSP (540.007 ± 0.003 µg/mL).

2.2. Compatibility Studies of RSP/RM-β-CD Inclusion Complex with Excipients

The compatibility of RSP with several pharmaceutical excipients, namely microcrystalline cellulose, anhydrous lactose, starch, sodium lauryl sulfate, and magnesium stearate has been assessed and drug–excipient incompatibility has been reported in the presence of magnesium stearate, lactose, and microcrystalline cellulose [44]. In order to evaluate the ability of RM-β-CD to prevent the incompatibilities between drug substance and the mentioned excipients, compatibility studies of inclusion complex and excipients were conducted using thermal and spectroscopic techniques.

2.2.1. Thermoanalytical Studies

The thermoanalytical TG/DTG/HF curves of the RSP/RM-β-CD inclusion complex and its mixture with pharmaceutical excipients, recorded in dynamic air atmosphere and heating rate of 10 °C·min^{-1} are presented in Figure 8a–c.

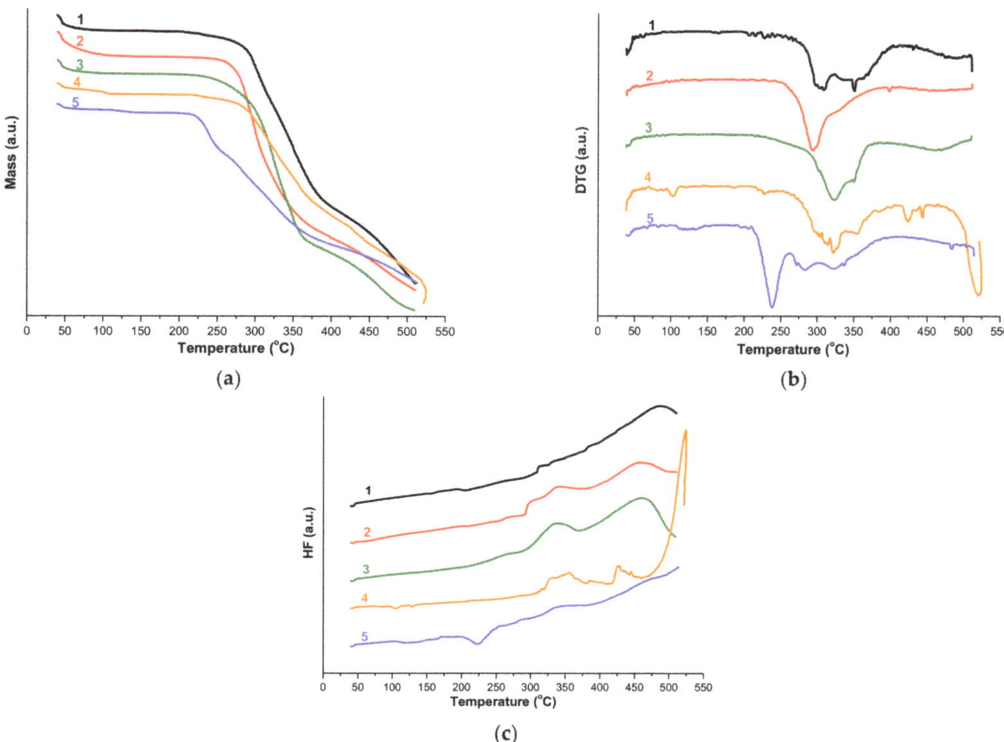

Figure 8. TG (**a**), DTG (**b**), and HF (**c**) curves of RSP/RM-β-CD inclusion complex (1) and its mixtures with excipients, as follows: RSP/RM-β-CD + STR (2); RSP/RM-β-CD + CEL (3); RSP/RM-β-CD + MgSTE (4); RSP/RM-β-CD + LCT (5).

The TG/DTG curves of RSP/RM-β-CD inclusion complex suggests its thermal degradation in the following temperature ranges: 35–86 °C (Δm = 5.2%, dehydration process), 197–338 °C (Δm = 32%, DTG_{peak} at 309 °C), 338–414 °C (Δm = 28.8%, DTG_{peak} at 351 °C) and 414–510 °C (Δm = 13.2%, DTG_{peak} at 490 °C). The HF curve of inclusion complex reveals two exothermic events, a small one at 313 °C and another one at 487 °C, corresponding to the decomposition of the complex.

Regarding the inclusion complex–excipients physical mixtures, in all situations, thermoanalytical TG/DTG curves show a mass loss process at temperatures lower than 110 °C due to the release of water from complex and/or excipient. For the mixtures with CEL and LCT, the dehydration occurs at temperatures below 100 °C, as follows: RSP/RM-β-CD + CEL has water loss up to 92 °C, with Δm = 4.47% and RSP/RM-β-CD + LCT has water loss up to 68 °C, with Δm = 1.71%. In the case of mixtures with STR and MgSTE, the dehydration takes place up to higher temperatures, as follows: RSP/RM-β-CD + STR reaches constant mass at 124 °C, after a mass loss Δm = 7.98%, while for RSP/RM-β-CD + MgSTE, the dehydration is complete at 106 °C, with a corresponding Δm = 4.31% (DTG process in 94–107 °C temperature range, DTG_{max} at 103 °C).

The stability profile of mixtures with excipients in an anhydrous state is good, as revealed by the thermoanalytical curves. Accordingly to this, the RSP/RM-β-CD + STR mixture shows no thermal events in the 124–214 °C, while with the increase of thermal stress, the decomposition begins. TG/DTG curves reveal a continuous mass loss process in the temperature range 214–510 °C (DTG_{max} at 292 °C), with a corresponding Δm = 77.78%. The HF curve does not reveal significant events up to 289 °C, when some exothermal processes are observed, due to the oxidative thermolysis of organic edifice (HF_{max} at

301 °C, 338 °C and 457 °C, respectively). In the case of the RSP/RM-β-CD + CEL mixture, thermoanalytical data suggest that the anhydrous mixture is stable in the 92–232 °C temperature range, since none of the three thermoanalytical curves reveal any process. In the temperature range 232–510 °C, a consistent mass loss is observed (Δm = 91.37%), which is sustained by the DTG profile (main process between 233 and 391 °C, DTG_{max} at 322 °C, shoulder at 349 °C), secondary process between 391 and 510 °C, DTG_{max} at 459 °C). The HF curve reveals oxidative thermodegradation at temperatures over 233 °C, with HF_{max} at 264, 334, and 459 °C, respectively. The mixture RSP/RM-β-CD + MgSTE shows the most complex thermoanalytical profile, due to complexity of excipient composition, being known that pharmaceutical-purity MgSTE is a mix of different fatty acid salts that may vary in proportion, and additionally, its properties heavily depend on its moisture content [53,54]. After the dehydration, the RSP/RM-β-CD + MgSTE shows stability in the 106–218 °C temperature range, without revealing any interactions between the components of this matrix. The main mass loss process takes place in the 218–510 °C (Δm = 81.64%), with several DTG peaks at 321, 354, 422, 444, and 519 °C, respectively), and it is accompanied by several HF peaks corresponding to an endothermic event—dehydration (96–111 °C, HF_{peak} at 104 °C), and with the increase of temperature with exothermic ones, in the 303–510 °C range: 327, 356, 428, and 524 °C, respectively. Last, for the RSP/RM-β-CD + LCT sample, the stability in anhydrous state is observed in the range 68–110 °C; then, a small mass loss takes place in the range 110–143 °C (Δm = 1.65%), which is followed by the main mass loss process that takes place in the range 196–510 °C (Δm = 66.18%). The DTG profile is more complex in this case, revealing peaks up to 210 °C at 63, 117, and 133 °C), and in the temperature range 200–510 °C at 238 (main), 272, 283, 322, and 482 °C, respectively). The HF curve reveals some endothermal events at 119 °C and 221 °C, while the thermal events associated with the thermooxidation of complex are no longer visible at temperatures over 300 °C. This behavior is a clear indication that some incompatibilities take place in this system, which are mainly due to the fact that they are facilitated by the presence of the melted excipient (LCT), which occurs in the range 203–243 °C (HF_{peak} at 221 °C). The stability profile of mixtures with excipients in an anhydrous state is good, as revealed by the thermoanalytical curves, and no interactions are revealed between the inclusion complex and three of the selected excipients (STR, CEL, and MgSTE), while for mixture with LCT, interactions are observed during the thermal treatment of the samples. For this last sample, a concrete evaluation of interactions can be realized solely by the implementation of two other investigational tools, namely UATR-FTIR and PXRD.

2.2.2. UATR-FTIR Studies

UATR-FTIR spectroscopy is commonly used as screening technique for assessing the potential physicochemical interaction between an API and the excipients employed in the pharmaceutical dosage forms. Usually, UATR-FTIR spectroscopy is used as a complementary tool along thermal analysis for the evaluation of compatibility/incompatibility between API and excipient, which is kept under ambient conditions.

The disappearance of an absorption band, a reduction of the band intensity, or the appearance of new bands reveal the existence of interactions between the API and the studied excipient [34,55–58]. This method provides information about the chemical groups to avoid in the excipients in order to develop stable pharmaceutical formulations [55].

The UATR-FTIR spectra of the inclusion complex RSP/RM-β-CD and its physical mixtures with selected excipients recorded at ambient temperature are shown in Figure 9.

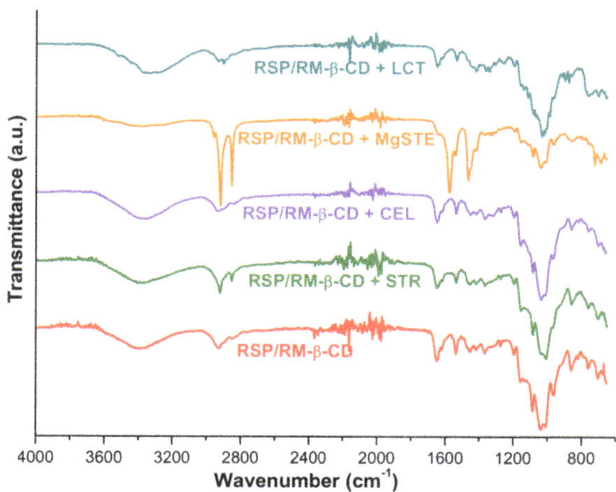

Figure 9. UATR-FTIR spectra of RSP/RM-β-CD inclusion complex and its physical mixtures with selected excipients.

The UATR-FTIR spectrum of RSP/RM-β-CD inclusion complex exhibits a broad band between 3500 and 3300 cm^{-1} (peak at 3387 cm^{-1}) related to hydroxyl groups (O-H stretching vibration) and other several bands noticed in Table 1; they are presented in Section 2.1.5.

Table 1. UATR-FTIR characteristic bands for RSP/RM-β-CD and its mixtures with excipients.

Sample	Analysis of UATR-FTIR Spectral Regions (cm^{-1})		
	3600–2700	1700–1000	1000–650
RSP/RM-β-CD	3387; 2924	1642; 1535; 1450; 1415; 1364; 1272; 1194; 1154; 1083; 1037; 1006	959; 857; 758
RSP/RM-β-CD + STR	3386; 2919; 2847	1649; 1535; 1459; 1416; 1364; 1272; 1195; 1152; 1083;1035; 1006	962; 857; 758
RSP/RM-β-CD + CEL	3386; 2930	1649; 1535; 1458; 1416; 1364; 1272; 1194; 1155; 1083; 1034; 1009	956; 858; 758
RSP/RM-β-CD + MgSTE	2957; 2916; 2850	1639; 1572; 1541; 1465; 1418; 1365; 1272; 1192; 1154; 1083; 1038; 1007	961; 857; 722
RSP/RM-β-CD + LCT	3338; 2945; 2931; 2901	1642; 1535; 1420; 1369; 1254; 1195; 1157; 1142; 1118; 1083; 1068; 1031; 1018	989; 968; 877; 761

The spectral data collected in Table 1 along with the spectra depicted in Figure 1 reveal that the characteristic bands of RSP/RM-β-CD are present in the mixture of the KP with excipients either at the same wavenumber as in inclusion complex or slightly shifted to different wavenumbers, except for the case of the physical mixture of KP with LCT, where the spectral bands form 1465 and 1006 cm^{-1} have disappeared from the mixture spectrum. This situation indicates an interaction between the inclusion complex and LCT even in ambient conditions, which are results that were previously suggested by the thermoanalytical methods.

2.2.3. PXRD Studies

X-ray diffractometry has been used as a complementary tool for evaluating the possible interactions between the RSP/RM-β-CD inclusion complex and the excipients, which correlate with changes in the crystallinity profile of the samples.

The X-ray diffraction patterns of the RSP/RM-β-CD inclusion complex, selected excipients, and their physical mixtures are depicted in Figure 10a–d.

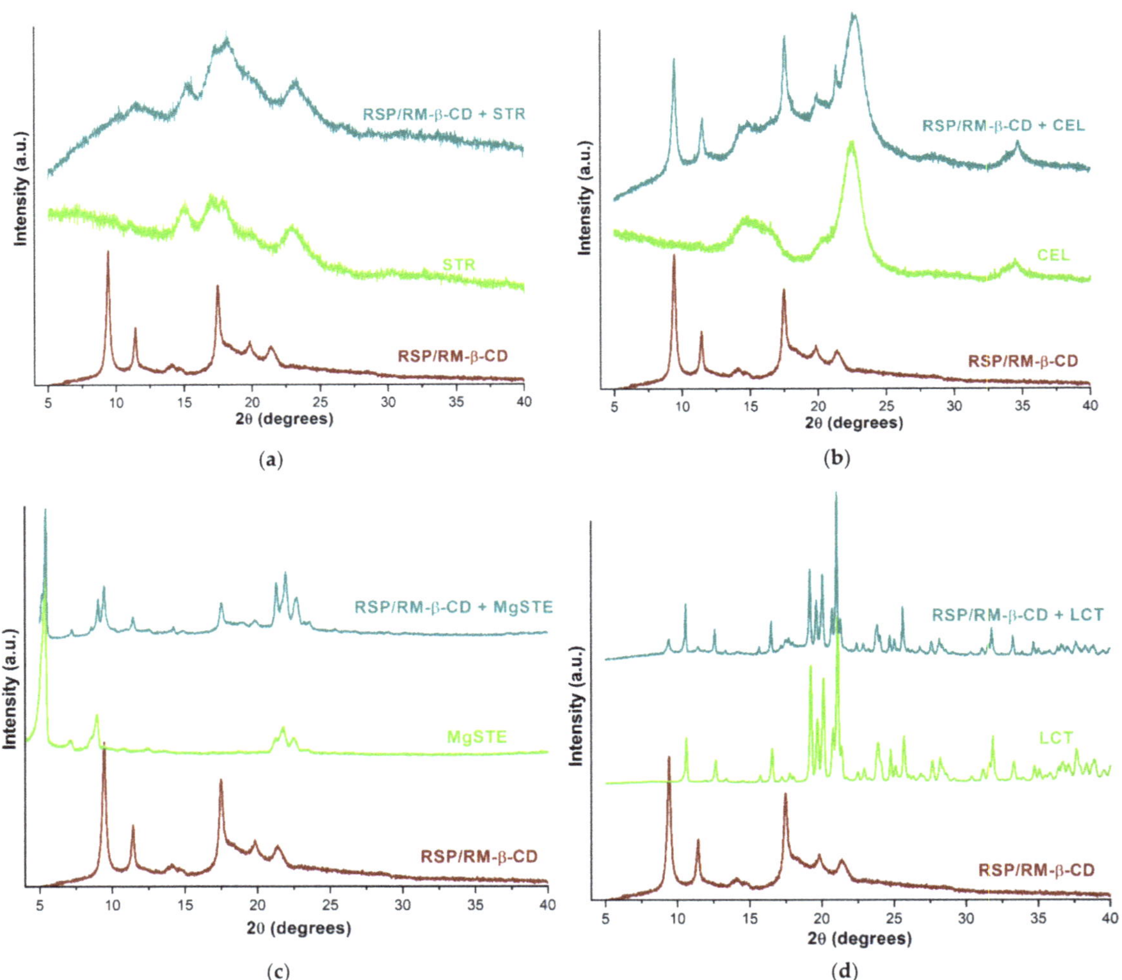

Figure 10. X-ray diffraction profiles of RSP/RM-β-CD inclusion complex, selected excipients, and their physical mixtures: starch (STR) (**a**); microcrystalline cellulose (CEL) (**b**); magnesium stearate (MgSTE) (**c**); lactose (LCT) (**d**).

The diffraction profile of the RSP/RM-β-CD inclusion complex presents three characteristic reflections of higher intensity at 9.42, 11.42, and 17.48 2θ and others of lower intensity at 14.17, 19.87, and 21.47 2θ. In the diffractograms of STR and CEL, there are several main broad peaks and numerous undefined ones, with low intensities indicating the amorphous nature of the excipients (Figure 10a,b). In the case of physical mixture of the RSP/RM-β-CD inclusion complex with STR, the diffraction pattern reveals the disappearance of the high intensity crystalline reflection of the inclusion complex from 9.42 2θ; the other two characteristic peaks of the inclusion complex can be observed at 11.64 and 14.47 2θ, while the rest of the bands are overlapped over the broad bands of excipient. The PXRD pattern of the inclusion complex–CEL physical mixture exhibits an overlap of characteristic peaks of both the RSP/RM-β-CD inclusion complex observed at 9.45, 11.50,

14.29, 17.52, 19.90, 21.32 2θ, and the excipient, indicating a lack of interaction between the components.

The diffractogram of MgSTE shows characteristic crystalline reflections at 2θ of 5.35, 7.20, 8.90, 21.78, and 22.53, suggesting its crystalline state [18] (Figure 10c). The diffraction profile of physical mixtures of the inclusion complex with MgSTE represents a sum of characteristic peaks of KP that appear at 2θ of 9.42, 11.45, 14.27, 17.47, 19.84, and 21.30, and excipient peaks that appear at 5.41, 7.23, 9.02, 21.93, and 22.71 2θ, highlighting no interaction between the inclusion complex and MgSTE.

The crystalline profile of LCT is demonstrated by the high intensity crystalline reflections at 19.22, 19.69, 20.11, 20.77, and 21.09 2θ and other peaks with lower intensity at 10.61, 12.62, 16.53, 23.85, 24.75, 25.67, 31.85, and 33.32 2θ, which are present in its diffraction pattern (Figure 10d). The diffractogram of the inclusion complex–LCT mixture shows all the characteristic crystalline reflections of excipient (at 10.57, 12.58, 16.48, 19.17, 19.63, 20.06, 20.72, 21.03, 23.84, 24.71, 25.61, 31.78, and 33.24 2θ), but some corresponding to RSP/RM-β-CD KP are greatly attenuated (at 14.14 and 21.30 2θ) or shifted (9.36, 11.36, and 17.17 2θ), confirming the interaction in solid state between the inclusion complex and LCT that was previously demonstrated by thermal analysis and UATR-FTIR spectroscopy.

3. Materials and Methods

3.1. Materials

Risperidone (as Pharmaceutical Secondary Standard) was acquired from Sigma-Aldrich, (Steinheim, Germany) and randomly methylated β-cyclodextrin (DS~12) was purchased from Cyclolab R&L Ltd. (Budapest, Hungary). The pharmaceutical grade excipients, namely starch, microcrystalline cellulose, magnesium stearate, anhydrous lactose, and methylcellulose were obtained from Sigma-Aldrich (Steinheim, Germany). All other reagents and chemicals used were of analytical purity.

3.2. Phase Solubility Studies

Phase solubility studies were performed according to the method reported by Higuchi and Connors [38] in 0.1 M phosphate buffer of pH 7.4, at 25 °C. An excessive amount of RSP (10 mg) was added to 3 mL of solution containing RM-β-CD in concentration of 0–50 mM. The obtained suspensions were vigorously shaken at 25 °C for 5 days. After the equilibrium was reached, the samples were filtered using a 0.45 μm nylon disk filter and suitably diluted. The RSP concentration in filtered solutions was evaluated using UV spectroscopic measurements at 277 nm. Spectronic Unicam—UV 320 UV-Visible double beam spectrophotometer (Spectronic Unicam, Cambridge, UK) with 1 cm matched quartz cells was used for all UV spectrophotometric measurements.

The apparent stability constant ($K_{1:1}$) was calculated from the phase solubility diagram, using the following equation:

$$K_{1:1} = \frac{Slope}{S_0\,(1 - Slope)} \qquad (2)$$

where S_0 is RSP solubility in phosphate buffer 0.1 M (pH 7.4) in the absence of RM-β-CD.

3.3. Molecular Docking Studies

Molecular docking studies were carried out to characterize the interaction between drug substance and CD. The RM-β-CD structure used in this work was generated from the curated coordinates of ligand 2QKH (X-ray diffraction, resolution 1.9 Å) downloaded from the Protein Data Bank database [59]. Methyl groups were manually added on free hydroxyl groups in order to obtain a degree of substitution equal with 12 (GaussView 5, Semichem Inc). Substituents were added on the β-cyclodextrin natural core, namely, 4-CH_3 on the O_2- position for the 2, 3, 4, and 6 glucopyranose units, 5 groups on the O_3- for the 1, 2, 4, 5, and 7 glucose residues, and finally, 3-CH_3 on 1, 5, and 7 glucopyranose units on the O_6- position. All dihedral angles of the methoxy groups were homogenized, the resulting conformations being compatible with an unhindered CD cavity. CD was optimized in the same manner with RSP (DFT/B3LYP/6-311G). Three-dimensional coordinates of RSP was generated using the Gaussian program suite at DFT/B3LYP/6-311G optimization.

The molecular docking analysis was carried out using the Autodock 4.2.6 software together with the AutoDockTools [41]. The docking between RSP and RM-β-CD involves adding all the polar hydrogens, computing the Gasteiger charge; a grid box was created using Autogrid 4 with $50 \times 50 \times 50$ Å in x, y and z directions with 0.375 Å spacing from CD center. All the calculations were performed in vacuum. For the docking process, we chose the Lamarckian genetic algorithm (Genetic Algorithm combined with a local search), with a population size of 150 and a number of 50 runs. In order to generate the molecular modeling figures, we exported all Autodock results in the PyMOL (The PyMOL Molecular Graphics System, Version 2.0 Schrödinger, LLC, New York, NY, USA) [43]. To validate the repeatability and reproducibility of the docking method, we performed redocking and then expressed the results as RMSD in Å using Discovery Studio software. All the calculations were performed in triplicate and expressed as an average.

3.4. Preparation of the Solid Inclusion Complex and Physical Mixtures with Excipients

The kneading method in a 1:1 molar ratio was employed to prepare the inclusion complex of RSP with RM-β-CD. For this purpose, 0.2851 g of RSP and 0.9149 g of CD were weighed, and the mixture was pulverized in an agate mortar and triturated with 1.2 g ethanol:HCl 0.1 M solution (1:1, m/m). Then, the thick slurry was kneaded for 45 min, and during the process, a few drops of solvent were added to maintain a suitable consistency. Thus, the product obtained was dried at ambient temperature and then in the oven at 40 °C for 24 h. The dried kneaded product was pulverized and passed through a 75-μm size sieve. In addition, a physical mixture of RSP with RM-β-CD in 1:1 molar ratio was obtained by mixing in the agate mortar and pestle for 10 min in a solvent-free manner.

The physical mixtures of RSP/RM-β-CD inclusion complex and each excipient were prepared by mixing in an agate mortar with pestle for approximately 5 min in the ratio of 1:1 (m/m).

3.5. Thermal Analysis

The pure RSP, RM-β-CD, and RSP/RM-β-CD physical mixture and kneaded product and also the mixtures of KP and selected excipients were analyzed using a Perkin-Elmer DIAMOND TG/DTA instrument. Samples with masses around 3–4 mg were weighed in aluminum crucibles and studied under air atmosphere at a flow rate of 100 mL/min, over the temperature range of 40–500 °C, with a heating rate of 10 °C/min.

3.6. Powder X-ray Diffractometry

PXRD studies were carried out using a Bruker D8 Advance powder X-ray diffractometer (Bruker AXS, Karlsruhe, Germany). The X-ray diffraction patterns were collected at ambient temperature, using CuKα radiation (40 kV, 40 mA) and a Ni filter, over the interval of 10–45° angular domain (2θ).

3.7. UATR-FTIR Spectroscopy

The UATR-FTIR spectroscopic analysis was performed using a Perkin Elmer SPEC-TRUM 100 device. The data were collected directly on solid samples in the spectral domain 4000–600 cm^{-1} on an UATR device. Spectra were built up after a number of 16 co-added acquisitions, with a spectral resolution of 4 cm^{-1}.

3.8. Solubility Profile of RSP/RM-β-CD Kneaded Product

The saturation shake-flask method was employed in order to evaluate the RSP solubility upon complexation with RM-β-CD. To this end, an excessive amount of drug substance and RSP/RM-β-CD kneaded product were added in 5 mL of phosphate buffer 0.1 M (pH 7.4), so that saturated solutions were obtained. The samples were shaken for

24 h at room temperature, and then the solutions were separated from the insoluble drug substance by filtration using a 0.45 μm nylon disk filter. After appropriate dilution, the filtrate was subjected to UV-spectrophotometric analysis at 277 nm. The RSP quantification was realized using a calibration curve. A set of RSP solutions in phosphate buffer 0.1 M with concentration in the range of 10–70 μg/mL were prepared, and their absorbance was recorded at 277 nm. The standard curve was obtained by plotting the absorbance (A) vs. the concentration (C); the equation of the RSP calibration curve is: $A = 0.02801 \cdot C + 0.00677$, with $R = 0.99981$.

4. Conclusions

This study investigates the encapsulation of antipsychotic drug RSP by RM-β-CD and the compatibility of the inclusion complex with several excipients commonly used in pharmaceutical dosage forms. The RSP/RM-β-CD binary product was obtained using the kneading method and was evaluated by experimental and theoretical approaches. The experimental results provided by thermal methods, powder X-ray diffractometry, and UATR-FTIR spectroscopy demonstrate the inclusion complex formation between RSP and RM-β-CD in 1:1 molar ratio as the solubility studies indicated. As a result of inclusion complex formation, the RSP solubility was increased by 2.58-fold as compared with free RSP, highlighting the solubilizing effect of CD.

Since the corroboration of thermoanalytical data suggested that the inclusion complex is compatible with three of the selected excipients (namely STR, CEL, and MgSTE), but incompatible with LCT, two complementary investigational tools were used, namely UATR-FTIR spectroscopy and PXRD. These last two instrumental techniques allow the evaluation of compatibility/incompatibility between the components of a complex matrix in ambient conditions. It was shown that interactions between the RSP/RM-β-CD and LCT occur in solid state even under ambient conditions, and they are accentuated by the thermal stress. In the development of new generic forms containing RSP formulated as a supramolecular adduct with RM-β-CD, precautions should be taken in the selection of excipients, without using lactose in the final product.

Author Contributions: Conceptualization, L.S., I.-M.T., A.L., O.S. and I.L.; data curation, I.-M.T.; formal analysis, A.L. and D.C.; investigation, L.S., I.-M.T., A.L., G.V., P.B., M.M., R.-M.V. and O.S.; methodology, L.S., G.V., R.-M.V. and I.L.; project administration, L.S.; resources, I.-M.T., D.C., G.V., P.B. and M.M.; software, R.-M.V.; supervision, A.L. and I.L.; validation, I.-M.T., A.L., O.S. and I.L.; visualization, A.L., D.C. and G.V.; writing—original draft, L.S., I.-M.T. and O.S.; writing—review and editing, L.S., A.L., O.S. and I.L. All authors have read and agreed to the published version of the manuscript.

Funding: This research received no external funding.

Institutional Review Board Statement: Not applicable.

Informed Consent Statement: Not applicable.

Data Availability Statement: Not applicable.

Conflicts of Interest: The authors declare no conflict of interest.

Sample Availability: Samples of the compounds are available in limited quantity from the authors.

References

1. Germann, D.; Kurylo, N.; Han, F. Risperidone. In *Profiles of Drug Substances, Excipients and Related Methodology*; Academic Press Elsevier: Amsterdam, The Netherlands, 2012; Volume 37, pp. 313–361, ISBN 9780123972200.
2. Yunusa, I.; El Helou, M.L. The Use of Risperidone in Behavioral and Psychological Symptoms of Dementia: A Review of Pharmacology, Clinical Evidence, Regulatory Approvals, and Off-Label Use. *Front. Pharmacol.* **2020**, *11*, 1–7. [CrossRef]
3. Love, R.C.; Nelson, M.W. Pharmacology and clinical experience with risperidone. *Expert Opin. Pharmacother.* **2000**, *1*, 1441–1453. [CrossRef] [PubMed]
4. Rahman, Z.; Zidan, A.S.; Khan, M.A. Risperidone solid dispersion for orally disintegrating tablet: Its formulation design and non-destructive methods of evaluation. *Int. J. Pharm.* **2010**, *400*, 49–58. [CrossRef]

5. Censi, R.; Di Martino, P. Polymorph Impact on the Bioavailability and Stability of Poorly Soluble Drugs. *Molecules* **2015**, *20*, 18759–18776. [CrossRef] [PubMed]
6. Crini, G.; Fourmentin, S.; Fenyvesi, Éva; Torri, G.; Fourmentin, M.; Morin-Crini, N. Cyclodextrins, from molecules to applications. *Environ. Chem. Lett.* **2018**, *16*, 1361–1375. [CrossRef]
7. Carneiro, S.B.; Duarte, F.; Ílary, C.; Heimfarth, L.; Quintans, J.D.S.S.; Quintans-Júnior, L.J.; Júnior, V.F.D.V.; De Lima, Á.A.N. Cyclodextrin–Drug Inclusion Complexes: In Vivo and In Vitro Approaches. *Int. J. Mol. Sci.* **2019**, *20*, 642. [CrossRef]
8. Usacheva, T.; Kabirov, D.; Beregova, D.; Gamov, G.; Sharnin, V.; Biondi, M.; Mayol, L.; D'Aria, F.; Giancola, C. Thermodynamics of complex formation between hydroxypropyl-β-cyclodextrin and quercetin in water–ethanol solvents at T = 298.15 K. *J. Therm. Anal. Calorim.* **2019**, *138*, 417–424. [CrossRef]
9. Han, D.; Han, Z.; Liu, L.; Wang, Y.; Xin, S.; Zhang, H.; Yu, Z. Han Solubility Enhancement of Myricetin by Inclusion Complexation with Heptakis-O-(2-Hydroxypropyl)-β-Cyclodextrin: A Joint Experimental and Theoretical Study. *Int. J. Mol. Sci.* **2020**, *21*, 766. [CrossRef]
10. Saokham, P.; Muankaew, C.; Jansook, P.; Loftsson, T. Solubility of Cyclodextrins and Drug/Cyclodextrin Complexes. *Molecules* **2018**, *23*, 1161. [CrossRef]
11. Sahakijpijarn, S.; Moon, C.; Koleng, J.; Christensen, D.; Williams, R. Development of Remdesivir as a Dry Powder for Inhalation by Thin Film Freezing. *Pharmaceutics* **2020**, *12*, 1002. [CrossRef] [PubMed]
12. Li, S.; Yuan, L.; Chen, Y.; Zhou, W.; Wang, X. Studies on the Inclusion Complexes of Daidzein with β-Cyclodextrin and Derivatives. *Molecules* **2017**, *22*, 2183. [CrossRef]
13. Saxena, K.; Kurian, S.; Saxena, J.; Goldberg, A.; Chen, E.; Simonetti, A. Mixed States in Early-Onset Bipolar Disorder. *Psychiatr. Clin. N. Am.* **2020**, *43*, 95–111. [CrossRef]
14. Braga, S.S.; El-Saleh, F.; Lysenko, K.; Paz, F.A.A. Inclusion Compound of Efavirenz and γ-Cyclodextrin: Solid State Studies and Effect on Solubility. *Molecules* **2021**, *26*, 519. [CrossRef]
15. Simsek, T.; Rasulev, B.; Mayer, C.; Simsek, S. Preparation and Characterization of Inclusion Complexes of β-Cyclodextrin and Phenolics from Wheat Bran by Combination of Experimental and Computational Techniques. *Molecules* **2020**, *25*, 4275. [CrossRef]
16. He, J.; Zheng, Z.-P.; Zhu, Q.; Guo, F.; Chen, J. Encapsulation Mechanism of Oxyresveratrol by β-Cyclodextrin and Hydroxypropyl-β-Cyclodextrin and Computational Analysis. *Molecules* **2017**, *22*, 1801. [CrossRef]
17. Du, F.; Pan, T.; Ji, X.; Hu, J.; Ren, T. Study on the preparation of geranyl acetone and β-cyclodextrin inclusion complex and its application in cigarette flavoring. *Sci. Rep.* **2020**, *10*, 1–10. [CrossRef] [PubMed]
18. Tănase, I.-M.; Sbârcea, L.; Ledeți, A.; Barvinschi, P.; Cîrcioban, D.; Vlase, G.; Văruț, R.-M.; Ledeți, I. Compatibility studies with pharmaceutical excipients for aripiprazole–heptakis (2,6-di-O-methyl)-β-cyclodextrin supramolecular adduct. *J. Therm. Anal. Calorim.* **2020**, *142*, 1963–1976. [CrossRef]
19. Tănase, I.-M.; Sbârcea, L.; Ledeți, A.; Vlase, G.; Barvinschi, P.; Văruț, R.-M.; Dragomirescu, A.; Axente, C.; Ledeți, I. Physicochemical characterization and molecular modeling study of host–guest systems of aripiprazole and functionalized cyclodextrins. *J. Therm. Anal. Calorim.* **2020**, *141*, 1027–1039. [CrossRef]
20. Sbârcea, L.; Ledeți, I.; DrĂgan, L.; Kurunczi, L.; Fuliaş, A.; Udrescu, L. Fosinopril sodium–hydroxypropyl-β-cyclodextrin inclusion complex. *J. Therm. Anal. Calorim.* **2015**, *120*, 981–990. [CrossRef]
21. Ferreira, E.B.; Júnior, W.F.D.S.; Pinheiro, J.G.D.O.; Da Fonseca, A.G.; Lemos, T.M.A.M.; Rocha, H.A.D.O.; De Azevedo, E.P.; Junior, F.J.B.M.; De Lima, Á.A.N. Characterization and Antiproliferative Activity of a Novel 2-Aminothiophene Derivative-β-Cyclodextrin Binary System. *Molecules* **2018**, *23*, 3130. [CrossRef]
22. Sbârcea, L.; Tănase, I.-M.; Ledeți, A.; Cîrcioban, D.; Vlase, G.; Barvinschi, P.; Miclău, M.; Văruț, R.-M.; Trandafirescu, C.; Ledeți, I. Encapsulation of Risperidone by Methylated β-Cyclodextrins: Physicochemical and Molecular Modeling Studies. *Molecules* **2020**, *25*, 5694. [CrossRef]
23. Pires, F.Q.; Pinho, L.A.; Freire, D.O.; Silva, I.C.R.; Sa-Barreto, L.L.; Cardozo-Filho, L.; Gratieri, T.; Gelfuso, G.M.; Cunha-Filho, M. Thermal analysis used to guide the production of thymol and Lippia origanoides essential oil inclusion complexes with cyclodextrin. *J. Therm. Anal. Calorim.* **2018**, *137*, 543–553. [CrossRef]
24. Júnior, F.J.D.L.R.; Da Silva, K.M.A.; Brandão, D.O.; Júnior, J.V.C.; Dos Santos, J.A.B.; De Andrade, F.H.D.; Batista, R.S.D.A.; Lins, T.B.; De Sousa, D.P.; Medeiros, A.C.D.; et al. Investigation of the thermal behavior of inclusion complexes with antifungal activity. *J. Therm. Anal. Calorim.* **2018**, *133*, 641–648. [CrossRef]
25. Stella, V.J.; He, Q. Cyclodextrins. *Toxicol. Pathol.* **2008**, *36*, 30–42. [CrossRef]
26. Gu, W.; Liu, Y. Characterization and stability of beta-acids/hydroxypropyl-β-cyclodextrin inclusion complex. *J. Mol. Struct.* **2020**, *1201*, 127159. [CrossRef]
27. Szente, L. Highly soluble cyclodextrin derivatives: Chemistry, properties, and trends in development. *Adv. Drug Deliv. Rev.* **1999**, *36*, 17–28. [CrossRef]
28. Brewster, M.E.; Loftsson, T. Cyclodextrins as pharmaceutical solubilizers. *Adv. Drug Deliv. Rev.* **2007**, *59*, 645–666. [CrossRef]
29. Shukla, D.; Chakraborty, S.; Singh, S.; Mishra, B. Preparation and in-vitro characterization of Risperidone-cyclodextrin inclusion complexes as a potential injectable product. *DARU J. Pharm. Sci.* **2009**, *17*, 226–235.
30. Jug, M.; Kos, I.; Bećirević-Laćan, M. The pH-dependent complexation between risperidone and hydroxypropyl-β-cyclodextrin. *J. Incl. Phenom. Macrocycl. Chem.* **2009**, *64*, 163–171. [CrossRef]

31. El-Barghouthi, M.I.; Masoud, N.A.; Al-Kafawein, J.K.; Zughul, M.B.; Badwan, A.A. Host–Guest Interactions of Risperidone with Natural and Modified Cyclodextrins: Phase Solubility, Thermodynamics and Molecular Modeling Studies. *J. Incl. Phenom. Macrocycl. Chem.* **2005**, *53*, 15–22. [CrossRef]
32. Kacso, I.; Rus, L.M.; Martin, F.; Miclaus, M.; Filip, X.; Dan, M. Solid-state compatibility studies of Ketoconazole-Fumaric acid co-crystal with tablet excipients. *J. Therm. Anal. Calorim.* **2021**, *143*, 3499–3506. [CrossRef]
33. Ledeți, I.; Romanescu, M.; Cîrcioban, D.; Ledeți, A.; Vlase, G.; Vlase, T.; Suciu, O.; Murariu, M.; Olariu, S.; Matusz, P.; et al. Stability and Compatibility Studies of Levothyroxine Sodium in Solid Binary Systems—Instrumental Screening. *Pharmaceutics* **2020**, *12*, 58. [CrossRef] [PubMed]
34. Cristea, M.; Baul, B.; Ledeți, I.; Ledeți, A.; Vlase, G.; Vlase, T.; Karolewicz, B.; Ştefănescu, O.; Dragomirescu, A.O.; Mureşan, C.; et al. Preformulation studies for atorvastatin calcium. *J. Therm. Anal. Calorim.* **2019**, *138*, 2799–2806. [CrossRef]
35. Zhang, L.; Luan, H.; Lu, W.; Wang, H. Preformulation Studies and Enabling Formulation Selection for an Insoluble Compound at Preclinical Stage—From In Vitro, In Silico to In Vivo. *J. Pharm. Sci.* **2020**, *109*, 950–958. [CrossRef]
36. Jiang, W.; Yu, L.X. Modern Pharmaceutical Quality Regulations. *Dev. Solid Oral Dos. Forms* **2009**, 885–901. [CrossRef]
37. U.S. Department of Health and Human Services Food and Drug Administration. ICH Q8(R2) Pharmaceutical Development. *Work. Qual. Des. Pharm.* **2009**, *8*, 28.
38. Higuchi, T.K.; Connors, A. Phase-solubility techniques. *Adv. Anal. Chem. Instrum.* **1965**, *4*, 117–211.
39. Yin, H.; Wang, C.; Yue, J.; Deng, Y.; Jiao, S.; Zhao, Y.; Zhou, J.; Cao, T. Optimization and characterization of 1,8-cineole/hydroxypropyl-β-cyclodextrin inclusion complex and study of its release kinetics. *Food Hydrocoll.* **2021**, *110*, 106159. [CrossRef]
40. Jiang, L.; Yang, J.; Wang, Q.; Ren, L.; Zhou, J. Physicochemical properties of catechin/β-cyclodextrin inclusion complex obtained via co-precipitation. *CyTA J. Food* **2019**, *17*, 544–551. [CrossRef]
41. Morris, G.M.; Huey, R.; Lindstrom, W.; Sanner, M.F.; Belew, R.K.; Goodsell, D.S.; Olson, A.J. AutoDock4 and AutoDockTools4: Automated docking with selective receptor flexibility. *J. Comput. Chem.* **2009**, *30*, 2785–2791. [CrossRef]
42. Huey, R.; Morris, G.M.; Olson, A.J.; Goodsell, D.S. A semiempirical free energy force field with charge-based desolvation. *J. Comput. Chem.* **2007**, *28*, 1145–1152. [CrossRef]
43. DeLano, W.L. Pymol: An open-source molecular graphics tool. *Newsl. Protein Crystallogr.* **2002**, *40*, 1–8.
44. Daniel, J.S.P.; Veronez, I.P.; Rodrigues, L.L.; Trevisan, M.G.; Garcia, J.S. Risperidone—Solid-state characterization and pharmaceutical compatibility using thermal and non-thermal techniques. *Thermochim. Acta* **2013**, *568*, 148–155. [CrossRef]
45. Rao, K.S.; Udgirkar, D.B.; Mule, D.D. Enhancement of dissolution rate and bioavailability of Aceclofenac by Complexation with cyclodextrin *Res. J. Pharm. Biol. Chem. Sci.* **2010**, *1*, 142–151.
46. Sbârcea, L.; Ledeți, A.; Udrescu, L.; Văruţ, R.-M.; Barvinschi, P.; Vlase, G.; Ledeți, I. Betulonic acid—cyclodextrins inclusion complexes. *J. Therm. Anal. Calorim.* **2019**, *138*, 2787–2797. [CrossRef]
47. Circioban, D.; Ledeti, I.; Suta, L.-M.; Vlase, G.; Ledeti, A.; Vlase, T.; Varut, R.; Sbarcea, L.; Trandafirescu, C.; Dehelean, C. Instrumental analysis and molecular modelling of inclusion complexes containing artesunate. *J. Therm. Anal. Calorim.* **2020**, *142*, 1951–1961. [CrossRef]
48. Sbarcea, L.; Udrescu, L.; Dragan, L.; Trandafirescu, C.; Sasca, V.; Barvinschi, P.; Bojita, M. Characterization of fosinopril natrium-hydroxypropyl-β-cyclodextrin inclusion complex. *Rev. Chim.* **2011**, *62*, 349–351.
49. Circioban, D.; Ledeti, A.; Vlase, G.; Coricovac, D.; Moaca, A.; Farcas, C.; Vlase, T.; Ledeti, I.; Dehelean, C. Guest–host interactions and complex formation for artemisinin with cyclodextrins: Instrumental analysis and evaluation of biological activity. *J. Therm. Anal. Calorim.* **2018**, *134*, 1375–1384. [CrossRef]
50. Sbârcea, L.; Udrescu, L.; DrĂgan, L.; Trandafirescu, C.; Szabadai, Z.; Bojiţă, M. Fosinopril-cyclodextrin inclusion complexes: Phase solubility and physicochemical analysis. *Die Pharm.* **2011**, *66*.
51. Circioban, D.; Ledeti, A.; Vlase, G.; Moaca, A.; Ledeti, I.; Farcaş, C.; Vlase, T.; Dehelean, C. Thermal degradation, kinetic analysis and evaluation of biological activity on human melanoma for artemisinin. *J. Therm. Anal. Calorim.* **2018**, *134*, 741–748. [CrossRef]
52. Udrescu, L.; Sbârcea, L.; Fuliaş, A.; Ledeți, I.; Vlase, G.; Barvinschi, P.; Kurunczi, L. Physicochemical Analysis and Molecular Modeling of the Fosinoprilβ-Cyclodextrin Inclusion Complex. *J. Spectrosc.* **2014**, *2014*, 1–14. [CrossRef]
53. Koivisto, M.; Jalonen, H.; Lehto, V.-P. Effect of temperature and humidity on vegetable grade magnesium stearate. *Powder Technol.* **2004**, *147*, 79–85. [CrossRef]
54. Craig, D.Q.M.; Reading, M. *Thermal Analysis of Pharmaceuticals*; CRC Press: Boca Raton, FL, USA, 2006; ISBN 9781420014891.
55. Segall, A.I. Preformulation: The use of FTIR in compatibility studies. *J. Innov. Appl. Pharm. Sci.* **2019**, *4*, 1–6.
56. da Silva, E.P.; Pereira, M.A.V.; Lima, I.P.D.B.; Lima, N.G.P.B.; Barbosa, E.G.; Aragão, C.F.S.; Gomes, A.P.B. Compatibility study between atorvastatin and excipients using DSC and FTIR. *J. Therm. Anal. Calorim.* **2016**, *123*, 933–939. [CrossRef]
57. Mohamed, A.I.; Abd-Motagaly, A.M.E.; Ahmed, O.A.A.; Amin, S.; Ali, A.I.M. Investigation of Drug–Polymer Compatibility Using Chemometric-Assisted UV-Spectrophotometry. *Pharmaceutics* **2017**, *9*, 7. [CrossRef]
58. Rus, L.M. Development of meloxicam oral lyophilisates: Role of thermal analysis and complementary techniques. *Farmacia* **2019**, *67*, 56–67. [CrossRef]
59. Protein DATA Bank. Available online: http://www.pdb.org/pdb/home/home.do (accessed on 10 January 2021).

Article

Cytotoxicity of β-Cyclodextrins in Retinal Explants for Intravitreal Drug Formulations

Manisha Prajapati [1], Gustav Christensen [2], François Paquet-Durand [2] and Thorsteinn Loftsson [1,*]

[1] Faculty of Pharmaceutical Sciences, University of Iceland, Hofsvallagata 53, IS-107 Reykjavik, Iceland; map52@hi.is

[2] Institute for Ophthalmic Research, University of Tübingen, Elfriede-Aulhorn-Strasse 5-7, 72076 Tübingen, Germany; gustav.christensen@uni-tuebingen.de (G.C.); francios.paquet-durand@klinikum.uni-tuebingen.de (F.P.-D.)

* Correspondence: thorstlo@hi.is; Tel.: +354-525-4464; Fax: +354-525-4071

Abstract: Cyclodextrins (CDs) have been widely used as pharmaceutical excipients for formulation purposes for different delivery systems. Recent studies have shown that CDs are able to form complexes with a variety of biomolecules, such as cholesterol. This has subsequently paved the way for the possibility of using CDs as drugs in certain retinal diseases, such as Stargardt disease and retinal artery occlusion, where CDs could absorb cholesterol lumps. However, studies on the retinal toxicity of CDs are limited. The purpose of this study was to examine the retinal toxicity of different beta-(β)CD derivatives and their localization within retinal tissues. To this end, we performed cytotoxicity studies with two different CDs—2-hydroxypropyl-βCD (HPβCD) and randomly methylated β-cyclodextrin (RMβCD)—using wild-type mouse retinal explants, the terminal deoxynucleotidyl transferase dUTP nick end labeling (TUNEL) assay, and fluorescence microscopy. RMβCD was found to be more toxic to retinal explants when compared to HPβCD, which the retina can safely tolerate at levels as high as 10 mM. Additionally, studies conducted with fluorescent forms of the same CDs showed that both CDs can penetrate deep into the inner nuclear layer of the retina, with some uptake by Müller cells. These results suggest that HPβCD is a safer option than RMβCD for retinal drug delivery and may advance the use of CDs in the development of drugs designed for intravitreal administration.

Keywords: cyclodextrin; retinal explant; cytotoxicity; uptake

Citation: Prajapati, M.; Christensen, G.; Paquet-Durand, F.; Loftsson, T. Cytotoxicity of β-Cyclodextrins in Retinal Explants for Intravitreal Drug Formulations. *Molecules* **2021**, *26*, 1492. https://doi.org/10.3390/molecules26051492

Academic Editor: Marina Isidori

Received: 15 February 2021
Accepted: 3 March 2021
Published: 9 March 2021

Publisher's Note: MDPI stays neutral with regard to jurisdictional claims in published maps and institutional affiliations.

1. Introduction

Cyclodextrins (CDs) are cyclic oligosaccharides consisting of (α-1,4-)-linked α-D-glucopyranose units with a hydrophobic central cavity and a hydrophilic outer surface. They can form water-soluble inclusion complexes with numerous lipophilic drugs provided that their structure (or part of it) fits in the CD cavity. No covalent bonds are formed or broken during the complexation and drug molecules in the complex are in rapid equilibria with free molecules in the aqueous complexation media [1]. The complexation affects many physicochemical properties of drugs, such as their aqueous solubility and chemical stability [2]. Natural CDs (i.e., αCD, βCD, and γCD) have a limited solubility in water and, thus, CD derivatives with an improved solubility have been synthesized and are currently used as solubilizing complexing agents in various marketed pharmaceutical products, particularly 2-hydroxypropyl-β-cyclodextrin (HPβCD) and sulfobutylether-β-cyclodextrin (SBEβCD), as well as randomly methylated β-cyclodextrin (RMβCD), although to a lesser extent [1]. CDs have also undergone extensive safety studies and have been approved by both the European Medicines Agency (EMA) and the Food and Drug Administration (FDA) for pharmaceutical use and in dietary supplements [3,4]. CDs have been the subject of numerous review publications [5–13].

Drugs for the treatment of retinal diseases are most often delivered via intravitreal injections or implants, where the drug is administered directly into the vitreous humor, which is the hydrogel-type fluid that occupies the space between the lens and the retina. The vitreous mainly consists (99%) of water in a network of collagen and hyaluronic acid. In humans, the volume of the vitreous humor is about 4 mL [14]. Drug molecules must be dissolved in the aqueous vitreous to permeate into the retinal tissue. After an intravitreal injection, hydrophilic and high molecular weight drugs (e.g., proteins and peptides) are known to be excreted via an anterior route to the aqueous humor, while small lipophilic drugs easily pass the retina and are removed via the posterior choroidal flow [15]. The half-life of a dissolved drug in the vitreous humor is typically less than 10 to 24 h, where small molecules have a shorter half-life than biomolecules such as proteins [16,17]. It is expected that the hydrophilic CD molecules (molecular weight between about 1000 and 2000 Da) are readily removed from the vitreous humor after an intravitreal injection. CDs might be able to enhance the retinal delivery of poorly soluble lipophilic drugs after intravitreal administration.

The ability of CDs to complex biomolecules depends upon the molecular structure and the CD binding constants and generally follows the order carbohydrates << nucleic acids << proteins < lipids [18]. Therefore, most biological effects of CDs are based on their interaction with membranes rich in lipids and their ability to extract lipids (e.g., phospholipids and cholesterol) from the plasma membrane [18]. In ophthalmic drug delivery, CDs have mainly been applied topically to the eye, with little or no reports on their intravitreal administration [19–21]. Although hydrophilic CDs, such as HPβCD and SBEβCD, have been shown to be well-tolerated when applied topically to the eye, with no detectable side effects, their parent βCD and RMβCD are known to extract cholesterol from cell membranes and form cholesterol/CD complexes [22–24]. Furthermore, Nociari et al. observed that βCD extracted lipofuscin bisretinoids from the retinal pigment epithelium (RPE) [25]. Based on these observations, it was proposed that CDs can be used to develop therapeutic candidates for many retinal degenerative diseases, such as Stargardt disease, which is an inherited form of macular degeneration causing central vision loss and sometimes referred to as juvenile macular degeneration. The cause of Stargardt disease is characterized by an abnormal accumulation of lipofuscin in the retina [26]. Similarly, the topical administration of HPβCD as eye drops over 3 months has shown a significant efficacy in reducing amyloid-beta and inflammation in aged mouse retina, consequently improving the retinal function by elevating retinal pigment epithelium-specific protein 65 (RPE 65), which is a key molecule in the visual cycle [27]. Even oral HPβCD treatment in mice resulted in a reduction in the retinal cholesterol content and changes in the retinal sterol, gene, and protein levels [28].

Our study aimed to examine the cytotoxicity of CDs in mouse retinal explants using βCD derivatives to explore their applicability for future ophthalmic formulations. Additionally, fluorophore-conjugation to the same CDs was used to trace the uptake and localization within the retina (Figure 1).

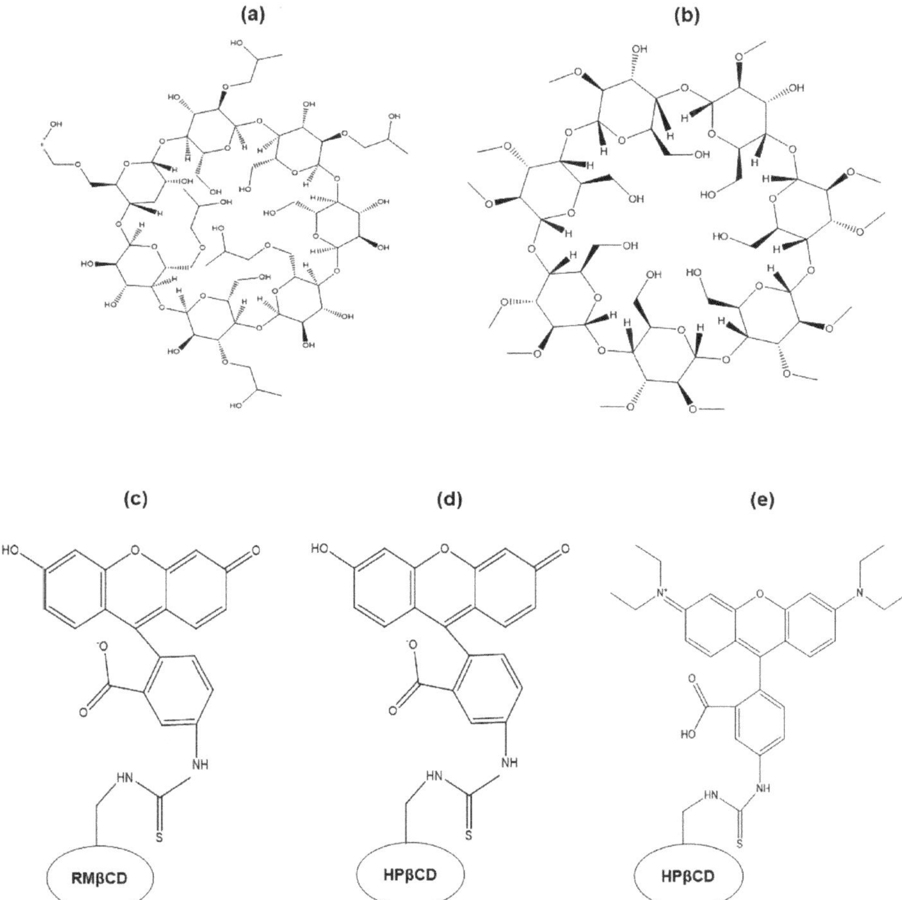

Figure 1. Chemical structures of (**a**) 2-hydroxypropyl-β-cyclodextrin (HPβCD), (**b**) randomly methylated β-cyclodextrir. (RMβCD), (**c**) 6-deoxy-6-[(5/6)-fluoresceinylthioureido]-HPβCD (FITC-RMβCD), (**d**) 6-deoxy-6-[(5/6)-fluoresceinylthioureido]-RMβCD(FITC-HPβCD), and (**e**) 6-deoxy-6-[(5/6)-rhodaminylthioureido]-HPβCD (RBITC-HPβCD).

2. Results and Discussion

2.1. Particle Size and TEM Data Analysis

The particle size of 100 mM CD aggregates was measured by Nano Sight and confirmed by Transmission Electron Microscopy (TEM) (Figure 2). CDs tend to form nano-sized aggregates when their concentration is increased [29]. Although the aggregation in pure aqueous CD solutions is generally low, the aggregation is frequently enhanced by the formation of drug/CD complexes. Furthermore, the aggregation and the size of the aggregates increase with an increasing CD concentration [30].

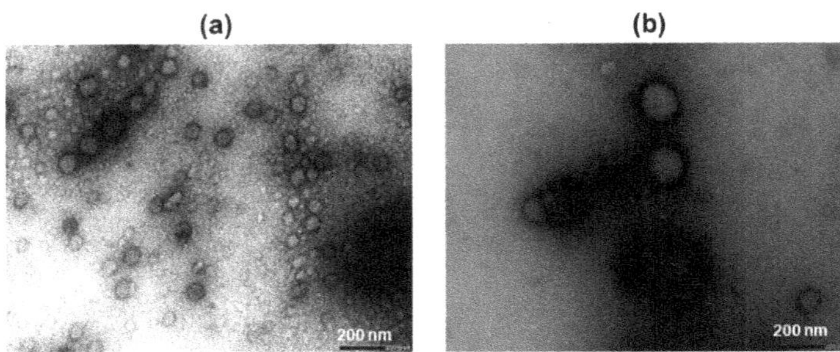

Figure 2. Transmission electron microscopy images of cyclodextrin (CD) aggregates at a magnitude of 30 K. (**a**) 100 mM HPβCD and (**b**) 100 mM RMβCD. The average diameter of RMβCD aggregates appeared to be around two times larger than that of HPβCD.

The concentration of CDs used in pharmaceutical formulations depends on the type of CD used, but due to their favorable toxicological and pharmacological profiles, the CD concentrations can be relatively high [31]. However, the limited solubility of the natural CDs, especially βCD, can limit their concentration [1]. Both HPβCD and RMβCD are very soluble in water (Section 3.1), so we used very high concentrations of these CD derivatives to observe their potential toxicity on retinal explants. Aggregates were observed in aqueous HPβCD and RMβCD solutions. The average diameter of the HPβCD aggregate, as determined by Nano Sight, was 148 nm and that of RMβCD was 305 nm. Larger aggregates (diameter > 100 nm) do not have a spherical shape like smaller ones (<100 nm). Instead, they look like clusters of smaller, spherically-shaped aggregates [32]. The aggregate diameters observed with TEM are smaller than those determined by Nano Sight. This can be explained by the TEM sample preparation, where the aggregate size or structure can change during sample preparation.

Generally, it is thought that only the free drug molecules, which have dissociated from the CD complex, are able to permeate cell membranes [1]. However, recent findings have revealed that CD molecules can enter the cells by endocytosis [33] and this may also be true for drug/CD complexes [34]. However, it appears highly unlikely that the large and hydrophilic CD aggregates can enter cells.

2.2. Cytotoxicity of β-Cyclodextrin Derivatives in Retinal Explant Cultures

The terminal deoxynucleotidyl transferase dUTP nick end labeling (TUNEL) assay was employed to quantify the number of dying cells in histological sections from retinal explant cultures incubated with CDs (Figure 3).

The relative cytotoxicity of the CDs was expressed as the percentage of TUNEL positive cells in the respective part of the retinal tissue section. They were counted in both the outer retina (outer nuclear layer, ONL) and inner retina (inner nuclear layer, INL) (Figure 3a). TUNEL positive cells were not counted in the ganglion cell layer (GCL) as most of the cells typically degenerate quickly after explant tissue preparation due to the cutting of the optic nerve.

10 and 100 mM CDs were applied on top of the retinal cultures, i.e., the side closest to the GCL, which is the route the CDs would naturally follow after an intravitreal injection. For 100 mM, the TUNEL positive cell values in the INL were about 0.5%, 5%, and 0.6% for the control, RMβCD, and HPβCD, respectively, while for the ONL, they were about 3%, 12%, and 6% for the control, RMβCD, and HPβCD, respectively (Figure 3b). Therefore, both types of CDs were toxic compared to the control when used in 100 mM concentrations. RMβCD was significantly more toxic compared to HPβCD, both in INL and ONL. HPβCD was predominantly toxic to ONL cells, i.e., where the cell bodies of photoreceptors are located. At the 10 mM concentration, RMβCD still exhibited significant toxicity and killed cells in both ONL and INL, while the number of TUNEL positive cells for HPβCD was

similar to the control. This demonstrated that the retina could safely tolerate levels as high as 10 mM HPβCD.

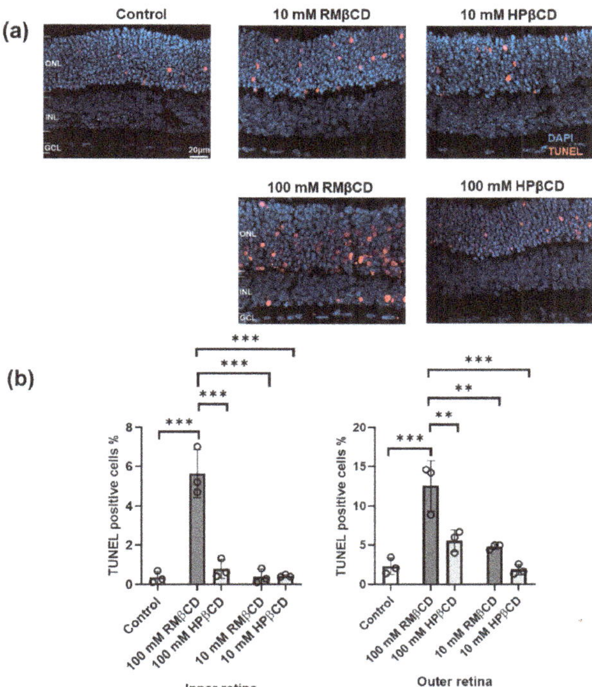

Figure 3. Cytotoxicity of β-derivatives of cyclodextrin in retina explant cultures. (**a**) Sections of retinal cultures to which different CD solutions were applied; saline solution was used as the control. The terminal deoxynucleotidyl transferase dUTP nick end labeling (TUNEL) assay (red) was used to detect dying cells. DAPI (blue) was used as a nuclear counterstain. Cultures were derived from wild-type mice at postnatal day (P) 13. CDs were added at P15 and incubated with the cultures for a duration of 48 h. ONL = outer nuclear layer, INL = inner nuclear layer, and GCL = ganglion cell layer. (**b**) Analysis of average TUNEL positive cells (%) in both INL and ONL from cultures with different CD solutions and the control. Results represent the mean ± SD for n = 3 explant cultures per group. Statistical analysis was performed using one-way ANOVA with Tukey's multiple comparisons test (α = 0.05) and asterisks represent the significant difference (** = $p \leq 0.01$ and *** = $p \leq 0.001$).

The higher toxicity of RMβCD at both concentrations compared to HPβCD may be due to various reasons. RMβCD is a modified βCD where about two-thirds of the hydroxy groups have been replaced by methoxy groups, while in the case of HPβCD, only a few of the CD-hydroxyl groups have been substituted by 2-hydroxypropyl groups, which makes RMβCD more lipophilic, with a logP value of −6 [12,35,36]. The logP value for HPβCD is about −11. The lipophilicity of the CDs clearly affects other properties, such as the solubilizing capacity, tissue irritating effect, hemolytic activity, and surface activity. The more lipophilic the compound, the easier it penetrates the cell layer, even though the permeability of CDs through biological membranes is negligible, as explained by Loftsson et al. [12], because of the size and hydrophilicity of the CD molecules.

Recently, it has been shown that RMβCD can penetrate the epidermis and dermis of the human skin [37]. In addition, relatively high amounts of HPβCD and dimethyl-β-cyclodextrin were absorbed via the rectum of rats and excreted into the urine, suggesting CD complexes may be absorbable through the rectal mucosa [38]. However, the latest findings regarding the endocytosis of CDs gave a whole new perspective of CDs being able to enter cells [33]. CD molecules can easily form complexes with natural hydrophobic molecules including the cellular components based on the host-guest interaction. Phospholipids are the preferred cellular target for αCD and cholesterol for βCD [39]. Because of this property, they have also been described to induce lysis of cell membranes by removal of membrane components such as cholesterol, phospholipids, and proteins [23,40,41].

Cholesterol is one of the major components of the cell membrane constituting 30% of total lipids and plays an important structural role in membrane stability [42]. Since βCD has an affinity for cholesterol, this CD can induce the release of cellular cholesterol directly affecting the cell and biological barriers [40]. Consequently, the cholesterol content of the membranes can decrease thus affecting the function of the cell membrane and disrupting the barrier function of the cell layers [39]. Additionally, it was found that cholesterol extraction caused the destabilization of tight-junction protein complexes, which are localized in lipid rafts [39]. Methylated β-cyclodextrins tend to interact strongly with lipids [43,44], and there is a correlation between the cytotoxic effect and the cholesterol complexation properties of βCD such that the higher complexation with cholesterol increases the toxicity to the cells [22].

In the retina, cholesterol represents >98% of total sterols [45]. The stronger interaction of CD with the lipids/cholesterol in retinal cells can aid in more vigorous cell membrane destabilization and more cell death. Likewise, there are similar reports on Müller glia cells where the cholesterol status plays an important role. Low cholesterol is harmful to the retinal cells; hence, more cholesterol extraction by RMβCD likely causes more toxicity [45]. This idea was further supported by our phase solubility studies of cholesterol with different β-cyclodextrin derivates (Figure 4).

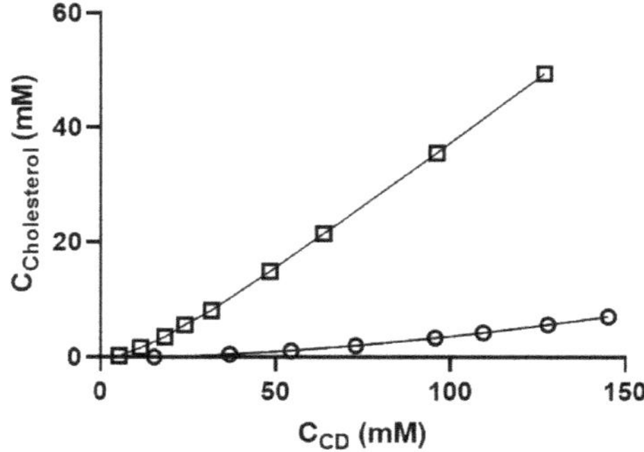

Figure 4. Phase-solubility diagrams of cholesterol in various aqueous β-cyclodextrin derivatives at room temperature. The diagram shows the CD concentration plotted against the cholesterol concentration. Overall, cholesterol was solubilized about five-fold more by RMβCD than by HPβCD. Each point represents the mean of triplicate experiments. Key: o = HPβCD and □ = RMβCD.

Cholesterol was solubilized five times more effectively by RMβCD compared to HPβCD (Figure 4). Cholesterol had the highest affinity for the lipophilic RMβCD but had a lower affinity for the very hydrophilic cyclodextrins like HPβCD. Cholesterol solubilization was also affected by the structure of the CD derivative, like the number and position of the methyl groups and the presence of ionic groups [22].

Similar results have been obtained when the toxicity of these CDs was tested on different cell lines. The cytotoxicity of methylated βCDs was found to be very high in intestinal Caco-2 cell cultures, while substitution of the βCD with hydroxyl groups drastically decreased the cytotoxicity [46]. However, in another study, HPβCD presented no cytotoxicity up to 200 mM on the same cell lines [22].

Furthermore, RMβCD possesses surface-active properties [47,48] and it even shows a detergent-like mechanism of lipid solubilization when interacting with lipid vesicles. Here, RMβCD was first adsorbed onto the vesicle surface, which was followed by RMβCD partitioning from the aqueous medium into the phospholipid bilayers forming lipid-RMβCD mixed assemblies and finally the lipid solubilization into micelle like aggregates [49]. The cells in the ONL are photoreceptor cells, and cholesterol is an important component of photoreceptor membranes, relevant for the cells function [50,51]. Hence, photoreceptors might suffer more from the cholesterol extraction capacity of the CDs, something that might be particularly relevant for the higher toxicity observed with RMβCD. Additionally, the cell death in the ONL might be exacerbated by an overall higher sensitivity of photoreceptors, when compared to INL cells.

2.3. Fluorescent Microscopy of Fluorescently-Labeled Cyclodextrin Derivates to Study Cellular Uptake in Retinal Cultures

Following this, we investigated the overall distribution of fluorescently-labeled CDs in the retina. To this end, RBITC-HPβCD, FITC-HPβCD, and FITC-RMβCD were used and the fluorescent intensity in the inner and outer retina was quantified (Figure 5) and compared to the control specimen to account for auto-fluorescence coming from the tissue itself. No difference between FITC-labeled CDs was observed for the inner and outer retina, indicating that the HPβCD and RMβCD distribute similarly within the retina. Therefore, the difference in cytotoxicity between the two compounds was likely due to their respective effect on the cells, as discussed above, and not because of differences in the overall tissue distribution.

Some subtle differences between FITC-labeled CDs were observed. For FITC-HPβCD, a string-like structure was seen, spanning across the inner and outer retina. This was not as prominent in cultures where FITC-RMβCD was applied (Figure 5a). This type of staining suggests that the CDs had partly been taken up by Müller glial cells.

When using a rhodamine-labeled CD, we observed an elevated signal in the retina. The signal-to-noise ratio was higher for RBITC-HPβCD than for FITC-HPβCD (Figure 5b,c). This allowed us to detect specific uptake in photoreceptors, which supported the result from the cytotoxicity analysis, where HPβCD mainly killed cells in the outer retina.

However, these results should be taken with a grain of salt as the dye labelling could also have affected cell uptake. First of all, we did not observe the same Müller cell uptake in retinas with RBITC-HPβCD, compared with retinas with FITC-HPβCD. Since the rhodamine molecule is positively charged, the rhodamine-conjugated CDs might bind to negatively charged cell membranes and extracellular matrix elements, possibly facilitating cellular uptake. FITC and RBITC derivatives behave differently and are internalized by different processes. The labelling increases the molecular weight and may alter the properties of the parental CDs, while these still retain high water solubility and cannot cross the cell membrane by passive diffusion. However, there are reports suggesting that endocytosis was observed in fluorescent CDs as well [34]. Müller cells have the capacity to assemble and secrete lipoproteins which can be utilized by photoreceptors or inner retinal neurons, serving as an intraretinal source of cholesterol [52]. This could explain the Müller cell uptake of HPβCD.

While the fluorescent CDs do not behave exactly as their non-fluorescent form, these studies enhance the understanding of the behavior of labelled CD derivatives at the tissue level [53]. Together with the toxicity analysis, these results may be employed in further the development of CDs as drugs or drug carriers for the treatment of retinal diseases.

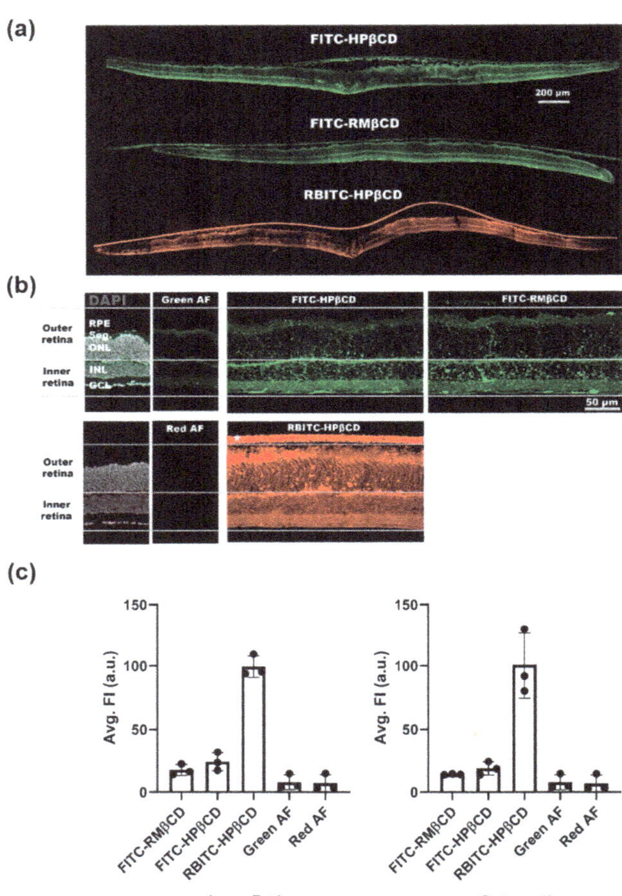

Figure 5. Uptake of fluorescently-labeled CDs in the retina explant culture. (**a**) Overview of tissue sections from wild type (WT) mice explant cultures to which three different fluorescently-labeled CDs were added to the side facing down in the image. (**b**) Close-up images of the sections were taken for an analysis of the fluorescent signal. Images from control sections (without added CDs) were used to measure the intensity of the red and green auto-fluorescence (AF) coming from the tissue itself. For this analysis, the outer retina is defined as the area from the outer nuclear layer (ONL) to the retinal pigment epithelium (RPE), while the inner retina encompasses the area from the ganglion cell layer (GCL) to the inner nuclear layer (INL). Seg. = segments of the photoreceptors (inner and outer segment). * membrane on which the retinal explant was cultured. (**c**) Analysis of the average fluorescent intensity coming from either the inner or outer retina. For both inner and outer retina, a high signal was detected from RBITC-labeled CDs, while the signal was lower and similar for FITC-labeled CDs. Results represent the mean $\pm$ SD for $n = 3$, where n is the number of animals.

3. Materials and Methods

3.1. Materials

Randomly methylated β-cyclodextrin (RMβCD, also referred to as RAMEB) with a degree of substitution (DS) of 12.6 (molecular weight, MW 1312) was purchased from Wacker Chemie (Munich, Germany), and cholesterol from Sigma Chemical Co. (St. Louis, MO, USA), while 2-hydroxypropyl-β-cyclodextrin (HPβCD) with DS 4.2 (MW 1380) was kindly provided by Janssen Pharmaceutica, Belgium. The solubility of HPβCD in water is >600 mg/mL, while that of RMβCD is >500 mg/mL [1].

6-deoxy-6-[(5/6)-fluoresceinylthioureido]-RMβCD, 6-deoxy-6-[(5/6)-fluoresceinyl thioureido]-HPβCD and 6-deoxy-6-[(5/6)-rhodaminylthioureido]-HPβCD were kindly provided by CycloLab, Budapest, Hungary. Milli-Q water and sterile saline solution were used for the preparation of CD solution and all other chemicals were commercially available products (Sigma-Aldrich, Saint Louis, MO, USA) of special reagent grade.

3.2. Methods

3.2.1. Animals

The C3H wild-type mouse line was used in all studies [54]. Animals were used irrespective of gender and were housed with free access to food and water under standard white light with 12 h light/dark cycles. They were sacrificed at postnatal day (P) 13 by CO_2 asphyxiation, followed by cervical dislocation. All procedures were performed in accordance with §4 of the German law on animal protection and approved by the animal protection committee of the University of Tübingen (*Einrichtung für Tierschutz, Tierärztlichen Dienst und Labortierkunde; Registration No. AK02/19M, 3 April 2019*).

3.2.2. Assessment of the Retinal Cytotoxicity of β-Cyclodextrin Derivatives

Culturing of Organotypic Retinal Explant Cultures

To study the cytotoxicity of CDs on the retinal tissue, retinas from mice were isolated for culturing for an extended period of time. The detailed protocol is described elsewhere [55], but will be summarized here. Immediately after animal sacrifice, the eyes were enucleated and incubated for 5 min at room temperature (RT) in R16 serum-free culture medium (Gibco, Carlsbad, CA, USA). To promote the removal of the sclera and choroid, the eyes were transferred to a preheated (37 °C) solution of 0.12% proteinase K (MP Biomedicals, Illkirch-Grafenstaden, France) and incubated for 15 min. Afterwards, the eyes were soaked in 1:4 mixtures of 10% FBS/medium to stop the protease reaction. The eyes were dissected under sterile conditions. The retina with the retinal pigment epithelium (RPE) attached was isolated and cultured on a Transwell membrane (polycarbonate, 0.4 μm pore size, COSTAR, NY) with the RPE side facing down in a 6-well plate. Then, 1 mL complete medium (CM, R16 medium with supplements; detailed under [55]) was added to each well. The explants were allowed to recover from the explantation procedure in a sterile incubator (37 °C, 5% CO_2) for 48 h. The CM was exchanged every second day by removing 0.7 mL of the CM in the plate and adding 0.9 mL fresh CM to account for evaporation and conserve neuroprotective agents produced by the retinal culture. At P15, CD and saline solution were applied to the cultures for a 48 h incubation period. Here, 20 μL of isotonic CD solutions (adjusted by NaCl) was carefully placed on the top of the retina to cover the whole tissue. Either 10 or 100 mM CD was used. For fluorescently-tagged CD, 5 mM was applied. Alternatively, a 0.9% NaCl solution was added as a control. All solutions were passed through sterile filters (PES, 0.22 μm, Merck Millipore, Ireland) before being introduced to the culture.

Preparation of Retinal Tissue Sections

After the culturing of organotypic retinal explants with CD or saline, the cultures were fixed in a 4% paraformaldehyde/PBS solution for at least 45 min. The explants were cryoprotected by introducing an incremental amount of sucrose to the well plate, i.e., 10% sucrose, 20% sucrose, and 30% sucrose for 10, 20, and 30 min (at RT), respectively. Afterwards, the area of the retina culture attached to the Transwell membrane was cut out with fine scissors, and the membrane piece was submerged in an embedding medium (Tissue-Tek O.C.T. Compound, Sakura Finetek Europe, Netherlands), followed by snap freezing in liquid nitrogen. The frozen specimens were sectioned on an NX50 cryostat (ThermoFisher, Waltham, MA) to produce 14 μm thick sections on Superfrost Plus object slides (R. Langenbrinck, Emmendingen, Germany) used for direct imaging or further staining.

Assessing Cell Death in Retinal Sections Using the TUNEL Assay

A terminal deoxynucleotidyl transferase dUTP nick end labeling (TUNEL) assay was used to stain the nuclei with damaged (nick-end) DNA [56]. This was done to quantify the number of dying cells in retinal cultures after CD or saline was applied to assess the cytotoxicity of the CDs. Firstly, microscopy slides with retinal sections were rehydrated with PBS. A proteinase K solution preheated to 37 °C (0.21 µg/mL, Tris-buffered saline (TBS) (Sigma-Aldrich, Saint Louis, MO, USA) was added and incubated for 5 min at 37 °C. After washing with TBS, a solution of ethanol/acetic acid was added and incubated for 5 min before washing. A blocking solution consisting of 1% bovine serum albumin, 10% normal goat serum, 3% Triton-X (Sigma-Aldrich, Saint Louis, MO, USA), and 2.5% fish gelatin in PBS was incubated on the sections (1 h, RT). A TUNEL reaction solution consisting of the enzyme solution and labeling solution from the In Situ Cell Death Detection Kit (TMR red, Product No. 12156792910, Sigma Aldrich) was prepared at a 1:9 ratio, diluted in blocking solution (1:1), and incubated with the slides at 37 °C for 1 h. The slides were washed with PBS, and a mounting medium with DAPI (Vectashield, Vector Laboratories, USA) was added. Samples were kept at 2–8 °C for at least 30 min before imaging with fluorescent microscopy (Axio Imager Z2 ApoTome, Carl Zeiss Microscopy GmbH), using a CCD camera with a 20× objective. Ex./Em. of 548/561 nm was used to detect TUNEL labeling at random locations of the section. Image acquisition was conducted by recording z-stacks, each with 10 images 1 µm apart. Values such as the exposure time for the red (TUNEL) channel, binning, and brightness/contrast of each image were kept consistent. To quantify the number of dying cells, in either the inner or outer nuclear layer in the tissue section, the following equation was used:

$$\text{TUNEL positive cells (\%)} = \frac{\text{\#TUNEL positive nuclei}}{\text{Area of layer / Average area of nuclei}} \cdot 100\%. \tag{1}$$

To analyze the statistical significance within the dataset, one-way ANOVA with Tukey's multiple comparisons test ($\alpha = 0.05$) was performed using GraphPad Prism 8.

Determining the Retinal Uptake of Fluorescently-Labeled β-Cyclodextrin

Fluorescently-labeled CD (FITC-HPβCD, RBITC-HPβCD, and FITC-RMβCD) was added to organotypic retinal explant cultures, as described previously. Slides with retinal tissue sections were rehydrated in PBS for 10 min before mounting medium with DAPI (Vectashield, Vector Laboratories, USA) was added. Images were recorded with fluorescent microscopy, as described above. A green (Ex./Em. 493/513) and red (Ex./Em. 558/575) channel were used to detect the fluorescein or rhodamine labeling, respectively. Image acquisition was conducted by recording z-stacks, each with 14 images 1 µm apart. Values such as the exposure time for the green and red channel, binning, and brightness/contrast of each image were kept consistent. Sections from the saline-treated retinas were imaged using the same parameters to determine the level of green and red auto-fluorescence from the cultures. Alternatively, tile pictures showing the entire retinal section were recorded by stitching together adjacent projected z-stacks in the image acquisition software (ZEN 2.6, Carl Zeiss Microscopy GmbH).

3.2.3. Particle Size Measurement

Nano Sight Wave

The laser-based light scattering analysis of CD particles was performed with Nano Sight NS300 (Malvern, Worcestershire, UK), fitted with an O-ring top-plate. Nanoparticle tracking analysis (NTA) software was used to capture images and process data, representing the concentration, size distribution, and intensity of particles in the sample. Sample measurement was done in static mode using a capture time of 60 s and five repeats. The camera level was adjusted to 11 so that all particles were visible. The same camera level was used for all the samples. A suitable detection level was selected for data analysis to limit the detection of non-particles and was between levels 4 and 12. The result for each sample was based on the average of five measurements obtained from the NTA and represented by the average particle concentration, average particle size (i.e., mean size), and mode size (i.e., the size that displays the highest peak).

Transmission Electron Microscopy (TEM) Analysis

The size and morphology of the CD in an aqueous solution were evaluated by using Model JEM-1400 Transmission Electron Microscopy (JEOL, Tokyo, Japan). The negative staining technique was utilized. Firstly, a small amount of clear CD solution (3 µL) was dropped on a 300-mesh coated grid and dried at 37–40 °C for one hour. Then, a drop of centrifuged 4% w/v uranyl acetate (26 µL) was added to the loaded grid. After 6 min of staining, the sample was dried overnight at room temperature. Finally, the stained grid was placed in the sample holder and inserted into the microscope.

3.2.4. Phase-Solubility Studies

The phase-solubility studies of cholesterol were performed by dissolving cholesterol (20 mg/mL) in methylene chloride, followed by solvent evaporation (0.3 mL per vial) under a flow of nitrogen in cylindrical vials. This left a very thin layer of cholesterol on the inner surface of the vials. Aqueous CD solutions were added to the vials, which were tightly sealed and heated in an autoclave (121 °C for 20 min) [57]. The vials were opened after equilibrating them at room temperature overnight and a small amount of solid cholesterol was added to each vial, which were again allowed to equilibrate at room temperature for an additional 6 days. The seeding with cholesterol was performed to reduce the possibility of complications caused by differences in the solid-state of cholesterol. Finally, the aqueous cholesterol suspensions were filtered through a 0.45 µm nylon membrane filter, and the filtrate was analyzed using high-pressure liquid chromatography (HPLC).

Quantitative determination of cholesterol was performed on a reversed-phase HPLC system from Merck-Hitachi (Darmstadt, Germany) consisting of an L 4250 UV-Vis detector operated at 203 nm, L 6200 A Intelligent pump, AS-2000A Autosampler, D-2500 Cromato-Integrator, and Phenomex Luna (Cheshire, UK) 5 µm C18 reversed-phase column (150 × 4.6 mm). The mobile phase consisted of methanol, acetonitrile, isopropyl alcohol, and tetrahydrofuran (50:25:25:0.1).

Author Contributions: Conceptualization, M.P.; methodology, M.P. and G.C.; data curation, M.P., T.L. and G.C.; formal analysis, M.P. and G.C.; investigation; M.P. and G.C.; writing—original draft preparation, M.P.; writing, reviewing, and editing, M.P., G.C., F.P.-D. and T.L.; supervision, F.P.-D. and T.L.; funding acquisition, F.P.-D. and T.L. All authors have read and agreed to the published version of the manuscript.

Funding: This work was financially supported by the European Union grant no. MSCA-ITN-2017-765441 (transMed); the German Research Council (DFG) grant no. PA1751/10-1; and the Faculty of Pharmaceutical Sciences, University of Iceland.

Data Availability Statement: No data reported.

Acknowledgments: The authors gratefully acknowledge CycloLab, Budapest, Hungary, for providing us with the fluorescent cyclodextrins used in the study.

Conflicts of Interest: The authors declare no conflict of interest.

Sample Availability: Samples of the compounds are not available from the authors.

References

1. Loftsson, T.; Jarho, P.; Másson, M.; Järvinen, T. Cyclodextrins in drug delivery. *Expert Opin. Drug Deliv.* **2005**, *2*, 335–351. [CrossRef]
2. Loftsson, T.; Björnsdóttir, S.; Pálsdóttir, G.; Bodor, N. The effects of 2-hydroxypropyl-β-cyclodextrin on the solubility and stability of chlorambucil and melphalan in aqueous solution. *Int. J. Pharm.* **1989**, *57*, 63–72. [CrossRef]
3. Stella, V.J.; He, Q. Cyclodextrins. *Toxicol. Pathol.* **2008**, *36*, 30–42. [CrossRef]
4. Questions and Answers on Cyclodextrins Used as Excipients in Medicinal Products for Human Use. Available online: https://www.ema.europa.eu/en/documents/report/cyclodextrins-used-excipients-report-published-support-questions-answers-cyclodextrins-used_en.pdf (accessed on 11 February 2021).
5. Zhang, D.; Lv, P.; Zhou, C.; Zhao, Y.; Liao, X.; Yang, B. Cyclodextrin-based delivery systems for cancer treatment. *Mater. Sci. Eng. C* **2019**, *96*, 872–886. [CrossRef] [PubMed]
6. Carneiro, S.B.; Duarte, F.; Ílary, C.; Heimfarth, L.; Quintans, J.D.S.S.; Quintans-Júnior, L.J.; Júnior, V.F.D.V.; Neves de Lima, Á.A. Cyclodextrin–Drug Inclusion Complexes: In Vivo and In Vitro Approaches. *Int. J. Mol. Sci.* **2019**, *20*, 642. [CrossRef] [PubMed]
7. Saokham, P.; Muankaew, C.; Jansook, P.; Loftsson, T. Solubility of Cyclodextrins and Drug/Cyclodextrin Complexes. *Molecules* **2018**, *23*, 1161. [CrossRef] [PubMed]
8. Conceicao, J.; Adeoye, O.; Cabral-Marques, H.M.; Lobo, J.M.S. Cyclodextrins as Drug Carriers in Pharmaceutical Technology: The State of the Art. *Curr. Pharm. Des.* **2018**, *24*, 1405–1433. [CrossRef]
9. Muankaew, C.; Loftsson, T. Cyclodextrin-Based Formulations: A Non-Invasive Platform for Targeted Drug Delivery. *Basic Clin. Pharmacol. Toxicol.* **2018**, *122*, 46–55. [CrossRef]
10. Loftsson, T.; Duchene, D. Cyclodextrins and their pharmaceutical applications. *Int. J. Pharm.* **2007**, *329*, 1–11. [CrossRef]
11. Saokham, P.; Loftsson, T. g-Cyclodextrin. *Int. J. Pharm.* **2017**, *516*, 278–292.
12. Jansook, P.; Ogawa, N.; Loftsson, T. Cyclodextrins: Structure, physicochemical properties and pharmaceutical applications. *Int. J. Pharm.* **2018**, *535*, 272–284. [CrossRef] [PubMed]
13. Stella, V.J.; Rajewski, R.A. Sulfobutylether-β-cyclodextrin. *Int. J. Pharm.* **2020**, *583*, 119396. [CrossRef]
14. Friedrich, S.; Cheng, Y.-L.; Saville, B. Drug distribution in the vitreous humor of the human eye: The effects of intravitreal injection position and volume. *Curr. Eye Res.* **1997**, *16*, 663–669. [CrossRef] [PubMed]
15. Varela-Fernández, R.; Díaz-Tomé, V.; Luaces-Rodríguez, A.; Conde-Penedo, A.; García-Otero, X.; Luzardo-Álvarez, A.; Fernández-Ferreiro, A.; Otero-Espinar, F.J. Drug Delivery to the Posterior Segment of the Eye: Biopharmaceutic and Pharmacokinetic Considerations. *Pharmaceutics* **2020**, *12*, 269. [CrossRef] [PubMed]
16. Ashton, P. *Intraocular Drug Delivery*; Jaffe, G.J., Ashton, P., Pearson, P.A., Eds.; Taylor & Francis: New York, NY, USA, 2006; pp. 1–25.
17. Chaudhari, P.; Ghate, V.M.; Lewis, S.A. Supramolecular cyclodextrin complex: Diversity, safety, and applications in ocular therapeutics. *Exp. Eye Res.* **2019**, *189*, 107829. [CrossRef] [PubMed]
18. Leclercq, L. Interactions between cyclodextrins and cellular components: Towards greener medical applications? *Beilsteinj. Org. Chem.* **2016**, *12*, 2644–2662. [CrossRef]
19. Loftsson, T.; Stefánsson, E. Cyclodextrins in ocular drug delivery: Theoretical basis with dexamethasone as a sample drug. *J. Drug Deliv. Sci. Technol.* **2007**, *17*, 3–9. [CrossRef]
20. Castro-Balado, A.; Mondelo-García, C.; Zarra-Ferro, I.; Fernández-Ferreiro, A. New ophthalmic drug delivery systems. *Farm. Hosp.* **2020**, *44*, 149–157. [CrossRef] [PubMed]
21. Moiseev, R.V.; Morrison, P.W.J.; Steele, F.; Khutoryanskiy, V.V. Penetration Enhancers in Ocular Drug Delivery. *Pharmaceutics* **2019**, *11*, 321. [CrossRef] [PubMed]
22. Kiss, T.; Fenyvesi, F.; Bácskay, I.; Váradi, J.; Fenyvesi, É.; Iványi, R.; Szente, L.; Tósaki, Á.; Vecsernyés, M. Evaluation of the cytotoxicity of beta-cyclodextrin derivatives: Evidence for the role of cholesterol extraction. *Eur. J. Pharm. Sci.* **2010**, *40*, 376–380. [CrossRef]
23. Irie, T.; Uekama, K. Pharmaceutical applications of cyclodextrins 3. Toxicological issues and safety evaluation. *J. Pharm. Sci.* **1997**, *86*, 147–162. [CrossRef] [PubMed]
24. Vecsernyés, M.; Fenyvesi, F.; Bácskay, I.; Deli, M.A.; Szente, L. Fenyvesi, Éva Cyclodextrins, Blood–Brain Barrier, and Treatment of Neurological Diseases. *Arch. Med. Res.* **2014**, *45*, 711–729. [CrossRef]
25. Nociari, M.M.; Lehmann, G.L.; Bay, A.E.P.; Radu, R.A.; Jiang, Z.; Goicochea, S.; Schreiner, R.; Warren, J.D.; Shan, J.; De Beaumais, S.A.; et al. Beta cyclodextrins bind, stabilize, and remove lipofuscin bisretinoids from retinal pigment epithelium. *Proc. Natl. Acad. Sci. USA* **2014**, *111*, E1402–E1408. [CrossRef] [PubMed]
26. Racz, B.; Varadi, A.; Kong, J.; Allikmets, R.; Pearson, P.G.; Johnson, G.; Cioffi, C.L.; Petrukhin, K. A non-retinoid antagonist of retinol-binding protein 4 rescues phenotype in a model of Stargardt disease without inhibiting the visual cycle. *J. Biol. Chem.* **2018**, *293*, 11574–11588. [CrossRef] [PubMed]
27. Kam, J.H.; Lynch, A.; Begum, R.; Cunea, A.; Jeffery, G. Topical cyclodextrin reduces amyloid beta and inflammation improving retinal function in ageing mice. *Exp. Eye Res.* **2015**, *135*, 59–66. [CrossRef]
28. El-Darzi, N.; Mast, N.; Petrov, A.M.; Pikuleva, I.A. 2-Hydroxypropyl-β-cyclodextrin reduces retinal cholesterol in wild-type and $Cyp27a1^{-/-}Cyp46a1^{-/-}$ mice with deficiency in the oxysterol production. *Br. J. Pharm.* **2020**, 1–15. [CrossRef] [PubMed]

29. Messner, M.; Kurkov, S.V.; Jansook, P.; Loftsson, T. Self-assembled cyclodextrin aggregates and nanoparticles. *Int. J. Pharm.* **2010**, *387*, 199–208. [CrossRef]
30. Loftsson, T.; Saokham, P.; Couto, A.R.S. Self-association of cyclodextrins and cyclodextrin complexes in aqueous solutions. *Int. J. Pharm.* **2019**, *560*, 228–234. [CrossRef]
31. Loftsson, T. Cyclodextrins in Parenteral Formulations. *J. Pharm. Sci.* **2021**, *110*, 654–664. [CrossRef]
32. Do, T.T.; Van Hooghten, R.; Mooter, G.V.D. A study of the aggregation of cyclodextrins: Determination of the critical aggregation concentration, size of aggregates and thermodynamics using isodesmic and K2–K models. *Int. J. Pharm.* **2017**, *521*, 318–326. [CrossRef]
33. Rosenbaum, A.I.; Zhang, G.; Warren, J.D.; Maxfield, F.R. Endocytosis of beta-cyclodextrins is responsible for cholesterol reduction in Niemann-Pick type C mutant cells. *Proc. Natl. Acad. Sci. USA* **2010**, *107*, 5477–5482. [CrossRef]
34. Réti-Nagy, K.; Malanga, M.; Fenyvesi, É.; Szente, L.; Vámosi, G.; Váradi, J.; Bácskay, I.; Fehér, P.; Ujhelyi, Z.; Róka, E.; et al. Endocytosis of fluorescent cyclodextrins by intestinal Caco-2 cells and its role in paclitaxel drug delivery. *Int. J. Pharm.* **2015**, *496*, 509–517. [CrossRef]
35. Řezanka, M. Synthesis of substituted cyclodextrins. *Environ. Chem. Lett.* **2018**, *17*, 49–63. [CrossRef]
36. Szente, L.; Szejtli, J. Highly soluble cyclodextrin derivatives: Chemistry, properties, and trends in development. *Adv. Drug Deliv. Rev.* **1999**, *36*, 17–28. [CrossRef]
37. Weisse, S.; Perly, B.; Creminon, C.; Ouvrard-Baraton, F.; Djedai'Ni-Pilard, F. Enhancement of vitamin A skin absorption by cyclodextrins. *J. Drug Deliv. Sci. Technol.* **2004**, *14*, 77–86. [CrossRef]
38. Matsuda, H.; Arima, H. Cyclodextrins in transdermal and rectal delivery. *Adv. Drug Deliv. Rev.* **1999**, *36*, 81–99. [CrossRef]
39. Haimhoffer, Á.; Rusznyák, Á.; Réti-Nagy, K.; Vasvári, G.; Váradi, J.; Vecsernyés, M.; Bácskay, I.; Fehér, P.; Ujhelyi, Z.; Fenyvesi, F.; et al. Cyclodextrins in Drug Delivery Systems and Their Effects on Biological Barriers. *Sci. Pharm.* **2019**, *87*, 33. [CrossRef]
40. Kilsdonk, E.P.C.; Yancey, P.G.; Stoudt, G.W.; Bangerter, F.W.; Johnson, W.J.; Phillips, M.C.; Rothblat, G.H. Cellular Cholesterol Efflux Mediated by Cyclodextrins. *J. Biol. Chem.* **1995**, *270*, 17250–17256. [CrossRef]
41. Ohtani, Y.; Irie, T.; Uekama, K.; Fukunaga, K.; Pitha, J. Differential effects of α-, β- and γ-cyclodextrins on human erythrocytes. *Eur. J. Biochem.* **1989**, *186*, 17–22. [CrossRef]
42. Krause, M.R.; Regen, S.L. The Structural Role of Cholesterol in Cell Membranes: From Condensed Bilayers to Lipid Rafts. *Acc. Chem. Res.* **2014**, *47*, 3512–3521. [CrossRef]
43. Yancey, P.G.; Rodrigueza, W.V.; Kilsdonk, E.P.C.; Stoudt, G.W.; Johnson, W.J.; Phillips, M.C.; Rothblat, G.H. Cellular Cholesterol Efflux Mediated by Cyclodextrins: DEMONSTRATION OF KINETIC POOLS AND MECHANISM OF EFFLUX*. *J. Biol. Chem.* **1996**, *271*, 16026–16034. [CrossRef] [PubMed]
44. Irie, T.; Wakamatsu, K.; Arima, H.; Aritomi, H.; Uekama, K. Enhancing effects of cyclodextrins on nasal absorption of insulin in rats. *Int. J. Pharm.* **1992**, *84*, 129–139. [CrossRef]
45. Lakk, M.; Yarishkin, O.; Baumann, J.M.; Iuso, A.; Križaj, D. Cholesterol regulates polymodal sensory transduction in Müller glia. *Glia* **2017**, *65*, 2038–2050. [CrossRef] [PubMed]
46. Roka, E.; Ujhelyi, Z.; Deli, M.; Bocsik, A.; Fenyvesi, E.; Szente, L.; Fenyvesi, F.; Vecsernyés, M.; Váradi, J.; Fehér, P.; et al. Evaluation of the Cytotoxicity of α-Cyclodextrin Derivatives on the Caco-2 Cell Line and Human Erythrocytes. *Molecules* **2015**, *20*, 20269–20285. [CrossRef] [PubMed]
47. Cserháti, T.; Szejtli, J. Surfactant activity of methylated b-cyclodextrins. *Tenside Deterg.* **1985**, *22*, 237–238.
48. Azarbayjani, A.F.; Lin, H.; Yap, C.W.; Chan, Y.W.; Chan, S.Y. Surface tension and wettability in transdermal delivery: A study on the in-vitro permeation of haloperidol with cyclodextrin across human epidermis. *J. Pharm. Pharmacol.* **2010**, *62*, 770–778. [CrossRef] [PubMed]
49. Boulmedarat, L.; Bochot, A.; Lesieur, S.; Fattal, E. Evaluation of buccal methyl-beta-cyclodextrin toxicity on human oral epithelial cell culture model. *J. Pharm. Sci.* **2005**, *94*, 1300–1309. [CrossRef]
50. Mauch, D.H. CNS Synaptogenesis Promoted by Glia-Derived Cholesterol. *Science* **2001**, *294*, 1354–1357. [CrossRef] [PubMed]
51. Fliesler, S.J.; Vaughan, D.K.; Jenewein, E.C.; Richards, M.J.; Nagel, B.A.; Peachey, N.S. Partial Rescue of Retinal Function and Sterol Steady-State in a Rat Model of Smith-Lemli-Opitz Syndrome. *Pediatr. Res.* **2007**, *61*, 273–278. [CrossRef]
52. Rao, S.R.; Fliesler, S.J. Cholesterol homeostasis in the vertebrate retina: Biology and pathobiology. *J. Lipid Res.* **2021**, 100057. [CrossRef]
53. Váradi, J.; Hermenean, A.; Gesztelyi, R.; Jeney, V.; Balogh, E.; Majoros, L.; Malanga, M.; Fenyvesi, É.; Szente, L.; Bácskay, I.; et al. Pharmacokinetic Properties of Fluorescently Labelled Hydroxypropyl-Beta-Cyclodextrin. *Biomolecules* **2019**, *9*, 509. [CrossRef]
54. Sanyal, S.; Bal, A.K. Comparative light and electron microscopic study of retinal histogenesis in normal and rd mutant mice. *Z. Anat. Entwick!.* **1973**, *142*, 219–238. [CrossRef]
55. Belhadj, S.; Tclone, A.; Christensen, G.; Das, S.; Chen, Y.; Paquet-Durand, F. Long-Term, Serum-Free Cultivation of Organotypic Mouse Retina Explants with Intact Retinal Pigment Epithelium. *J. Vis. Exp.* **2020**, *165*, e61868. [CrossRef]
56. Loo, D.T. In Situ Detection of Apoptosis by the TUNEL Assay: An Overview of Techniques. *Pericytes* **2010**, *682*, 3–13. [CrossRef]
57. Loftsson, T.; Másson, M.; Sigurjónsdóttir, J.F. Methods to enhance the complexation efficiency of cylodextrins. *S.T.P. Pharma Sci.* **1999**, *9*, 237–242.

Article

Effect of Water Microsolvation on the Excited-State Proton Transfer of 3-Hydroxyflavone Enclosed in γ-Cyclodextrin

Khanittha Kerdpol [1], Rathawat Daengngern [2,3], Chanchai Sattayanon [4], Supawadee Namuangruk [4], Thanyada Rungrotmongkol [5,6,7], Peter Wolschann [8,9], Nawee Kungwan [10,11,*] and Supot Hannongbua [1,7,*]

1. Center of Excellence in Computational Chemistry (CECC), Department of Chemistry, Faculty of Science, Chulalongkorn University, Bangkok 10330, Thailand; khanittha.view@gmail.com
2. Department of Chemistry, Faculty of Science, King Mongkut's Institute of Technology Ladkrabang, Bangkok 10520, Thailand; rathawat.da@kmitl.ac.th
3. Integrated Applied Chemistry Research Unit, Faculty of Science, King Mongkut's Institute of Technology Ladkrabang, Bangkok 10520, Thailand
4. National Nanotechnology Center (NANOTEC), NSTDA, 111 Thailand Science Park, Pahonyothin Road, Klong Luang, Pathum Thani 12120, Thailand; c.sattayanon@gmail.com (C.S.); supawadee@nanotec.or.th (S.N.)
5. Structural and Computational Biology Research Unit, Department of Biochemistry, Faculty of Science, Chulalongkorn University, Bangkok 10330, Thailand; t.rungrotmongkol@gmail.com
6. Program in Bioinformatics and Computational Biology, Graduate School, Chulalongkorn University, Bangkok 10330, Thailand
7. Molecular Sensory Science Center, Faculty of Science, Chulalongkorn University, Bangkok 10330, Thailand
8. Department of Pharmaceutical Chemistry, University of Vienna, 1090 Vienna, Austria; karl.peter.wolschann@univie.ac.at
9. Institute of Theoretical Chemistry, University of Vienna, 1090 Vienna, Austria
10. Department of Chemistry, Faculty of Science, Chiang Mai University, Chiang Mai 50200, Thailand
11. Center of Excellence in Materials Science and Technology, Chiang Mai University, Chiang Mai 50200, Thailand
* Correspondence: naweekung@gmail.com (N.K.); supot.h@chula.ac.th (S.H.); Fax: +66-53-892277 (N.K.); +66-22-187598 (S.H.)

Citation: Kerdpol, K.; Daengngern, R.; Sattayanon, C.; Namuangruk, S.; Rungrotmongkol, T.; Wolschann, P.; Kungwan, N.; Hannongbua, S. Effect of Water Microsolvation on the Excited-State Proton Transfer of 3-Hydroxyflavone Enclosed in γ-Cyclodextrin. *Molecules* 2021, 26, 843. https://doi.org/10.3390/molecules26040843

Academic Editors: Marina Isidori, Margherita Lavorgna and Rosa Iacovino

Received: 10 January 2021
Accepted: 2 February 2021
Published: 5 February 2021

Abstract: The effect of microsolvation on excited-state proton transfer (ESPT) reaction of 3-hydroxyflavone (3HF) and its inclusion complex with γ-cyclodextrin (γ-CD) was studied using computational approaches. From molecular dynamics simulations, two possible inclusion complexes formed by the chromone ring (C-ring, Form I) and the phenyl ring (P-ring, Form II) of 3HF insertion to γ-CD were observed. Form II is likely more stable because of lower fluctuation of 3HF inside the hydrophobic cavity and lower water accessibility to the encapsulated 3HF. Next, the conformation analysis of these models in the ground (S_0) and the first excited (S_1) states was carried out by density functional theory (DFT) and time-dependent DFT (TD-DFT) calculations, respectively, to reveal the photophysical properties of 3HF influenced by the γ-CD. The results show that the intermolecular hydrogen bonding (interHB) between 3HF and γ-CD, and intramolecular hydrogen bonding (intraHB) within 3HF are strengthened in the S_1 state confirmed by the shorter interHB and intraHB distances and the red-shift of O–H vibrational modes involving in the ESPT process. The simulated absorption and emission spectra are in good agreement with the experimental data. Significantly, in the S_1 state, the keto form of 3HF is stabilized by γ-CD, explaining the increased quantum yield of keto emission of 3HF when complexing with γ-CD in the experiment. In the other word, ESPT of 3HF is more favorable in the γ-CD hydrophobic cavity than in aqueous solution.

Keywords: 3-hydroxyflavone (3HF); γ-cyclodextrin (γ-CD); excited-state proton transfer (ESPT); molecular dynamics (MD); density functional theory (DFT)

1. Introduction

Fluorescent organic molecules with strong intramolecular hydrogen bonds (intraHBs) connected by proton donating and accepting groups have gained considerable attention in recent years owing to their unique fluorescent emission in term of large Stokes shift without self-absorption [1]. Their unique properties harnessing from the excited state intramolecular proton transfer (ESIntraPT) could be typically described by the characteristic four-level photocycle. Initially, the molecule in an enol form (E) in the ground (S_0) state absorbing light in the shorter wavelength region results in the photoexcitation process from the S_0 state into the excited (S_1) state. The intraHB of E is strengthened in the S_1 state because of the charge redistribution upon photoexcitation, leading to the transfer of proton from the donor (D: $-NH_2$, $-OH$) to the acceptor (A: C=O, --N=), which changes the enol form (E*) to the keto form (K*) in the S_1 state. After that, the K* emits the fluorescence in the remarkably longer wavelength than the absorption and relaxes to the S_0 state, resulting in the notably large Stokes shift (the difference between positions of absorption and emission peaks). Then the K changes to the E through the back proton transfer (BPT) process spontaneously in the S_0 state due to the PT barrierless and high exothermic reaction. Generally, their photophysical properties can be easily modulated using many strategies [2,3] such as introducing electron-donating and withdrawing substituents, the heteroatom substitution, and the π-conjugation into the main core structure, to give the desirable absorption and emission spectra as well as large Stokes shift. The ESIntraPT molecules with such tunable photophysical properties including derivatives of salicylates [4–6], salicylideneanilines [7–9], flavones [10–13], benzazoles [14–18], and chalcones [19–21] have been reported and widely used in various applications ranging from chemical sensing to light-emitting diodes [22–27].

Among various ESIntraPT molecules, 3-hydroxyflavone (3HF), which consists of a chromone ring (C-ring) and a phenyl ring (P-ring), is one of the best-known molecular systems. 3HF exhibits ESIntraPT and gives dual fluorescence corresponding to its E* and K* forms with a large Stokes shift and photostability [28–30]. Thus, 3HF has been used as a prototype for the ESIntraPT processes and as sensitive fluorescence probes for discovering binding sites in various bio-relevant targets such as DNA, protein, and biomembranes [31–33]. The photophysical properties and ESPT processes of 3HF in organic solvents have been extensively studied [34–42]. In non-polar solvent, only the K* emission peak of 3HF in toluene was observed at 530 nm with large Stokes shift [33] because the ESIntraPT process effectively occurs, giving only K* form. However, in a protic solvent, the dual emission peaks from E* and K* of 3HF in methanol were observed at 409 and 528 nm, respectively [33], because the IntraHB of 3HF is disrupted and the intermolecular hydrogen bonds (interHBs) between 3HF and protic solvents is formed depending on the nature of solvents and the arrangement of protic solvent around 3HF. This favorable formation of interHBs could reduce the formation of K*, resulting in the low quantum yield in protic solvents or aqueous solution [40–44].

One of the strategies to enhance the K* emission intensity of 3HF in an aqueous solution is disrupting the interHBs of 3HF-water cage-like network using cyclodextrin (CD) [45–47]. The reduction of polarity and restricted environment inside CD's cavity is essential for many aspects of photophysical phenomena by inclusion complexes between 3HF and CD [45–47]. CDs are the cyclic oligosaccharides consisting of the crucial α-D-glucose unit, which exhibit the conical shape and the α-D-glucose units of 6, 7, and 8 are represented as α-, β-, and γ-cyclodextrins (α-, β-, and γ-CDs), respectively [48]. S. Das and N. Chattopadhyay experimentally studied the fluorescence anisotropy of the inclusion complexes of 3HF in α-, β-, and γ-CDs compared to 3HF in the aqueous medium. The fluorescence anisotropy of these probes decreased following the order α-, β-, and γ-CD, which is attributed to the disruption of the 3HF-water network in an aqueous medium [46]. It can be stated that the micro-environment of 3HF derivatives was able to alter and prevent the self-aggregation effectively by using CDs, especially γ-CD. From the investigation on ESPT processes of 3HF in β-, and γ-CDs [45,47], the intensity of K* of 3HF/γ-CD inclusion complex is higher than that of 3HF/β-CD inclusion complex. Consequently, the encapsulation in CDs should be a possible method to tune the fluorescence emission of 3HF derivatives and other hydrophobic compounds [45–47,49–52].

From the studies of the multi-spectroscopic approaches and molecular docking of the encapsulation of 3HF in different small-ring CDs, the 3HF/γ-CD complex showed the strongest interaction, and it provided a higher fluorescent yield than that in water medium [45–47]. Understanding the role of CD in enhancement of the fluorescent yield at atomic level might help us to be able to adjust the fluorescent wavelength to fit for fluorescence probes for bio-labeling in aqueous medium. However, to the best of our knowledge, the detailed information of ESPT reaction in the S_0 and S_1 states of 3HF in γ-CD at atomic level has not been reported. In this work, we aimed to systematically investigate the effect of a water molecule on the photophysical properties and ESPT reactions of an isolated 3HF and its encapsulation. All-atom molecular dynamics (MD) simulations for 300 ns were performed to study the structure and dynamics properties of the two possible 3HF/γ-CD complexes. The detailed information of each complex both in the S_0 and S_1 states was then investigated using density functional theory (DFT) and time-dependent DFT (TD-DFT) methods. The important distances and simulated infrared (IR) vibrational spectra from the optimized structures as well as the topology analysis were used to describe the hydrogen-bonded strength. The frontier molecular orbitals (MOs) were analyzed to provide the charge distribution of the complex. The simulated absorption and emission spectra were calculated and compared with the experimental data. Moreover, the energies of E and K forms of 3HF in each system at the S_0 and S_1 states were discussed to explain why the fluorescent yield of K* in aqueous medium increases when encapsulating 3HF into γ-CD.

2. Results and Discussion

2.1. Possible Inclusion Complexes

From the docking study, 3HF could form two possible inclusion complexes with γ-CD through its chromone ring (C-ring, Form I) or phenyl ring (P-ring, Form II) insertion into the hydrophobic cavity as depicted in Figure 1B. Although a higher occurrence (65%) was found for Form I, the interaction energies of both forms were likely comparable (Form I: −21.42 kcal/mol and Form II: −20.78 kcal/mol). Additionally, the previous docking studies [46,47] suggested that Form II was more stable. Thus, in the present work the 3HF/γ-CD complexes in both forms were further studied by MD simulations and DFT calculations.

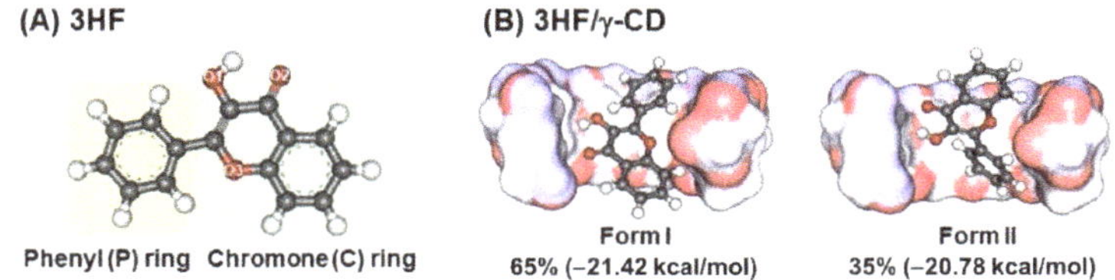

Figure 1. (**A**) Chemical structure of 3-Hydroxyflavone, 3HF. (**B**) Docked structures of the two possible 3HF/γ-CD complexes, Form I and Form II, where their percentage of occurrence and the lowest interaction energy retrieved from 100 independent docking runs are also given.

2.2. 3HF Mobility in γ-CD Cavity and Water Accessibility

To study the inclusion complexes in solution, the four different initial structures of Form I and Form II obtained from molecular docking and QM calculation were simulated by 300-ns MD simulations (MD1-MD4). All trajectories were analyzed and discussed as follows. The plots of RMSD and R_{gyr} of complex in Supplemental Figure S1 suggested that in Form I 3HF spontaneously released from the γ-CD cavity at ~68 ns, ~54 ns, and ~198 ns for MD1, MD2 and MD3. Interestingly, it feasibly moved back to form a complexation with γ-CD, resulting in Form I (MD3 ~201 ns) and Form II (MD1 ~70 ns and MD2 ~57 ns) as considerably seen by the plot of distance between the center of mass (C_m) of each 3HF ring and the C_m of the primary rim of the γ-CD in Figure 2, and the plot of distance between the C_m of each 3HF ring and the C_m of the secondary rim of the γ-CD in Supplementary Materials Figure S2. This is in contrast for Form II, in which the 3HF was well encapsulated inside the hydrophobic cavity in all MD1-MD4 systems throughout the simulation time (RMSD of 1.2–4.0 Å, and R_{gyr} of 6.0−7.0 Å). Moreover, it can be noticeable that the both rings of 3HF in Form I (MD3 and MD4) were fluctuated higher than those in Form II suggesting that the complex in Form II was found to be more stable in aqueous solution; in other words, the P-ring insertion is the suitable binding mode of 3HF for encapsulation with γ-CD in consistent with the lower water accessibility to the encapsulated 3HF ($n(r)$ of 2.0 ± 0.7 and 2.5 ± 1.2 at O1 and O2, respectively, in Figure 3). In Form I, the hydroxyl oxygen O1 of 3HF positioned closer to the wider rim of γ-CD had a significantly higher interaction with waters (4.4 ± 0.8), and in vice versa less waters (2.0 ± 0.2) can access to the carbonyl oxygen O2. No peak detected within ~2.8 Å of the O3 of 3HF, suggesting that this atom had a very weak hydration interaction as found in some flavonoids/CD complexes [53,54]. It is worth noting that such accessible water molecules at O1 and O2 sites may involve into proton transfer processes either blocking ESIntraPT or assisting ESInterPT in 3HF/γ-CD inclusion complex. To study the water assisted PT in 3HF/γ-CD either ground state or excited state, the model of 3HF/γ-CD with a water molecule placed between these two sites of 3HF was further investigated by DFT and TD-DFT calculations and discussed in the next sections.

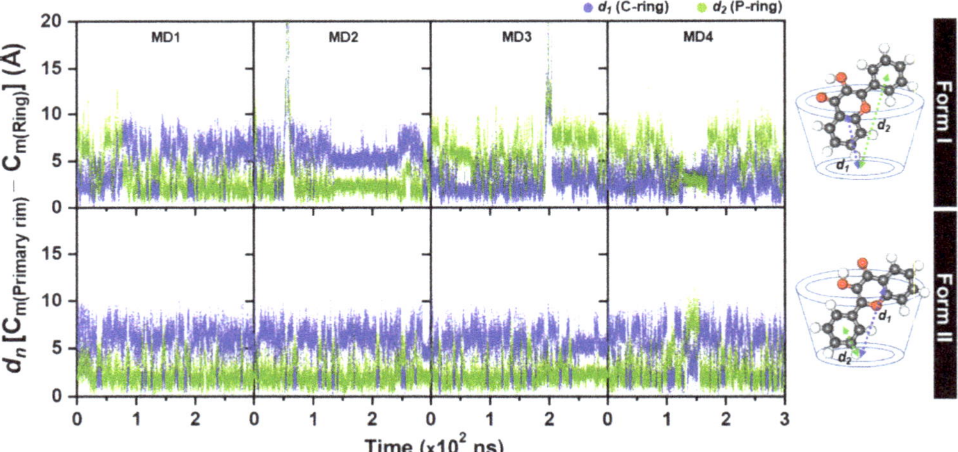

Figure 2. The plots of distance measured from the C_m of each 3HF ring to the C_m of the primary rim of γ-CD (all 7 O6 atoms) for the four MD simulations MD1-MD4 with different initial structures of complexes in Form I and Form II.

2.3. Structural Optimizations

All optimized E forms of 3HF, 3HFW, and 3HF encapsulated in the γ-CD cavity (Form I, Form II, Form I-W and Form II-W) with important labeled atoms and distances (O−H covalent bonds, intraHB, and interHBs) in the S_0 state are shown in Figure 4, where the measured distances in the S_0 and S_1 states are summarized in Table 1.

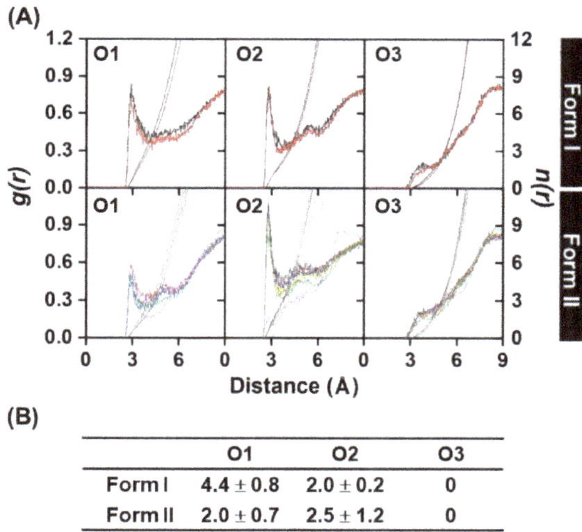

(B)

	O1	O2	O3
Form I	4.4 ± 0.8	2.0 ± 0.2	0
Form II	2.0 ± 0.7	2.5 ± 1.2	0

Figure 3. (**A**) Radial distribution function (RDF or $g(r)$) of water oxygen atoms around the O1-O3 atoms of 3HF encapsulated in the γ-CD cavity over the last 50-ns MD simulations for Form I (MD3-MD4) and Form II (MD1-MD4 with initial structures in this form and additional MD1-MD2 from re-encapsulation process found in Form I). (**B**) Average integration number, $n(r)$, up to the first minimum derived from RDF plots, corresponding to the number of water molecules pointing toward the focused oxygens of 3HF.

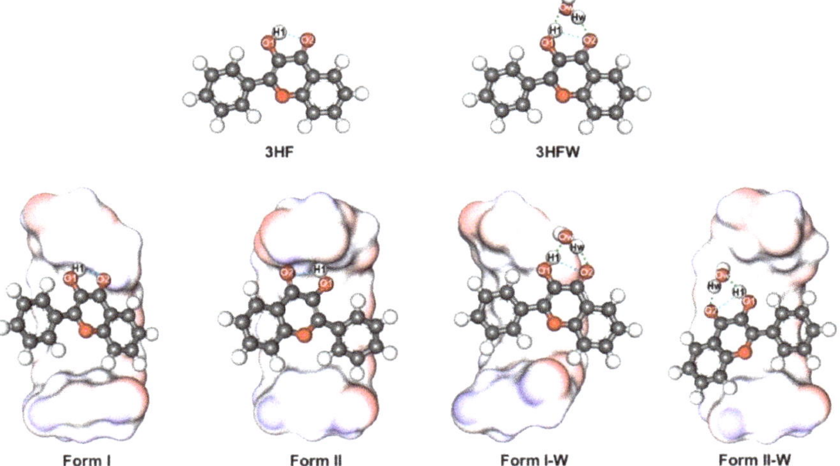

Figure 4. All S_0 optimized structures of 3HF, 3HFW, and the different conformations of the 3HF/γ-CD inclusion complexes (Form I and Form II) as well as its inclusion complex with a water molecule (Form I-W and Form II-W) computed by PBE0/def2-SVP level of theory. The blue and green dot lines represent intraHB in 3HF and interHBs between 3HF and a water molecule, respectively.

From Table 1, for the compounds without a water molecule (3HF, Form I, and Form II), the O1–H1 covalent bond at the 3HF hydroxyl group slightly increases from the S_0 to S_1 states around 0.015–0.034 Å. Whereas, the length of O2$\cdots$H1 intraHB between the hydroxyl group and carbonyl group of 3HF significantly decreases in the S_1 state at 0.214,

0.174, and 0.097 Å for 3HF, Form I, and Form II, respectively. The length of $O2\cdots H1$ intraHB of Form I and Form II is longer than that of the isolated 3HF because the O2 of 3HF forms a stronger interHB with a hydroxy group at the primary rim of γ-CD (O6–H) for Form I and at the secondary rim of γ-CD (O2–H) for Form II in the S_1 state. Overall, these results indicate that the strength of an intraHB in the S_1 state of 3HF, Form I, and Form II is stronger than that in the S_0 state. Consequently, the ESIntraPT process might easily occur in the S_1 state.

Table 1. A summary of the important bonds and distances involving PT process of enol form for all complexes in the S_0 and S_1 states.

Compound	Important Bond Distance (Å)									
	S_0 State					S_1 State				
	O1–H1	O2$\cdots$H1	Ow$\cdots$H1	Ow–Hw	O2$\cdots$Hw	O1–H1	O2$\cdots$H1	Ow$\cdots$H1	Ow–Hw	O2$\cdots$Hw
3HF	0.983	1.920				1.017	1.706			
3HFW	1.000	2.387	1.630	0.984	1.723	1.069	2.439	1.394	1.008	1.572
Form I	0.979	1.978				1.002	1.804			
Form II	0.981	2.177				0.996	2.080			
Form I-W	1.009	2.390	1.585	0.977	1.808	1.030	2.415	1.503	0.978	1.903
Form II-W	1.004	2.429	1.612	0.983	1.733	1.078	2.452	1.375	1.006	1.581

For the compounds with a water molecule (3HFW, Form I-W, and Form II-W), a water molecule forms interHBs with 3HF, therefore an $O2\cdots H1$ intraHB of these complexes is dramatically longer than that of the compounds without a water molecule. It can be stated that a water molecule might block ESIntraPT process and support ESInterPT process. The O1–H1 and Ow–Hw covalent bonds of 3HFW/Form II-W are slightly increased from the S_0 to S_1 states at 0.069/0.074 Å for O1–H1 and 0.024/0.023 Å for Ow–Hw covalent bonds. While the lengths of $Ow\cdots H1$ and $O2\cdots Hw$ interHBs of 3HFW/Form II-W are significantly decreased from the S_0 to S_1 states around 0.236/0.237 Å for $Ow\cdots H1$ and 0.151/0.152 Å for $O2\cdots Hw$ interHBs, so ESInterPT may easily take place more than ESIntraPT for 3HFW and Form II-W due to the strong interHBs in the S_1 state. For Form I-W, the length of $Ow\cdots H1$ interHB decreases (0.082 Å), while the length of $O2\cdots Hw$ interHB increases (0.095 Å) in the S_1 state because Hw of a water molecule forms a stronger interHB with O6 at a primary rim of γ-CD instead of O2 acceptor of 3HF in the S_1 state. Consequently, ESInterPT of Form I-W is quite difficult to occur because the $O2\cdots Hw$ interHB between 3HF and a water molecule is weaker (elongated length) in the S_1 state. Furthermore, the strength of intraHBs and interHBs in the S_1 state will be further discussed in the next section.

2.4. O–H Stretching and Topology Analysis

In general, the strength of the intraHBs and interHBs in the S_1 state could be revealed based on monitoring the red-shift of vibrational modes involving the hydrogen-bonded formation and the topology analysis at bond critical points (BCPs) using the QTAIM method. The calculated IR spectra of all studied compounds in the conjunct vibrational regions of the O–H stretching modes related with PT process both in the S_0 and S_1 states are listed in Table 2. These O–H stretching modes can be classified into the O1–H1 stretching mode of 3HF and Ow–Hw stretching mode of a water molecule.

For 3HF, the O1–H1 stretching vibrational mode of 3HF is located at 3544 cm^{-1} and 3003 cm^{-1} for the S_0 and S_1 states, respectively, giving a large red-shift of 541 cm^{-1}. Moreover, the large red-shift of the O1–H1 stretching vibrational mode is also observed for Form I (367 cm^{-1}) and Form II (244 cm^{-1}). Therefore, the strength of the $O2\cdots H1$ intraHB for these compounds is increased in the S_1 state providing ESIntraPT process.

For the compounds with a water molecule, the O1–H1 stretching vibrational modes of 3HFW, Form I-W, and Form II-W are located around 2967–3129 cm^{-1} in the S_0 state. Note that these vibrational modes changed to be around 2575–2599 cm^{-1} in the S_1 state, which evidently demonstrates that the red-shift is induced by the strengthening of the O1–H1···Ow interHB after photoexcitation. Similarly, the Ow–Hw stretching modes of these compounds are also red-shifted. In addition, the red-shift value of 3HFW and Form II-W is larger than that of Form I-W, indicating that the O1–H1···Ow and Ow–Hw···O2 intermolecular hydrogen-bonded strength of 3HFW and Form II-W in the S_1 state is stronger than that of Form I-W. Overall, these results show that O–H stretching vibrational frequencies shift to lower frequencies in the S_1 state compared with the S_0 state, which confirms that the hydrogen bonding interaction is stronger in the S_1 state.

Table 2. A summary of the values of the O1–H1 and the Ow–Hw stretching vibrational modes of enol form for all compounds in both S_0 and S_1 states and their spectral shifts (Δv in cm^{-1}).

| Compound | Wavenumber (cm^{-1}) | | | | | |
| | O1–H1 | | | Ow–Hw | | |
	S_0	S_1	Δv	S_0	S_1	Δv
3HF	3544	3002	541			
3HFW	3129	2599	530	3510	3074	436
Form I	3617	3250	367			
Form II	3574	3330	244			
Form I-W	2967	2575	392	3662	3499	163
Form II-W	3063	2577	486	3526	3111	415

A topology analysis of the electron density was used to further determine the strength of the intraHB and interHBs in the excited-state structures (E* form) of all compounds. The following parameters at BCPs were analyzed: the electron density $\rho(r)$, the potential energy density V(r), the Laplacian of the electron density $\nabla^2\rho(r)$, the Lagrangian kinetic energy G(r), the Hamiltonian kinetic energy density H(r), and the electron delocalization index (DI) between the proton acceptor and transferred proton, which are an O2···H1 intraHB, and the Ow···H1 and O2···Hw interHBs for all studied compounds. Additionally, the hydrogen-bonded energy (E_{HB}) can be calculated by using the Espinosa's equation: $E_{HB} = \frac{1}{2}|V(r_{BCP})|$. These results are summarized in Table S1 in the Supplementary Materials. From Table S1, E_{HB} of O2···H1 intraHB of 3HF (0.0224 a.u.) is the highest among the compounds without a water molecule. Then, E_{HB} is slightly decreased for Form I (0.0175 a.u.) and Form II (0.0111 a.u.). These results indicate that the intraHB of isolated 3HF is stronger than that the inclusion complexes. Nevertheless, this intraHB of all compounds can facilitate ESIntraPT processes. For the compounds with a water molecule, E_{HB} of Ow···H1 and O2···Hw of 3HFW/Form II-W are 0.0696/0.0649 and 0.0306/0.0317 a.u., respectively. The Ow···H1 interHB of 3HFW/Form II-W is stronger than the O2···Hw interHB, indicating that the proton might transfer via a water molecule from O1–H1 bond-breaking before Ow–Hw bond-breaking. However, only E_{HB} of Ow···H1 of Form I-W is obviously dropped (0.0253 a.u.). The result implies that the ESInterPT process might occur in 3HFW/Form II-W better than Form I-W. Overall, it can be observed that the intraHB and interHBs are strengthened in the S_1 state confirmed by the shorter distances of important bonds involving in the ESPT process, the red-shift observed by IR vibrational spectral calculations, and a high value of E_{HB} from topology analysis of the electron density.

2.5. Frontier MOs Analysis and Simulated Spectra

To further investigate behaviors of charge distribution and charge transfer in the S_1 state, the frontier MOs of the highest occupied molecular orbitals (HOMO) and the lowest unoccupied molecular orbitals (LUMO) of all studied compounds were analyzed because the main electronic transition is only related with these orbitals in range of 98% (HOMO $\rightarrow$ LUMO), which is assigned as π to π^* characters, and illustrated in Figure 5. It can be noted that electron density of both HOMO and LUMO is fully localized on the 3HF moieties and no electron density is located on water or γ-CD, indicating that no intramolecular charge transfer within 3HF and no intermolecular charge transfer between 3HF and water or γ-CD. Moreover, the HOMO and LUMO are localized on different parts of 3HF. For the HOMO orbitals, the electron density is distributed more on P-ring and partially on C-ring of 3HF. Whereas, that of LUMO is distributed completely on the whole molecule of 3HF.

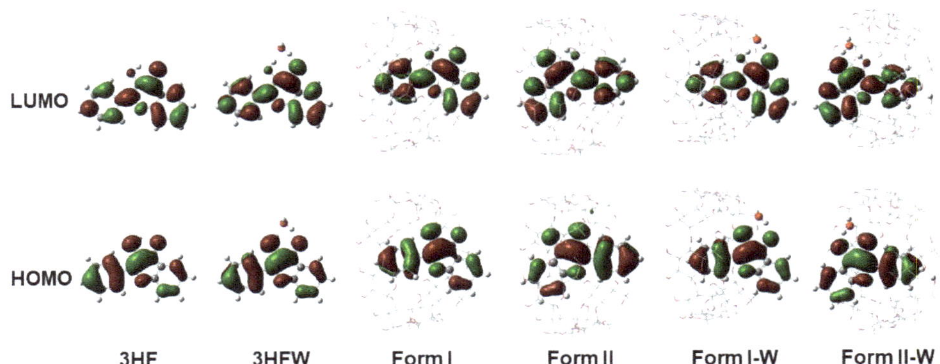

Figure 5. Frontier MOs of all studied compounds.

The UV-Vis absorption and emission spectra of all studied compounds were simulated based on their optimized S_0 and S_1 structures, respectively and then plotted in Figure S3 Furthermore, the absorption band maxima of E form ($\lambda_{abs\ of\ E}$), emission band maxima of E* ($\lambda_{emis\ of\ E*}$) and K* forms ($\lambda_{emis\ of\ K*}$), the excitation energy (E_{ex}), and the oscillator strength (f) as well as the major MOs contribution (%) of the absorption band for all compounds are reported in Table 3. From the detailed information in Table 3, the simulated absorption peaks of the complexes without and with a water molecule are around 341–345 nm and 350–353 nm, respectively, which are in good agreement with the experimental value of 341 nm for 3HF in water and 340 nm for 3HF in γ-CD [46,47]. Moreover, the predicted maximum wavelength for dual emission spectra of all studied compounds is also consistent with the experimental data [46,47], in which the $\lambda_{emis\ of\ E*}$ in water and in γ-CD are reported at 410 and 404 nm, respectively, while the $\lambda_{emis\ of\ K*}$ in water and in γ-CD are 511 and 538 nm. The deviations from the experimental data around 59–69 nm (0.52–0.61 eV) indicate that the chosen method at TD-PBE0/def2-SVP level of theory is adequate to describe the electronic spectra and provide the insight understanding of the ESPT process.

Table 3. UV/Vis absorption band maxima of enol form ($\lambda_{abs\ of\ E}$), emission band maxima of enol ($\lambda_{emis\ of\ E}$) and emission band maxima of keto forms ($\lambda_{emis\ of\ K}$), the excitation energy (E_{ex}), and the oscillator strength (f) as well as their major contribution (%) calculated by TD-PBE0/def2-SVP level of theory.

Compounds	UV/Vis Absorption				Emission					
	$\lambda_{abs\ of\ E}$ (nm)	E_{ex} (eV)	f	MOs (%contribution)	$\lambda_{emis\ of\ E*}$ (nm)	E_{ex} (eV)	f	$\lambda_{emis\ of\ K*}$ (nm)	E_{ex} (eV)	f
3HF	341	3.63	0.4907	HOMO→LUMO (98%)	390	3.18	0.5454	514	2.41	0.4632
3HFW	351	3.53	0.4857	HOMO→LUMO (98%)	408	3.04	0.5335	504	2.46	0.4619
Form I	345	3.59	0.4618	HOMO→LUMO (98%)	392	3.16	0.5123	524	2.36	0.3692
Form II	345	3.59	0.4081	HOMO→LUMO (98%)	393	3.15	0.4689	528	2.35	0.3792
Form I-W	353	3.51	0.4288	HOMO→LUMO (98%)	399	3.11	0.5213	529	2.34	0.3708
Form II-W	350	3.54	0.4174	HOMO→LUMO (98%)	412	3.01	0.4575	508	2.44	0.3899

The relative energy of E and K forms of all studied complexes at the S_0 and S_1 states was investigated to explain the ESPT phenomena as illustrated in Table 4. The results show that the E form is more stable than the K form in the S_0 state for all complexes with the energy differences at 6.63–16.85 kcal/mol. Moreover, most of K forms were stabilized in the γ-CD cavity except only Form I. However, in the S_1 state, the K* form is more stable than the E* form for all complexes. It is predicted that both ESIntraPT and ESInterPT processes are favorable in the S_1 state but not in the S_0 state. In case of the complexes without a water molecule (Form I and Form II), K* of Form II is more stable than Form I. So, the ESIntraPT process of Form II may be more effective than that of Form I, related to the MD results that Form II is favorably more stable. For the complexes with a water molecule, the ESInterPT via interHBs network is feasible to occur in both Form I-W and Form II-W especially Form II-W because of the slightly higher oscillator strength and the lower energy of K*. In addition, from our previous work, the ESInterPT of 3HFW is hard to occur due to the higher ESInterPT barrier and the rearrangement of a water molecule surrounding 3HF [40]. Therefore, it can be predicted that the encapsulating 3HF into γ-CD assists the disruption of the 3HF-water network in aqueous solution leading to an increment of the fluorescent yield of K* in aqueous solution from the ESIntraPT process.

Table 4. The relative energy and computed energy differences between the E and K forms ($\Delta E = E_{\text{enol}} - E_{\text{keto}}$) in the S_0 and S_1 states for all complexes.

Complex	Relative Energy (kcal/mol)				ΔE at S_0	ΔE at S_1
	S_0		S_1			
	E	K	E*	K*		
3HF	0	10.74	75.19	65.81	10.74	−9.38
3HFW	0	10.27	74.92	63.41	10.27	−11.51
Form I	0	16.85	74.79	71.16	16.85	−3.63
Form II	0	8.47	72.39	64.68	8.47	−7.71
Form I-W	0	6.63	71.24	63.43	6.63	−7.81
Form II-W	0	8.35	73.12	64.72	8.35	−8.40

3. Methodology

3.1. Model Preparation

The 3D structure of 3HF (Figure 1A) was generated by GaussView version 6.0 [55]. To investigate the effect of water on PT process, a water molecule near the PT site of 3HF was placed between the H1 and O2 atoms to model 3HF with a water molecule (3HFW). The 3HF and 3HFW structures were fully optimized with DFT and TD-DFT methods using Gaussian 16, Revision C.01 [56]. The γ-CD structure extracted from the co-crystal structure of cyclo/maltodextrin-binding protein in complex with γ-CD (PDB code: 2ZYK) was used in this study. To obtain the inclusion complex, encapsulation of the optimized 3HF into γ-CD's cavity was performed by molecular docking with 100 runs using the CDOCKER module in the Accelrys Discovery Studio 2.5 (Accelrys Software Inc., San Diego, CA, USA). The three docked inclusion complexes with the lowest CDOCKER interaction energy for the chromone ring (C-ring, Form I) and phenyl ring (P-ring, Form II) insertion into the hydrophobic cavity of the γ-CD were selected for MD simulations (MD1-MD3 for Form I and Form II). The most stable 3HF/γ-CD inclusion complexes in Form I and Form II from a docking study were fully optimized at PBE0/def2-SVP level of theory for investigating PT reaction. Note that, these Form I and Form II were used as the additional initial structures for MD simulations, MD4. To study the effect of water associated in PT for 3HF encapsulated in γ-CD, a water molecule was added to form interHBs with 3HF both in Form I and Form II namely Form I-W and Form II-W, respectively. Then, Form I-W and Form II-W were fully optimized with DFT and TD-DFT methods.

3.2. Molecular Dynamics Simulations

MD simulations of the 3HF/γ-CD complexes in Form I and Form II using the four different initial structures (eight complexes in total) were carried out using the AMBER16 program package [57]. The partial atomic charges and parameters of 3HF were generated in accordance with the previous standard procedures [58–60]. The general AMBER force field [61] and the Glycam06j carbohydrate force field [62] were applied for 3HF and γ-CD, respectively. The models were solvated by a truncated octahedral box of TIP3P water molecules with a spacing distance of 15 Å from complex. Afterward, all added water molecules were minimized using 1000 steps of steepest descent (SD) and continued by 3000 steps of conjugated gradient (CG). Next, the minimizations with the same steps were performed on the whole system. All studied models were heated up from 10 K to 298 K with a constant volume ensemble (NVT) for 100 ps and followed by all-atom MD simulations with a constant pressure ensemble (NPT) at 1 atm and 298 K for 300 ns. All chemical bonds involving hydrogen were constrained using the SHAKE algorithm [63]. The particle mesh Ewald's method [64] was performed for the treatment of the long-range interactions. The cpptraj module of AMBER16 program was used to calculate the root-mean-square displacement (RMSD), the radius of gyration (R_{gyr}), and the distance between guest and host molecules as well as the radial distribution function (RDF).

3.3. Quantum Chemical Calculations

All E and K forms of 3HF, 3HFW, and 3HF/γ-CD inclusion complexes in Form I and Form II in the S_0 and S_1 states were studied by PBE0 and TD-PBE0 methods, respectively with def2-SVP basis set by using Gaussian 16, Revision C.01 [56]. The solvation effect was taken into account by means of the non-equilibrium implementation of the conductor polarized continuum model (C-PCM) framework [65,66], so-called PCM-LR [67]. To confirm that the optimized structures (E, K, E*, and K*) are located at the local minimum, no imaginary frequency from vibrational calculations was found for all optimized structures both in S_0 and S_1 states. A hydrogen-bonded strength was determined by the important distance parameters involving ESIntraPT and ESInterPT processes namely the covalent O–H bond of 3HF and a water molecule, the intraHB between proton donor and proton acceptor of 3HF, and the interHBs between 3HF and a water molecule for the case study of water assisted effect. The strength of all intraHB and interHBs was further confirmed by the red-shift values of the O–H stretching vibrational frequencies between the S_0 and S_1 states from the simulated IR spectra, together with the topology analysis at bond critical point (BCP) from quantum theory of atoms in molecules (QTAIM) performed by Multiwfn [68], which was employed in previous studies [69–71]. The electronic spectra and frontier MOs were also calculated. Additionally, the absorption and emission spectra of all complexes were simulated to investigate the photophysical properties.

4. Conclusions

The effect of water microsolvation on the strength of hydrogen bonding, ESIntraPT and ESInterPT reactions as well as photophysical behaviors of 3HF and its inclusion complexes in γ-CD has been systematically studied using MD simulations and DFT/TD-DFT at PBE0/TD-PBE0 with def2-SVP basis set. Two possible 3HF/γ-CD inclusion complexes; C-ring and P-ring insertions (Form I and Form II), were observed from molecular docking. From MD results, a lower 3HF mobility in the hydrophobic cavity and a lower water accessibility to the encapsulated 3HF in Form II suggest that this form is favorably more stable. From the static calculation results, the strength of hydrogen bonding of all studied compounds increases upon photoexcitation into the S_1 state leading to being easier deprotonation, confirmed by the change of important bond lengths (the increasing of the covalent O–H bond length of proton donor, together with decreasing of the O$\cdots$H intraHBs and interHBs), the red shift of the O–H stretching modes, and the bond energy from the topology analysis. In addition, frontier MOs confirm that the main contribution for vertical S_0 to S_1 transition is π to π^* attributed from HOMO (π) to LUMO (π^*). For simulated spectra, the $\lambda_{abs\ of\ E}$, the $\lambda_{emis\ of\ E^*}$ and the $\lambda_{emis\ of\ K^*}$ are in good agreement with the experimental data (in the range of 0.52–0.61 eV relative differences) indicating that the present method is adequate to provide the information on their spectra and the possibility of ESPT processes. Besides, the ESIntraPT processes of the inclusion complexes (Form I and Form II) can occur easily like in the case of 3HF in aprotic solvents. Furthermore, the ESInterPT processes via interHBs of Form I-W and Form II-W inclusion complexes are feasible to take place. In addition, K* of Form II/Form II-W is more stable than that of Form I/Form I-W due to the lower energy and the higher oscillator strength. Consequently, the ESIntraPT and ESInterPT might be likely to occur in P-ring insertion in accordance with the MD results, in which P-ring insertion is the majority of the 3HF/γ-CD inclusion complexes with lower water accessibility. However, it is already known that the ESInterPT of 3HF in aqueous solution is hard to occur due to the higher ESInterPT barrier and the higher fluctuation of the water-rearrangement surrounding 3HF, which leads to a decrease of the fluorescence intensity. Thus, from the present work, we found that 3HF is stable inside the γ-CD hydrophobic cavity and promotes ESIntraPT by suppressing the 3HF-water network. This leads to the increment of fluorescent intensity. In the other word, the fluorescence intensity of K* could be efficiently tuned via host-guest complexation.

Supplementary Materials: The following are available online. Figure S1: RMSD plots for all atoms and R_{gyr} of the two orientations of C-ring insertion (Form I), and P-ring insertion (Form II) inclusion complexes for four different MD runs of each system, Figure S2: The plots of distance measured from the C_m of each 3HF ring to the C_m of the secondary rim of γ-CD (all 7 O2 atoms) for the four MD simulations MD1-MD4 with different initial structures of complexes in Form I and Form II, Figure S3: The simulated absorption spectra of E, and the simulated emission spectra of E* and K* for all studied compounds computed at TD-PBE0/def2-SVP level of theory, Table S1: Electron density $\rho(r)$, the Lagrangian kinetic energy G(r), potential energy density V(r), the Hamiltonian kinetic energy density H(r), the Laplacian of the electron density $\nabla^2\rho(r)$, the electron delocalization index (DI), and hydrogen-bonded energy (E_{HB}) at selected BCPs in the S_1 state (a.u.) for all compounds.

Author Contributions: Conceptualization, N.K., T.R., and S.H.; methodology, K.K.; validation, N.K., T.R., and S.N.; formal analysis, K.K., and C.S.; data curation, K.K.; writing—original draft preparation, K.K.; writing—review and editing, R.D., N.K., T.R., and S.N.; visualization, K.K.; supervision, P.W., N.K., T.R., S.N., and S.H.; project administration, N.K., and S.H.; funding acquisition, N.K., T.R., and S.H. All authors have read and agreed to the published version of the manuscript.

Funding: This research is supported by Ratchadapisek Somphot Fund for Postdoctoral Fellowship, Chulalongkorn University (CU).

Acknowledgments: The authors gratefully acknowledge the financial support from the Center of Excellence in Computational Chemistry (CECC) from Chulalongkorn University, the Center of Excellence in Materials Science and Technology, Chiang Mai University and the Office of National Higher Education Science Research and Innovation Policy Council (NXPO) in Global Partnership Project. N.K. thanks the Thailand Research Fund (Grant No. RSA6180044). Computational resources are provided by the Center of Excellence in Computational Chemistry (CECC), Computational Chemistry Laboratory Chiang Mai University (CCL-CMU), and NSTDA Supercomputer center (ThaiSC).

Conflicts of Interest: The authors declare no conflict of interest.

Sample Availability: Samples of the compounds are not available from the authors.

References

1. Zhou, P.; Han, K. Unraveling the detailed mechanism of excited-state proton transfer. *Acc. Chem. Res.* **2018**, *51*, 1681–1690. [CrossRef]
2. Uzhinov, B.M.; Khimich, M.N. Conformational effects in excited state intramolecular proton transfer of organic compounds. *Russ. Chem. Rev.* **2011**, *80*, 553–577. [CrossRef]
3. Padalkar, V.S.; Seki, S. Excited-state intramolecular proton-transfer (ESIPT)-inspired solid state emitters. *Chem. Soc. Rev.* **2016**, *45*, 169–202. [CrossRef]
4. Weller, A. Innermolekularer protonenübergang im angeregten zustand. *Z. Elektrochem. Ber. Bunsenges. Phys. Chem.* **1956**, *60*, 1144–1147.
5. Goodman, J.; Brus, L.E. Proton transfer and tautomerism in an excited state of methyl salicylate. *J. Am. Chem. Soc.* **1978**, *100*, 7472–7474. [CrossRef]
6. Yoon, M.; Kim, M.; Kim, M.H.; Kang, J.-G.; Sohn, Y.; Kim, I.T. Synthesis and photophysical properties of S, N and Se-modified methyl salicylate derivatives. *Inorg. Chim. Acta* **2019**, *495*, 119008. [CrossRef]
7. Lee, J.; Kwon, J.E.; You, Y.; Park, S.Y. Wholly π-Conjugated low-molecular-weight organogelator that displays triple-channel responses to fluoride ions. *Langmuir* **2014**, *30*, 2842–2851. [CrossRef]

8.   Mitra, S.; Tamai, N. Dynamics of photochromism in salicylideneaniline: A femtosecond spectroscopic study. *Phys. Chem. Chem Phys* **2003**, *5*, 4647–4652. [CrossRef]

9.   Borbone, F.; Tuzi, A.; Panunzi, B.; Piotto, S.; Concilio, S.; Shikler, R.; Nabha, S.; Centore, R. On–Off Mechano-responsive Switching of ESIPT Luminescence in Polymorphic N-Salicylidene-4-amino-2-methylbenzotriazole. *Cryst. Growth Des.* **2017**, *17*, 5517–5523 [CrossRef]

10.   Liu, B.; Wang, J.; Zhang, G.; Bai, R.; Pang, Y. Flavone-based ESIPT ratiometric chemodosimeter for detection of cysteine in living cells. *ACS Appl. Mater. Interfaces* **2014**, *6*, 4402–4407. [CrossRef] [PubMed]

11.   Chou, P.-T.; Chen, Y.-C.; Yu, W.-S.; Cheng, Y.-M. Spectroscopy and dynamics of excited-state intramolecular proton-transfer reaction in 5-hydroxyflavone. *Chem. Phys. Lett.* **2001**, *340*, 89–97. [CrossRef]

12.   Yang, Y.; Zhao, J.; Li, Y. Theoretical study of the ESIPT process for a new natural product quercetin. *Sci. Rep.* **2016**, *6*, 32152 [CrossRef]

13.   Zhang, Y.; Deng, Y.; Ji, N.; Zhang, J.; Fan, C.; Ding, T.; Cao, Z.; Li, Y.; Fang, Y. A rationally designed flavone-based ESIPT fluorescent chemodosimeter for highly selective recognition towards fluoride and its application in live-cell imaging. *Dyes Pigm.* **2019**, *166*, 473–479. [CrossRef]

14.   Henary, M.M.; Fahrni, C.J. Excited state intramolecular proton transfer and metal ion complexation of 2-(2′-Hydroxyphenyl)benzaz in aqueous solution. *J. Phys. Chem. A* **2002**, *106*, 5210–5220. [CrossRef]

15.   Das, S.; Chattopadhyay, N. Heteroatom controlled probe-water cluster formation of a series of ESIPT probes: An exploration with fluorescence anisotropy. *Chem. Phys. Lett.* **2018**, *708*, 37–41. [CrossRef]

16.   Liu, C.; Wang, F.; Xiao, T.; Chi, B.; Wu, Y.; Zhu, D.; Chen, X. The ESIPT fluorescent probes for N2H4 based on benzothiazol and their applications for gas sensing and bioimaging. *Sens. Actuators B* **2018**, *256*, 55–62. [CrossRef]

17.   Mai, V.T.N.; Shukla, A.; Mamada, M.; Maedera, S.; Shaw, P.E.; Sobus, J.; Allison, I.; Adachi, C.; Namdas, E.B.; Lo, S.-C Low amplified spontaneous emission threshold and efficient electroluminescence from a carbazole derivatized excited-state intramolecular proton transfer dye. *ACS Photonics* **2018**, *5*, 4447–4455. [CrossRef]

18.   Munch, M.; Curtil, M.; Vérité, P.M.; Jacquemin, D.; Massue, J.; Ulrich, G. Ethynyl-Tolyl Extended 2-(2′-Hydroxyphenyl)benzoxazole Dyes: Solution and Solid-state Excited-State Intramolecular Proton Transfer (ESIPT) Emitters. *Eur. J. Org. Chem.* **2019**, *2019*, 1134–1144. [CrossRef]

19.   Song, Z.; Kwok, R.T.K.; Zhao, E.; He, Z.; Hong, Y.; Lam, J.W.Y.; Liu, B.; Tang, B.Z. A ratiometric fluorescent probe based on ESIPT and AIE processes for alkaline phosphatase activity assay and visualization in living cells. *ACS Appl. Mater. Interfaces* **2014**, *6*, 17245–17254. [CrossRef]

20.   Dommett, M.; Crespo-Otero, R. Excited state proton transfer in 2′-hydroxychalcone derivatives. *Phys. Chem. Chem. Phys.* **2017**, *19*, 2409–2416. [CrossRef]

21.   Luo, Z.; Liu, B.; Qin, T.; Zhu, K.; Zhao, C.; Pan, C.; Wang, L. Cyclization of chalcone enables ratiometric fluorescence determination of hydrazine with a high selectivity. *Sens. Actuators B* **2018**, *263*, 229–236. [CrossRef]

22.   Kwon, J.E.; Park, S.Y. Advanced organic optoelectronic materials: Harnessing excited-state intramolecular proton transfer (ESIPT) process. *Adv. Mater.* **2011**, *23*, 3615–3642. [CrossRef]

23.   Ashton, T.D.; Jolliffe, K.A.; Pfeffer, F.M. Luminescent probes for the bioimaging of small anionic species in vitro and in vivo. *Chem. Soc. Rev.* **2015**, *44*, 4547–4595. [CrossRef] [PubMed]

24.   Lee, M.H.; Kim, J.S.; Sessler, J.L. Small molecule-based ratiometric fluorescence probes for cations, anions, and biomolecules *Chem. Soc. Rev.* **2015**, *44*, 4185–4191. [CrossRef] [PubMed]

25.   Azarias, C.; Budzák, Š.; Laurent, A.D.; Ulrich, G.; Jacquemin, D. Tuning ESIPT fluorophores into dual emitters. *Chem. Sci.* **2016**, *7*, 3763–3774. [CrossRef] [PubMed]

26.   Sedgwick, A.C.; Wu, L.; Han, H.-H.; Bull, S.D.; He, X.-P.; James, T.D.; Sessler, J.L.; Tang, B.Z.; Tian, H.; Yoon, J. Excited-state intramolecular proton-transfer (ESIPT) based fluorescence sensors and imaging agents. *Chem. Soc. Rev.* **2018**, *47*, 8842–8880. [CrossRef]

27.   Massue, J.; Pariat, T.M.; Vérité, P.; Jacquemin, D.; Durko, M.; Chtouki, T.; Sznitko, L.; Mysliwiec, J.; Ulrich, G. Natural born laser dyes: Excited-State Intramolecular Proton Transfer (ESIPT) Emitters and their use in random lasing studies. *Nanomaterials* **2019**, *9*, 1093. [CrossRef] [PubMed]

28.   Gunduz, S.; Goren, A.C.; Ozturk, T. Facile Syntheses of 3-Hydroxyflavones. *Org. Lett.* **2012**, *14*, 1576–1579. [CrossRef] [PubMed]

29.   Sengupta, P.K.; Kasha, M. Excited state proton-transfer spectroscopy of 3-hydroxyflavone and quercetin. *Chem. Phys. Lett.* **1979**, *68*, 382–385. [CrossRef]

30.   Kasha, M. Proton-transfer spectroscopy. Perturbation of the tautomerization potential. *J. Chem. Soc. Faraday Trans. 2* **1986**, *82*, 2379–2392. [CrossRef]

31.   Ash, S.; De, S.P.; Pyne, S.; Misra, A. Excited state intramolecular proton transfer in 3-hydroxy flavone and 5-hydroxy flavone: A DFT based comparative study. *J. Mol. Model.* **2010**, *16*, 831–839. [CrossRef] [PubMed]

32.   Das, S.; Chakrabarty, S.; Chattopadhyay, N. Origin of unusually high fluorescence anisotropy of 3-hydroxyflavone in water: Formation of probe–solvent cage-like cluster. *J. Phys. Chem. B* **2020**, *124*, 173–180. [CrossRef]

33.   Lazzaroni, S.; Dondi, D.; Mezzetti, A.; Protti, S. Role of solute-solvent hydrogen bonds on the ground state and the excited state proton transfer in 3-hydroxyflavone. A systematic spectrophotometric study. *Photochem. Photobiol. Sci.* **2018**, *17*, 923–933. [CrossRef] [PubMed]

34. Klymchenko, A.S.; Demchenko, A.P. Multiparametric probing of intermolecular interactions with fluorescent dye exhibiting excited state intramolecular proton transfer. *Phys. Chem. Chem. Phys* **2003**, *5*, 461–468. [CrossRef]
35. Klymchenko, A.S.; Pivovarenko, V.G.; Ozturk, T.; Demchenko, A.P. Modulation of the solvent-dependent dual emission in 3-hydroxychromones by substituents. *New J. Chem.* **2003**, *27*, 1336–1343. [CrossRef]
36. Klymchenko, A.S.; Kenfack, C.; Duportail, G.; Mély, Y. Effects of polar protic solvents on dual emissions of 3-hydroxychromones. *J. Chem. Sci.* **2007**, *119*, 83–89. [CrossRef]
37. Klymchenko, A.S.; Demchenko, A.P. Chapter 3 multiparametric probing of microenvironment with solvatochromic fluorescent dyes. *Meth. Enzymol.* **2008**, *450*, 37–58.
38. Ghosh, D.; Pradhan, A.K.; Mondal, S.; Begum, N.A.; Mandal, D. Proton transfer reactions of 4′-chloro substituted 3-hydroxyflavone in solvents and aqueous micelle solutions. *Phys. Chem. Chem. Phys* **2014**, *16*, 8594–8607. [CrossRef] [PubMed]
39. Ghosh, D.; Batuta, S.; Das, S.; Begum, N.A.; Mandal, D. Proton transfer dynamics of 4′-N,N-dimethylamino-3-hydroxyflavone observed in hydrogen-bonding solvents and aqueous micelles. *J. Phys. Chem. B* **2015**, *119*, 5650–5661. [CrossRef]
40. Salaeh, R.; Prommin, C.; Chansen, W.; Kerdpol, K.; Daengngern, R.; Kungwan, N. The effect of protic solvents on the excited state proton transfer of 3-hydroxyflavone: A TD-DFT static and molecular dynamics study. *J. Mol. Liq.* **2018**, *252*, 428–438. [CrossRef]
41. Zhao, J.; Ji, S.; Chen, Y.; Guo, H.; Yang, P. Excited state intramolecular proton transfer (ESIPT): From principal photophysics to the development of new chromophores and applications in fluorescent molecular probes and luminescent materials. *Phys. Chem. Chem. Phys.* **2012**, *14*, 8803–8817. [CrossRef]
42. Santos, F.S.; Ramasamy, E.; Ramamurthy, V.; Rodembusch, F.S. Excited state chemistry of flavone derivatives in a confined medium: ESIPT emission in aqueous media. *Photochem. Photobiol. Sci.* **2014**, *13*, 992–996. [CrossRef] [PubMed]
43. Bartl, K.; Funk, A.; Gerhards, M. IR/UV spectroscopy on jet cooled 3-hydroxyflavone (H2O)n (n = 1,2) clusters along proton transfer coordinates in the electronic ground and excited states. *J. Chem. Phys.* **2008**, *129*, 234306. [CrossRef] [PubMed]
44. Stamm, A.; Maué, D.; Gerhards, M. Structural rearrangement by isomer-specific infrared excitation in the neutral isolated dihydrated cluster of 3-hydroxyflavone. *J. Phys. Chem. Lett.* **2018**, *9*, 4360–4366. [CrossRef]
45. Banerjee, A.; Sengupta, P.K. Encapsulation of 3-hydroxyflavone and fisetin in β-cyclodextrins: Excited state proton transfer fluorescence and molecular mechanics studies. *Chem. Phys. Lett.* **2006**, *424*, 379–386. [CrossRef]
46. Das, S.; Chattopadhyay, N. Supramolecular inclusion-assisted disruption of probe-solvent network. *ChemistrySelect* **2017**, *2*, 6078–6081. [CrossRef]
47. Pahari, B.; Chakraborty, S.; Sengupta, P.K. Encapsulation of 3-hydroxyflavone in γ-cyclodextrin nanocavities: Excited state proton transfer fluorescence and molecular docking studies. *J. Mol. Struct.* **2011**, *1006*, 483–488. [CrossRef]
48. Szejtli, J. Introduction and general overview of cyclodextrin chemistry. *Chem. Rev.* **1998**, *98*, 1743–1754. [CrossRef]
49. Arunkumar, E.; Forbes, C.C.; Smith, B.D. Improving the properties of organic dyes by molecular encapsulation. *Eur. J. Org. Chem.* **2005**, *2005*, 4031. [CrossRef]
50. Hou, X.; Ke, C.; Bruns, C.J.; McGonigal, P.R.; Pettman, R.B.; Stoddart, J.F. Tunable solid-state fluorescent materials for supramolecular encryption. *Nat. Commun.* **2015**, *6*, 6884. [CrossRef]
51. Pahari, B.; Sengupta, B.; Chakraborty, S.; Thomas, B.; McGowan, D.; Sengupta, P.K. Contrasting binding of fisetin and daidzein in γ-cyclodextrin nanocavity. *J. Photochem. Photobiol. B* **2013**, *118*, 33–41. [CrossRef]
52. Douhal, A. Ultrafast guest dynamics in cyclodextrin nanocavities. *Chem. Rev.* **2004**, *104*, 1955–1976. [CrossRef]
53. Kicuntod, J.; Khuntawee, W.; Wolschann, P.; Pongsawasdi, P.; Chavasiri, W.; Kungwan, N.; Rungrotmongkol, T. Inclusion complexation of pinostrobin with various cyclodextrin derivatives. *J. Mol. Graph. Model.* **2016**, *63*, 91–98. [CrossRef] [PubMed]
54. Nutho, B.; Khuntawee, W.; Rungnim, C.; Pongsawasdi, P.; Wolschann, P.; Karpfen, A.; Kungwan, N.; Rungrotmongkol, T. Binding mode and free energy prediction of fisetin/β-cyclodextrin inclusion complexes. *Beilstein J. Org. Chem.* **2014**, *10*, 2789–2799. [CrossRef]
55. Dennington, R.; Keith, T.A.; Millam, J.M. (Eds.) *Gauss View*; Version 6; Semichem Inc.: Shawnee Mission, KS, USA, 2016.
56. *Gaussian 16, Revision, C.01*; Frisch, M.J.; Trucks, G.W.; Schlegel, H.B.; Scuseria, G.E.; Robb, M.A.; Cheeseman, J.R.; Scalmani, G.; Barone, V.; Petersson, G.A. (Eds.) Gaussian, Inc.: Wallingford, CT, USA, 2016.
57. Case, D.A.; Betz, R.M.; Cerutti, D.S.; Cheatham, T.E., III; Darden, T.A.; Duke, R.E.; Giese, T.J.; Gohlke, H.; Goetz, A.W.; Homeyer, N.; et al. *AMBER*; University of California: San Francisco, CA, USA, 2016.
58. Kaiyawet, N.; Rungrotmongkol, T.; Hannongbua, S. Effect of Halogen substitutions on dUMP to stability of thymidylate synthase/dUMP/mTHF ternary complex using molecular dynamics simulation. *J. Chem. Inf. Model.* **2013**, *53*, 1315–1323. [CrossRef]
59. Mahalapbutr, P.; Chusuth, P.; Kungwan, N.; Chavasiri, W.; Wolschann, P.; Rungrotmongkol, T. Molecular recognition of naphthoquinone-containing compounds against human DNA topoisomerase IIα ATPase domain: A molecular modeling study. *J. Mol. Liq.* **2017**, *247*, 374–385. [CrossRef]
60. Mahalapbutr, P.; Wonganan, P.; Charoenwongpaiboon, T.; Prousoontorn, M.; Chavasiri, W.; Rungrotmongkol, T. Enhanced solubility and anticancer potential of mansonone G By β-cyclodextrin-based host-guest complexation: A computational and experimental study. *Biomolecules* **2019**, *9*, 545. [CrossRef] [PubMed]
61. Wang, J.; Wolf, R.M.; Caldwell, J.W.; Kollman, P.A.; Case, D.A. Development and testing of a general amber force field. *J. Comput. Chem.* **2004**, *25*, 1157–1174. [CrossRef] [PubMed]

62. Kirschner, K.N.; Yongye, A.B.; Tschampel, S.M.; González-Outeiriño, J.; Daniels, C.R.; Foley, B.L.; Woods, R.J. GLYCAM06: A generalizable biomolecular force field. Carbohydrates. *J. Comput. Chem.* **2008**, *29*, 622–655. [CrossRef]
63. Ryckaert, J.-P.; Ciccotti, G.; Berendsen, H.J.C. Numerical integration of the cartesian equations of motion of a system with constraints: Molecular dynamics of n-alkanes. *J. Comput. Phys.* **1977**, *23*, 327–341. [CrossRef]
64. Luty, B.A.; van Gunsteren, W.F. Calculating electrostatic interactions using the particle−particle particle−mesh method with nonperiodic long-range interactions. *J. Phys. Chem.* **1996**, *100*, 2581–2587. [CrossRef]
65. Jacquemin, D.; Perpète, E.A.; Scalmani, G.; Frisch, M.J.; Assfeld, X.; Ciofini, I.; Adamo, C. Time-dependent density functional theory investigation of the absorption, fluorescence, and phosphorescence spectra of solvated coumarins. *J. Chem. Phys.* **2006**, *125*, 164324. [CrossRef] [PubMed]
66. Corni, S.; Cammi, R.; Mennucci, B.; Tomasi, J. Electronic excitation energies of molecules in solution within continuum solvation models: Investigating the discrepancy between state-specific and linear-response methods. *J. Chem. Phys.* **2005**, *123*, 134512. [CrossRef]
67. Jacquemin, D.; Mennucci, B.; Adamo, C. Excited-state calculations with TD-DFT: From benchmarks to simulations in complex environments. *Phys. Chem. Chem. Phys* **2011**, *13*, 16987–16998. [CrossRef]
68. Lu, T.; Chen, F. Multiwfn: A multifunctional wavefunction analyzer. *J. Comput. Chem.* **2012**, *33*, 580–592. [CrossRef] [PubMed]
69. Li, Y.; Wen, K.; Feng, S.; Yuan, H.; An, B.; Zhu, Q.; Guo, X.; Zhang, J. Tunable excited-state intramolecular proton transfer reactions with NH or OH as a proton donor: A theoretical investigation. *Spectrochim. Acta A Mol. Biomol. Spectrosc.* **2017**, *187*, 9–14. [CrossRef] [PubMed]
70. Prommin, C.; Kerdpol, K.; Saelee, T.; Kungwan, N. Effects of π-expansion, an additional hydroxyl group, and substitution on the excited state single and double proton transfer of 2-hydroxybenzaldehyde and its relative compounds: TD-DFT static and dynamic study. *New J. Chem.* **2019**, *43*, 19107–19119. [CrossRef]
71. Sun, C.; Su, X.; Zhou, Q.; Shi, Y. Regular tuning of the ESIPT reaction of 3-hydroxychromone-based derivatives by substitution of functional groups. *Org. Chem. Front.* **2019**, *6*, 3093–3100. [CrossRef]

Article

Inclusion Compound of Efavirenz and γ-Cyclodextrin: Solid State Studies and Effect on Solubility

Susana Santos Braga [1,*], Firas El-Saleh [2], Karyna Lysenko [1,3] and Filipe A. Almeida Paz [3]

[1] LAQV-REQUIMTE, Department of Chemistry, University of Aveiro, 3810-193 Aveiro, Portugal; karynalysenko@ua.pt
[2] Ashland Specialty Ingredients, Paul-Thomas Strasse, 56, D-40599 Düsseldorf, Germany; FElSaleh@ashland.com
[3] Department of Chemistry, CICECO—Aveiro Institute of Materials, University of Aveiro, 3810-193 Aveiro, Portugal; filipe.paz@ua.pt
* Correspondence: sbraga@ua.pt

Abstract: Efavirenz is an antiretroviral drug of widespread use in the management of infections with human immunodeficiency virus type 1 (HIV-1). Efavirenz is also used in paediatrics, but due to its very poor aqueous solubility the liquid formulations available resort to oil-based excipients. In this report we describe the interaction of γ-cyclodextrin with efavirenz in solution and in the solid state. In aqueous solution, the preferential host–guest stoichiometry was determined by the continuous variation method using ^{1}H NMR, which indicated a 3:2 host-to-guest proportion. Following, the solid inclusion compound was prepared at different stoichiometries by co-dissolution and freeze-drying. Solid-state characterisation of the products using FT-IR, ^{13}C{^{1}H} CP-MAS NMR, thermogravimetry, and X-ray powder diffraction has confirmed that the 3:2 stoichiometry is the adequate starting condition to isolate a solid inclusion compound in the pure form. The effect of γ-cyclodextrin on the solubility of efavirenz is studied by the isotherm method.

Keywords: cyclodextrin inclusion; solution-phase; solid state; antiretroviral

Citation: Braga, S.S.; El-Saleh, F.; Lysenko, K.; Paz, F.A.A. Inclusion Compound of Efavirenz and γ-Cyclodextrin: Solid State Studies and Effect on Solubility. *Molecules* **2021**, *26*, 519. https://doi.org/10.3390/molecules26030519

Academic Editors: Marina Isidori, Margherita Lavorgna and Rosa Iacovino

Received: 30 December 2020
Accepted: 16 January 2021
Published: 20 January 2021

Publisher's Note: MDPI stays neutral with regard to jurisdictional claims in published maps and institutional affiliations.

1. Introduction

Efavirenz (EFV) is a potent antiretroviral of widespread use as first-line therapy for patients with HIV infection. Administered in the form of a tablet and requiring only one dose per day [1], efavirenz is a practical therapeutic option. It is classified by the Biopharmaceutical Classification System (BCS) as a class II drug, that is, it is poorly water soluble and highly permeable [2]. Several approaches to improve the solubility of efavirenz are reported, from the manipulation of polymorphs [2] and blending with superdisintegrants [3] or into solid dispersions [4,5], to a broad variety of encapsulation strategies, which include liposomes [6], micelles [7,8], lipidic nanoparticles [9,10], and polymeric nanoparticles [11].

Molecular encapsulation of efavirenz with cyclodextrins is a less explored but quite promising solution for the amelioration of the physicochemical and organoleptic properties of this drug. Besides having low aqueous solubility, efavirenz presents a very bitter taste, and for this reason it may benefit from the taste-masking effect resulting from inclusion of each efavirenz molecule into the cavity of the cyclodextrin host. Cyclodextrins are cyclic oligosaccharides formed by natural or biotechnological enzymatic action on starch. The most abundant native cyclodextrins occur with six (α-CD), seven (β-CD), or eight (γ-CD, Figure 1a) D-glucose units, linked together by α-1,4 glycosidic bonds. The unique molecular geometry of cyclodextrins, with the shape of a truncated cone having the secondary hydroxyl groups facing the wider rim and the primary hydroxyls at the narrower rim, while the cavity is lined with protons, gives them hydrophobicity at the cavity in tandem with good aqueous solubility [12,13]. Cyclodextrins are thus used as biocompatible, solubilising, stabilising, and taste-masking agents for a variety of drugs [14], with highlight to the antiviral remdesivir [15] and to ozalin, a recently approved paediatric sedative and the only known commercial drug to contain γ-CD in its composition, acting as a solubiliser and taste-masking agent [16]. Cyclodextrins are gaining increased interest in medicinal applications, being employed as co-adjuvants in vaccines [17] or even as experimental drugs in disorders associated with lipid buildup, as Niemann–Pick disease or focal segmental glomerulosclerosis [18].

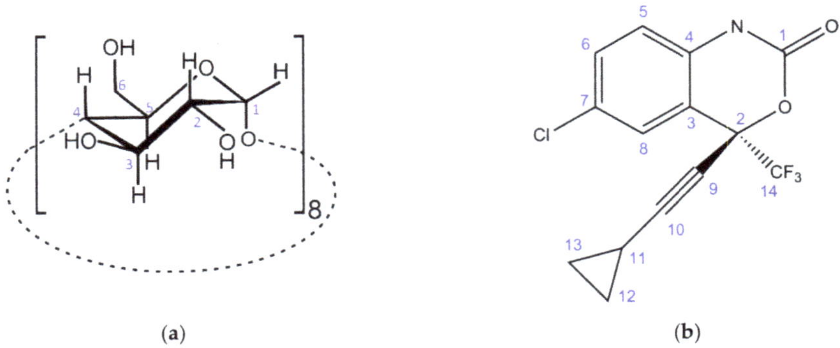

(a) (b)

Figure 1. Chemical structure and atom labelling of (a) γ-cyclodextrin; (b) efavirenz.

Efavirenz is, thus far, only known to form inclusion complexes with two chemical derivatives of beta-cyclodextrin: HPβCD (randomly (2-hydroxy)propylated β-CD) and RAMEB (randomly methylated β-CD); inclusion into the native β-CD was also investigated but the formation of an authentic inclusion complex in the solid state was not demonstrated [19]. The study, conducted by Sathigari et al., further showed that the freeze-dried adducts with RAMEB and HPβCD increased efavirenz solubility in water (at 180 min) from 5.64 ± 1.149% (pure drug) to 54.25 ± 1.031% and 43.13 ± 0.331%, respectively. In a preliminary approach to the inclusion of efavirenz into native cyclodextrins, we investigated the possibility of using β-CD and γ-CD as hosts for efavirenz [20]. Efavirenz and each cyclodextrin were co-dissolved in a water/alcohol solution and the mixture was co-precipitated by cooling; in the batch of efavirenz with β-CD, the precipitate consisted of crystals of the two separate components [20], which confirmed that β-CD does not have adequate cavity size for efavirenz. Our results were, therefore, in good agreement with the findings of Sathigari et al. [19]. In the batch of efavirenz and γ-CD, a new, low-crystalline phase was formed, indicating the formation of an inclusion complex [20]. The preliminary findings on the formation of γ-CD·EFC prompted further investigation on this host-guest system, which is herein presented. Studies were conducted both in the aqueous solution phase and in the solid state. Results indicate the presence of a complex with 3:2 host–guest stoichiometry, in both the solution and solid phase.

2. Results and Discussion

2.1. Stoichiometry in a Water–Methanol Solution

Inclusion stoichiometry in solution was first assessed as a means to understand the preferences of this host–guest system. For this, ¹H spectra were collected for series of solutions of EFV and γ-CD, prepared by the continuous variation method (see Section 3.3 for details).

The plot of χ(γ-CD) × Δδ against the χ(γ-CD), for the two protons of γ-CD located at the inner cavity (H3 and H5), is represented in the Figure 2. The maximum of the experimental points, indicative of the host-to-guest proportion in the inclusion complex, is found at the χ(γ-CD) value of 0.6, thus indicating that the preferred stoichiometry is 3:2, that is, the most abundant species in this liquid medium is (γ-CD)$_3$·(EFV)$_2$. Note that the presence of other species, namely the 1:1 complex, i.e., (γ-CD)·(EFV), albeit in lower abundance, cannot be excluded [21–23].

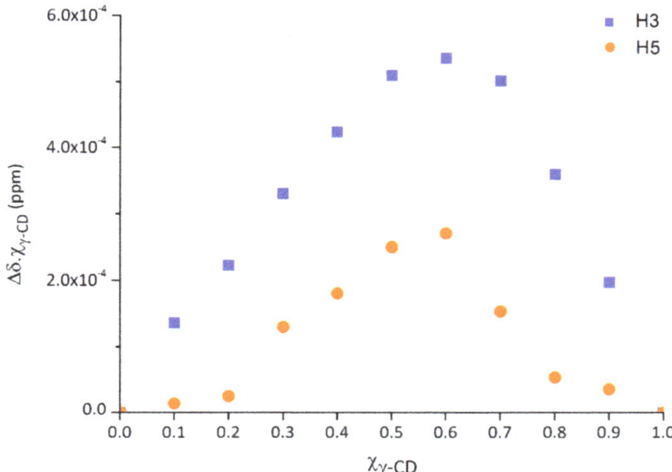

Figure 2. Job plot of the host protons H3 and H5 for the inclusion of efavirenz into γ-CD, measured in a mixed solution of D_2O and CD_3OD (1:1).

2.2. Solid-State Studies

Procedures for the preparation of inclusion complexes of efavirenz and γ-CD as solid materials comprised combining two separate solutions of each component. Efavirenz, dissolved in ethanol, was mixed with γ-CD, dissolved in ultrapure water, to obtain a clear mixed solution that was subsequently subjected to snap-freezing and freeze-drying. In agreement with the information obtained from the studies in solution, two different host-to-guest stoichiometries were tested, 3:2 and 1:1.

2.2.1. FT-IR Spectroscopy

Infrared spectroscopy can provide a quick insight into the formation of inclusion complexes, in particular for guests that contain, like efavirenz, oscillators sensitive to the hydrophobic environment of the host cavity and located in a spectral area free from host bands (which could eventually superimpose with it). The carbonyl (C=O) group of efavirenz is thus an excellent probe for inclusion. The carbonyl stretching frequency occurs at 1741 cm^{-1} in the spectrum of pure efavirenz, appearing blueshifted and with maxima centred at 1751 and 1746 cm^{-1} in the spectra of the 3:2 and the 1:1 products, respectively (Figure 3). A blueshift, i.e., an increase in the stretching energy of an oscillator, can be the result of the lower polarisation of the C=O group, with increased electron density in the double bond. It is frequently the result of an apolar environment around the affected oscillator, such as that caused by inclusion into the cavity of γ-CD. For the 1:1 product, the smaller blueshift can be interpreted as the combined result of two contributions, one from the inclusion complex and a second one originating from a small contamination with non-included efavirenz.

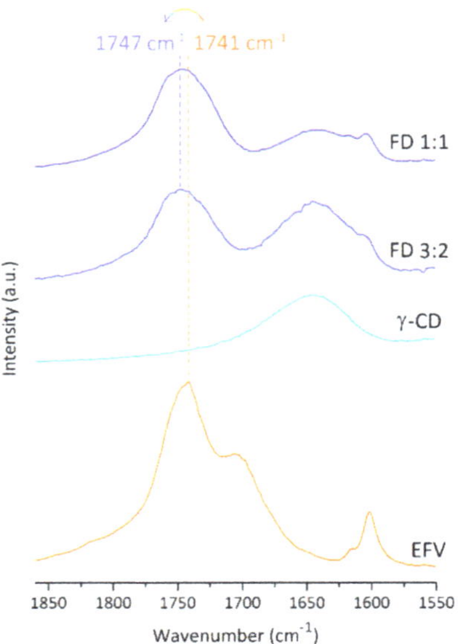

Figure 3. Fourier-transform infrared spectra of efavirenz (EFV), γ-CD and the two freeze-dried solid products starting from mixed solutions with γ-CD:EFV proportions of 3:2 and 1:1 (FD).

2.2.2. Powder X-ray Diffraction

Powder X-ray diffraction (PXRD) is a very useful tool in identifying the formation of inclusion complexes. It also contributes to investigate the presence of any eventual impurities that may appear, such as crystallites of the host or the guest. PXRD data is presented in Figure 4. Efavirenz presents a diffractogram with several well-resolved peaks that are indicative of its high crystallinity. The most intense reflections occur at 6.3, 10.5, 11.4, 11.9, 12.6, 19.3, 20.0, 20.5, 20.8, 21.2, 21.6, 23.0, 24.7, 25.0 and 27.6 degrees of 2θ.

A first attempt at collecting diffractograms of the two freeze-dried products revealed them to be mostly amorphous (results not shown), which was expected as a result of the preparation method. Restoration of the hydration waters in these samples was performed as a means to increase their crystallinity. The process consisted in placing the bulk materials at ambient temperature in a water-saturated atmosphere during ca. 16 h. PXRD patterns of the rehydrated freeze-dried compounds present overall quasi-similar diffraction patterns and they comprise essentially a new phase, with no traces of crystallites of γ-CD heptahydrate. The diffractogram of the sample with 3:2 stoichiometry exhibits reflections peaking at 5.3, 6.1, 7.5, 8.6, 10.7, 11.5, 12.3, 13.7, 14.2, 15.0, 15.7, 16.1, 16.6, 16.9, 18.9, 20.1, 20.9, 21.6, 22.2 and 22.6 degrees of 2θ. For the compound prepared with a starting stoichiometry of 1:1 (sample FD 1:1 in the Figure 4), the diffractogram is less well resolved, with the main reflections presenting some broadening and peaking at 7.6, 11.5, 11.9, 12.3, 13.7, 15.0, 16.0 and 16.9 degrees of 2θ. It should be noted that the peak centred around 11.9 degrees in the latter sample, albeit poorly resolved, coincides with the most intense reflection of pure efavirenz, thus indicating some degree of contamination with non-included efavirenz (similarly to the observations made from FT-IR data).

Figure 4 also shows the calculated diffractogram of γ-CD·12-crown-4-ether [24], which is herein used as a representative model for the only known isostructural series of γ-CD inclusion complexes [25]. The diffractogram of the 3:2 freeze-dried sample shows an overall diffraction envelope that is similar to that of the model complex, which suggests it belongs to this isostructural series [25]. It is, thus, fair to assume the inclusion complex of γ-CD and EFV should present the host molecules stacked in infinite channels, similarly to the host organisation reported for the complexes of this isostructural series and herein exemplified for γ-CD·12-crown-4-ether (inset in Figure 4).

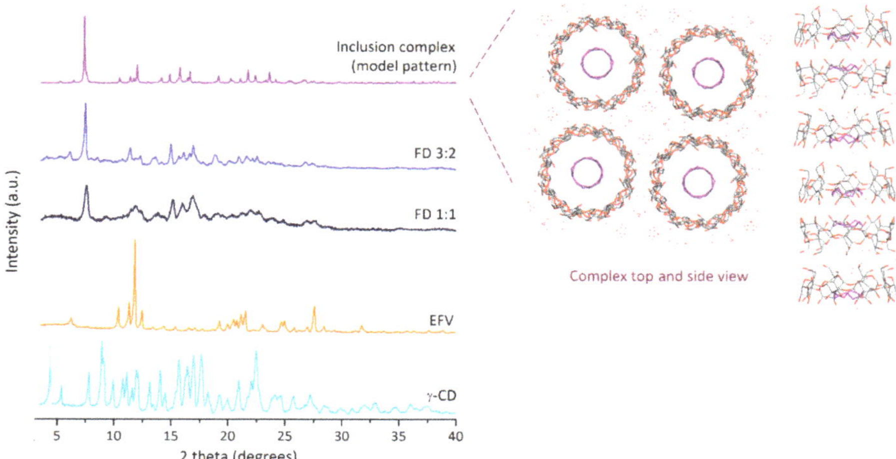

Figure 4. Experimental powder X-ray diffractograms of γ-CD, efavirenz (EFV), and the freeze-dried samples (FD) with γ-CD:EFV stoichiometries of 1:1 and 3:2 (rehydrated prior to data collection to increase their crystallinity). For comparison it also shows the trace of the inclusion complex γ-CD·12-crown-4-ether [24], calculated from its atomic coordinates using Mercury 3.5.1 (Copyright CCDC 2001–2014). The inset depicts the structure of γ-CD·12-crown-4-ether, as viewed from the top (crystallographic *c* axis) and from the side (*a* axis); the molecules of the crown ether guest are represented in purple for differentiation from those of the γ-CD macrocycle.

2.2.3. $^{13}C\{^1H\}$ CP-MAS NMR

The solid-state NMR spectra of efavirenz, γ-CD heptahydrate and the two freeze-dried products of γ-CD with EFV are depicted in Figure 5.

The spectrum of efavirenz presents a set of well-resolved resonances, with multiple signals observed for its carbons with the exception of C_2. Signal multiplicity for EFV carbons was previously reported by Rodrigues de Sousa et al., having been attributed to the presence of more than one polymorphs of efavirenz and to different molecular conformations [26]. The host, γ-CD, exhibits multiple sharp resonances for each type of carbon atom, which is ascribed, for C_1 and C_4 carbons, to differences in the conformation about the α-1,4 bonds, and, for carbons located closer to the rims, as is the case of C_6, to ambient changes in the hydrogen-bonding network and the varying number of hydration water molecules [27,28]. In the spectra of $(γ\text{-CD})_3 \cdot (EFV)_2$ and γ-CD·EFV, the host carbons appear as single broad resonances, thus indicating symmetrisation of the γ-CD as a result of inclusion of EFV and of the spatial organisation into channels. Regarding the guest signals, and as previously noted for FT-IR results, the 1:1 sample, γ-CD·EFV, shows contamination with pure efavirenz. In $^{13}C\{^1H\}$ CP-MAS NMR, this is particularly evident when one observes the carbonyl region. The spectrum of γ-CD·EFV exhibits two resonances for the carbonyl (C_1, see Figure 1) with chemical shift similar to those of pure efavirenz, thus indicating the presence of non-included guest. In turn, the spectrum of $(γ\text{-CD})_3 \cdot (EFV)_2$, bears only one resonance for the carbonyl. This indicates the presence of the pure inclusion complex, in which inclusion into the cavity of γ-CD resulted in a more symmetrical chemical environment for the C=O.

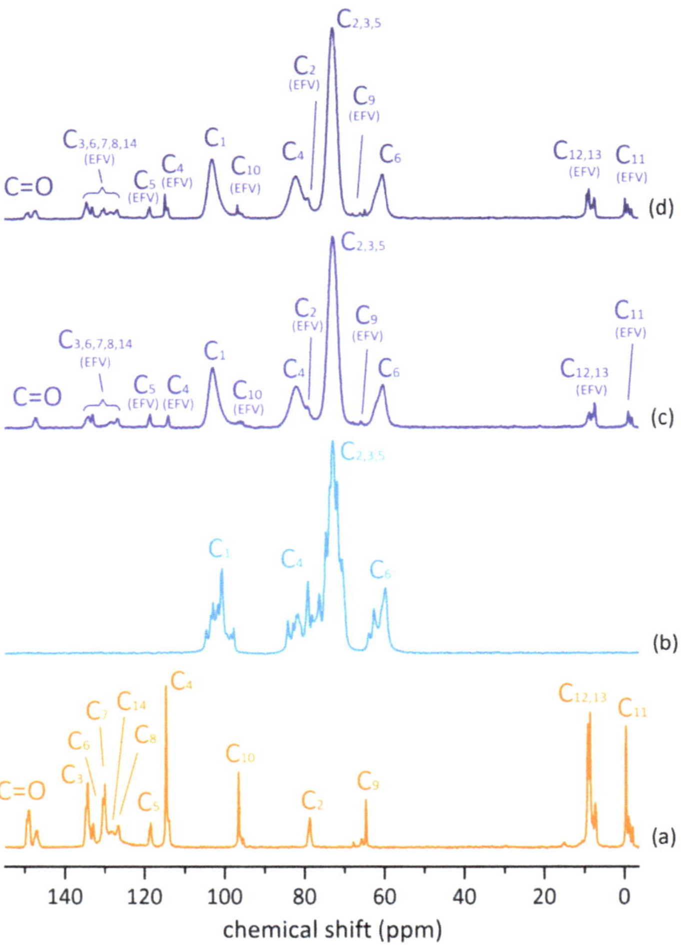

Figure 5. ^{13}C{^{1}H} CP-MAS NMR spectra of (**a**) EFV, (**b**) γ-CD, (**c**) (γ-CD)$_3$·(EFV)$_2$, and (**d**) γ-CD·EFV (see labelling in Figure 1). Efavirenz carbons were assigned as in Rodrigues de Sousa et al. [26].

2.2.4. Thermogravimetric Analysis

The thermograms of efavirenz, γ-CD heptahydrate and the two freeze-dried products, (γ-CD)$_3$·(EFV)$_2$ and γ-CD·EFV, are represented in Figure 6. The thermogram of pure efavirenz shows no mass losses from ambient temperature until about 185 °C, temperature that marks the onset of its decomposition. The absence of mass losses at temperatures lower than 100–130 °C indicates the absence of hydration waters, which is expectable due to the apolar nature of this compound. In turn, the thermogram of γ-CD heptahydrate is marked by an initial dehydration step that starts at ambient temperature and proceeds up to 95 °C. This step features a mass loss of 9% that translates into seven hydration waters, being thus coherent with the original specifications.

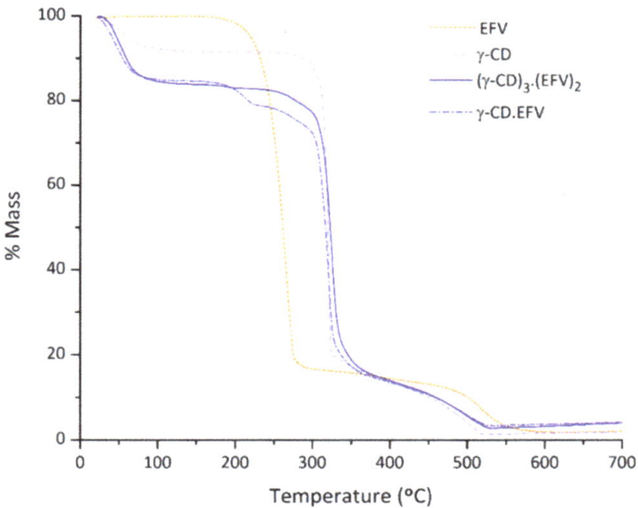

Figure 6. Thermogravimetry traces for efavirenz, γ-CD, and the freeze-dried adducts (γ-CD)$_3$·(EFV)$_2$ and γ-CD·EFV.

The thermogravimetric traces of the two freeze-dried products are marked by an increase in the number of hydration waters. This is evidenced by a more intense dehydration step, rounding 15.5%. For the 3:2 complex, the results allow inferring a general formula of (γ-CD)$_3$·(EFV)$_2$·(H$_2$O)$_{39}$. The increase in the number of hydration waters in comparison to those exhibited by the host is a characteristic of γ-CD inclusion complexes and it results from their distinctive tetragonal symmetry. As described in the section referring to PXRD, γ-CD molecules pack, with the guest molecules lodged inside the channel cavity and also with the formation of wide inter-channel spaces that are able to retain a large number of water molecules (see inset in Figure 4). Literature examples of γ-CD complexes with a large number of hydration waters include γ-CD·quercetin·(H$_2$O)$_{17}$ [29], γ-CD·fisetin·(H$_2$O)$_{17}$ [30], and (γ-CD)$_3$·(resveratrol)$_4$·(H$_2$O)$_{62}$ [31], to name only a few.

The presence of a strong dehydration step in the trace corresponding to the 1:1 sample is also indicative of the presence of an inclusion complex, however, the mass loss observed between ca. 180 and 220 °C denotes contamination with some amount of pure efavirenz that decomposes at this temperature. It is important to highlight that the step associated with efavirenz thermal decomposition is absent from the thermogram of (γ-CD)$_3$·(EFV)$_2$, which provides definitive corroboration of the presence of the pure inclusion complex in this sample.

2.3. Effect of γ-CD on Efavirenz Solubility

The solubilising effect of medicinal applications of γ-CD on efavirenz was evaluated by collecting the solubility isotherm for this API while in the presence of increasing concentrations of the host. Results are shown in Figure 7. The aqueous solubility reported for efavirenz is very low, with values between 0.0127 mM [32] and 0.0263 mM [33]. The isotherm data herein shown demonstrates that the presence of 37.7 mM γ-CD in solution increases the solubility of efavirenz to 0.124 mM, that is, by at least 5-fold. Efavirenz solubility increases further with the increase in concentration of γ-CD in aqueous solution, with 0.565 mM of EFV dissolved when γ-CD concentration is 75.0 and 1.424 mM of EFV dissolved for the γ-CD concentration of 112.5 mM. After this point, a plateau is reached, indicating the formation of aggregates containing only γ-CD molecules. Self-aggregation of γ-CD molecules is well known, and it can be attributed to their symmetry and extensive intermolecular hydrogen bonding.

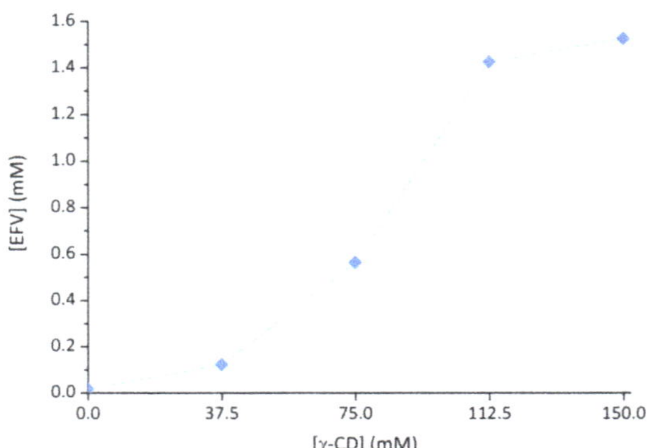

Figure 7. Phase solubility isotherm for efavirenz (EFV) in the presence of γ-CD.

For a brief discussion of these results, it is worth comparing them with those of β-CD, HPβCD, RAMEB and HPγCD, previously reported as solubilisers for EFV. The effect of RAMEB on EFV solubility was equivalent to that of γ-CD, with 80 mM of RAMEB solubilising roughly 0.5 mM of this guest [19]. Nevertheless, RAMEB is only approved for topical use (at the nasal and ocular mucosa), which limits its interest as a solubilising agent [14]. Regarding HPβCD, the literature shows contradictory data, one study depicting it as a good solubiliser, with 60 mM increasing EFV solubility to roughly 1 mM [19], whereas our previous report, in which the isotherm data were collected under the same conditions as those of the present study, revealed it to perform worse than γ-CD, since a concentration of 125 mM of HPβCD was required to solubilise ca. 0.5 mM of EFV [20]. Our study also evaluated HPγCD, which had an isotherm similar to that of HPβCD up to 150 mM and performed slightly worse at higher concentrations [20].

3. Materials and Methods

3.1. Materials

Pharmaceutical-grade γ-CD (Cavamax W8 Pharma) from Wacker-Chemie was kindly donated by Ashland Specialty Ingredients (Düsseldorf, Germany). Efavirenz was obtained from Smillax Pharma (Heiderabad, India).

Ultrapure water was used for the inclusion procedures. All organic solvents were of analytical grade, except otherwise specified.

3.2. Equipment

Laboratory powder XRD data were collected at ambient temperature on an Empyrean PANalytical diffractometer (Cu K$\alpha_{1,2}$ X-radiation, λ_1 = 1.540598 Å; λ_2 = 1.544426 Å) equipped with an PIXcel 1D detector and with the sealed tube operating at 45 kV and 40 mA (Bruker AXS, Karlsruhe, Germany). Intensity data were collected by the step-counting method (step 0.01°), in continuous mode, in the ca. $3.5 \leq 2\theta \leq 50°$ range.

Solution-phase ^{1}H nuclear magnetic resonance (NMR) spectra were recorded on an Avance 300 spectrometer (Bruker Biospin, Rheinstetten, Germany) at 300.13 MHz, at ambient temperature. A 50:50 solution of deuterated water and deuterated methanol was used as solvent, with the residual proton signal of methanol (^{1}H 3.31 ppm) and tetramethylsilane (TMS) being used as internal references. The chemical shifts are quoted in parts per million.

^{13}C{^{1}H}CP/MAS NMR spectra were recorded at 100.62 MHz on a (9.4 T) Avance III 400 spectrometer (Bruker Biospin), with an optimised $\pi/2$ pulse for ^{1}H of 4.5 µs, 3 ms contact time, a spinning rate of 12 kHz, and 4 s recycle delays. The chemical shifts are quoted in parts per million from TMS.

Infrared spectra were obtained as KBr pellets in a 7000 FTIR spectrometer (Mattson, Oakland, CA, USA) (resolution 2.0 cm^{-1}; 128 scans per spectrum).

TGA studies were performed on a Shimadzu TGA-50 thermogravimetric analyser (Kyoto, Japan), using a heating rate of 5 C min^{-1}, under air atmosphere, with a flow rate of 20 mL min^{-1}. The sample holder was a 5 mm ø platinum plate and the sample mass was about 5 mg.

UV-Vis data for the solubility isotherms were collected on a spectrophotometer Analytik Jena Specord 200 Plus.

3.3. Continuous Variation Method

The continuous variation method reported by Job [34] provides an estimate of the preferred stoichiometry in solution based on the measured changes in a physical parameter (in the present case, the chemical shift of sample protons). A series of solutions were prepared using as solvent a 1:1 mixture of deuterated water and deuterated methanol. In each solution, the ratio of host and guest, r, varied in steps of 0.1 while their sum ($[\gamma\text{-CD}]_0$ + $[\text{EFV}]_0$) was kept constant, at a value of 0.01 M. For the host, γ-CD, r is thus defined as:

$$r_{\gamma\text{-CD}} = [\gamma\text{-CD}]_0 / \{[\gamma\text{-CD}]_0 + [\text{EFV}]_0\}) \tag{1}$$

3.4. Preparation of the Inclusion Complexes as Solid Materials

3.4.1. γ-CD with EFV in the 3:2 Stoichiometry

A solution of γ-CD (213.1 mg, 0.15 mmol) in ultrapure water (1.5 mL) at 40 °C was treated with another solution of EFV (31.5 mg, 0.10 mmol) in ethanol (0.2 mL). The mixed solution was stirred for 3 min and then subjected to snap-freezing in liquid nitrogen. Solvents were subsequently removed by freeze-drying to obtain a white solid.

FT-IR: ν(tilde) (cm^{-1}) = 3405 s, 2934 m, 2894 sh, 2254 w, 1751 m, 1647 m, 1502 m, 1456 sh, 1417 m, 1385 m, 1373 m, 1337 m, 1302 m, 1252 m, 1200 m, 1161 s, 1102 sh, 1080 s, 1054 sh, 1027 vs, 1002 s, 944 m, 933 m, 864 w, 830 vw, 760 m, 743 w, 707 m, 693 w, 656 w, 610 w, 585 m, 531 w, 482 w.

^{13}C{^{1}H} CP-MAS NMR: δ (ppm) = 147.3 (EFV C_1), 134.4 (EFV C_3), 133.1 (EFV C_6), 130.0 (EFV C_7), 128.4 (EFV C_{14}), 126.8 (EFV C_8), 118.7 (EFV C_5), 114.2 (EFV C_4), 103.1 (γ-CD C_1), 96.1 (EFV C_{10}), 82.2 (γ-CD C_4), 79.3 (EFV C_2), 73.0 (γ-CD $C_{2,3,5}$), 66.0 (EFV C_9), 60.7 (γ-CD C_6), 8.8, 7.3 (EFV $C_{12,13}$), −0.3, −0.9, −1.7 (EFV C_{11}) ppm.

3.4.2. γ-CD with EFV in the 1:1 Stoichiometry

A solution of γ-CD (142.0 mg, 0.10 mmol) in ultrapure water (1.5 mL) at 40 °C was treated with another solution of EFV (31.5 mg, 0.10 mmol) in ethanol (0.2 mL). The mixed solution was stirred for 3 min and then subjected to snap-freezing in liquid nitrogen. Solvents were subsequently removed by freeze-drying to obtain a white solid.

FT-IR: ν(tilde) (cm^{-1}) = 3390 s, 2932 m, 2893 sh, 2252 w, 1746 m, 1637 m, 1499 m, 1458 m, 1414 m, 1383 m, 1372 m, 1335 m, 1302 m, 1249 m, 1198 m, 1187 m, 1159 s, 1099 sh, 1079 s, 1052 sh, 1025 vs, 1000 s, 942 m, 931 sh, 861 w, 838 vw, 760 m, 742 w, 706 m, 691 w, 674 vw, 671 vw, 655 w, 608 w, 580 m, 566 sh, 528 w, 480 w

^{13}C{^{1}H} CP-MAS NMR: δ (ppm) = 149.2, 147.3 (EFV C_1), 134.5 (EFV C_3), 133.1 (EFV C_6), 130.2 (EFV C_7), 128.4 (EFV C_{14}), 126.8 (EFV C_8), 118.7 (EFV C_5), 114.9, 114.2 (EFV C_4), 103.1 (γ-CD C_1), 96.8, 95.6 (EFV C_{10}), 82.2 (γ-CD, C), 79.1 (EFV C_2), 73.0 (γ-CD $C_{2,3,5}$), 66.1, 64.9 (EFV C_9), 60.8 (γ-CD C), 8.8, 7.5 (EFV $C_{12,13}$), −0.1, −0.8, −1.7 (EFV C_{11}) ppm.

For comparison, the data for efavirenz is as follows:

FT-IR data of efavirenz: ν(tilde) (cm^{-1}) = 3255 s, 3183 s, 3094 m, 3020 w, 2951 w, 2876 w, 2253 s, 1894 vw, 1741 vs, 1705 s, 1602 s, 1498 vs, 1457 m, 1428 w, 1403 m, 1134 m, 1060 m, 1333 vs, 1280 s, 1251 vs, 1184 vs, 1168 vs, 1096 s, 1073 s, 1028 s, 976 s, 948 s, 931 s, 884 m, 866 m, 835 m, 824 s, 755 m, 742 m, 711 s, 689 s, 668 w, 656 m, 568 m, 556 m, 541 m, 518 w, 484 m, 460 w, 417 w.

$^{13}C\{^1H\}$ CP-MAS NMR of efavirenz: δ (ppm) = 149.3, 147.5 (C_1), 134.5 (C_3), 133.2 (C_6) 130.7, 130.2 (C_7), 128.4 (C_{14}), 126.8 (C_8), 118.8 (C_5), 114.8, 114.2 (C_4), 96.8, 95.5 (C_{10}), 79.0 (C_2), 66.0, 64.9 (C_9), 9.4, 8.9, 8.1, 7.5 (EFV $C_{12,13}$), −0.1, −0.8, −1.0, −1.8 (EFV C_{11}).

3.5. Solubility Isotherms

Excess amounts of EFV were added to 20 mL of unbuffered aqueous solutions of increasing concentrations of γ-CD. Solutions were stirred for 48 h at room temperature. Aliquots were filtered (0.22 μm), diluted with a 1:1 mixture of water and isopropanol and their absorbance was measured at 293 nm.

4. Conclusions

The results described in the present report demonstrated that inclusion of efavirenz into γ-CD occurred both in solution and in the solid state, forming a complex with 3:2 stoichiometry, that is, $(γ\text{-}CD)_3 \cdot (EFV)_2$. The bulky nature of efavirenz, with a quasi-planar central bicyclic benzoxazin-2-one ring, and two substituents, a cyclopropylethynyl and a trifluoromethyl, protruding laterally from the main plane of the rings, implies that a host with a large cavity is required for molecular encapsulation. It is noteworthy that, even though we employed γ-CD, a host with a wide cavity diameter, inclusion required 1.5 host units per each molecule of efavirenz—the host-to-guest stoichiometry of 3:2 is quite rare for γ-CD inclusion complexes.

Powder X-ray diffraction further evidenced that the complex belongs to the isostructural series of γ-CD inclusion complexes with tetragonal symmetry in which the molecules of γ-CD are stacked in infinite channels with the guest molecules located inside. This contributed to the symmetrisation of the environment around the carbons of efavirenz, particularly C1, which is part of a carbonyl group, and that was observed as a single resonance in solid-state NMR and as a blueshifted vibrational band in FT-IR. The solubilising effect of γ-CD over the efavirenz guest was evaluated by collection of the solubility isotherm for this host–guest system to reveal a B_s-type diagram [35], that is, the association with γ-CD increases EFV solubility but only to a certain point, which is followed by a plateau. Besides the solubilising action of γ-CD on efavirenz, its ability to mask the bitter taste of this drug is another attractive application for the inclusion complex that warrants demonstration in future studies.

Author Contributions: Conceptualisation, S.S.B.; methodology, S.S.B.; data curation, K.L., F.E.-S., S.S.B.; formal analysis, K.L., F.E.-S., S.S.B.; investigation, K.L., F.E.-S., S.S.B.; writing—original draft preparation, S.S.B.; writing—review and editing, K.L., F.E.-S., S.S.B., F.A.A.P.; supervision, S.S.B., F.A.A.P.; funding acquisition, S.S.B., F.E.-S., F.A.A.P. All authors have read and agreed to the published version of the manuscript.

Funding: We acknowledge University of Aveiro and FCT/MCTES (Fundação para a Ciência e a Tecnologia, Ministério da Ciência, da Tecnologia e do Ensino Superior) for financial support for the QOPNA research Unit (FCT UID/QUI/00062/2019), LAQV-REQUIMTE (Ref. UIDB/50006/2020) and CICECO—Aveiro Institute of Materials (UIDB/50011/2020 and UIDP/50011/2020), through national founds and, where applicable, co-financed by the FEDER, within the PT2020 Partnership Agreement. The NMR spectrometers are part of the National NMR Network (PTNMR) and are partially supported by Infrastructure Project N° 022161 (co-financed by FEDER through the operational programme COMPETE 2020, POCI and PORL and FCT through PIDDAC).

Data Availability Statement: Solid-state characterisation data is available upon request from the NMR and diffraction services of the University of Aveiro.

Conflicts of Interest: The authors declare no conflict of interest.

Abbreviations

API	Active pharmaceutical ingredient
CP-MAS	Cross-polarisation with magic angle spinning (a solid-state NMR method)
FT-IR	Fourier-transform infrared spectroscopy
γ-CD	Gamma-cyclodextrin
HPβCD	(2-hydroxy)propyl-beta-cyclodextrin
HPγCD	(2-hydroxy)propyl-gamma-cyclodextrin
EFV	Efavirenz
NMR	Nuclear magnetic resonance
ppm	Parts per million
PXRD	Powder X-ray diffraction
RAMEB	Randomly methylated beta-cyclodextrin
TGA	Thermogravimetric analysis
TMS	Tetramethylsilane

References

1. Best, B.M.; Goicoechea, M. Efavirenz—Still first line king? *Expert Opin. Drug Metab. Toxicol.* **2018**, *4*, 965–972. [CrossRef] [PubMed]
2. Fandaruff, C.; Rauber, G.S.; Araya-Sibaja, A.M.; Pereira, R.N.; de Campos, C.E.M.; Rocha, H.V.A.; Monti, G.A.; Malaspina, T.; Silva, M.A.S.; Cuffini, S.L. Polymorphism of anti-HIV drug efavirenz: Investigations on thermodynamic and dissolution properties. *Cryst. Growth Des.* **2014**, *14*, 4968–4974. [CrossRef]
3. Rajesh, Y.V.; Balasubramaniam, J.; Bindu, K.; Sridevi, R.; Swetha, M.; Rao, V.U. Impact of superdisintegrants on efavirenz release from tablet formulations. *Acta Pharm.* **2010**, *60*, 185–195. [CrossRef] [PubMed]
4. Yang, J.; Grey, K.; Doney, J. An improved kinetics approach to describe the physical stability of amorphous solid dispersions. *Int. J. Pharm.* **2010**, *384*, 24–31. [CrossRef] [PubMed]
5. Madhavi, B.B.; Kusum, B.; Charanya, C.K.; Madhu, M.N.; Harsa, V.S.; Banji, D. Dissolution enhancement of efavirenz by solid dispersion and PEGylation techniques. *Int. J. Pharm. Investig.* **2011**, *1*, 29–34. [CrossRef]
6. Okafor, N.I.; Nkanga, C.I.; Walker, R.B.; Noundou, X.S.; Krause, R.W.M. Encapsulation and physicochemical evaluation of efavirenz in liposomes. *J. Pharm. Investig.* **2019**, *50*, 201–208. [CrossRef]
7. Chiappetta, D.A.; Facorro, G.; de Celis, E.R.; Sosnik, A. Synergistic encapsulation of the anti-HIV agent efavirenz within mixed poloxamine/poloxamer polymeric micelles. *Nanomed. Nanotech. Biol. Med.* **2011**, *7*, 624–637. [CrossRef]
8. Chiappetta, D.A.; Hocht, C.; Taira, C.; Sosnik, A. Efavirenz-loaded polymeric micelles for pediatric anti-HIV pharmacotherapy with significantly higher oral bioavailaibility. *Nanomedicine* **2010**, *5*, 11–23. [CrossRef]
9. Varshosaz, J.; Taymouri, S.; Jahanian-Najafabadi, A.; Alizadeh, A. Efavirenz oral delivery via lipid nanocapsules: Formulation, optimisation, and ex-vivo gut permeation study. *IET Nanobiotechnol.* **2018**, *12*, 795–806. [CrossRef]
10. Pokharkar, V.; Patil-Gadhe, A.; Palla, P. Efavirenz loaded nanostructured lipid carrier engineered for brain targeting through intranasal route: In-vivo pharmacokinetic and toxicity study. *Biomed. Pharmacother.* **2017**, *94*, 150–164. [CrossRef]
11. Tshweu, L.; Katata, L.; Kalombo, L.; Chiappetta, D.A.; Hocht, C.; Sosnik, A.; Swai, H. Enhanced oral bioavailability of the antiretroviral efavirenz encapsulated in poly(epsilon-caprolactone) nanoparticles by a spray-drying method. *Nanomedicine* **2014**, *9*, 1821–1833 [CrossRef] [PubMed]
12. Kim, D.-H.; Lee, S.-E.; Pyo, Y.-C.; Tran, P.; Park, J.-S. Solubility enhancement and application of cyclodextrins in local drug delivery. *J. Pharm. Investig.* **2020**, *50*, 17–27. [CrossRef]
13. Pereira, A.B.; Braga, S.S. Cyclodextrin Inclusion of Nutraceuticals, from the Bench to your Table. In *Cyclodextrins: Synthesis, Chemical Applications and Role in Drug Delivery*, 1st ed.; Ramirez, F.G., Ed.; NovaSience: Hauppage, NY, USA, 2015.
14. Background Review for Cyclodextrins Used as Excipients; European Medicines Agency. 2014. Available online: http://www.ema.europa.eu/docs/en_GB/document_library/Report/2014/12/WC500177936.pdf (accessed on 9 December 2020).
15. Veklury (Remdesivir) Prescribing Information. Available online: https://www.gilead.com/-/media/files/pdfs/medicines/covid-19/veklury/veklury_pi.pdf (accessed on 9 December 2020).
16. Lyseng-Williamson, K.A. Midazolam oral solution (Ozalin®): A profile of its use for procedural sedation or premedication before anaesthesia in children. *Drug Ther. Persp.* **2019**, *35*, 255–262. [CrossRef]
17. Phase 1 Study of Hydroxypropyl-Beta-Cyclodextrin(HP-beta-CyD)-Adjuvanted Influenza Split Vaccine. Available online: https://rctportal.niph.go.jp/en/detail?trial_id=UMIN000028530 (accessed on 2 August 2019).
18. Braga, S.S. Cyclodextrins: Emerging medicines of the new millennium. *Biomolecules* **2019**, *9*, 801. [CrossRef]
19. Sathigari, S.; Chadha, G.; Lee, Y.L.; Wright, N.; Parsons, D.L.; Rangari, V.K.; Fasina, R.; Babu, J. Physicochemical characterization of efavirenz cyclodextrin inclusion complexes. *AAPS Pharm. Sci. Tech.* **2009**, *10*, 81–87. [CrossRef]
20. Braga, S.S.; Lysenko, K.; El-Saleh, F.; Paz, F.A.A. Cyclodextrin-efavirenz complexes investigated by solid state and solubility studies. *Sciforum* **2020**. [CrossRef]

21. Braga, S.S.; Aree, T.; Immamura, K.; Vertut, P.; Boal-Palheiros, I.; Sänger, W.; Teixeira-Dias, J.J.C. Structure of the β-cyclodextrin *p*-hydroxybenzaldehyde inclusion complex in aqueous solution and in the crystalline state. *J. Incl. Phenom. Macrocycl. Chem.* **2002**, *43*, 115–125. [CrossRef]

22. Fernandes, J.A.; Paz, F.A.A.; Braga, S.S.; Ribeiro-Claro, P.; Rocha, J. Inclusion of potassium 4,40-biphenyldicarboxylate into b-cyclodextrin: The design and synthesis of an organic secondary building unit. *New J. Chem.* **2011**, *35*, 1280–1290. [CrossRef]

23. Silva, R.N.; Costa, C.C.; Santos, M.J.G.; Alves, M.Q.; Braga, S.S.; Vieira, S.I.; Rocha, J.; Silva, A.M.S.; Guieu, S. Fluorescent light-up probe for the detection of protein aggregates. *Chem. Asian J.* **2019**, *14*, 859–863. [CrossRef]

24. Kamitori, S.; Hirotsu, K.; Higuchi, T. Crystal and molecular structure of the γ-cyclodextrin–12-crown-4 1: 1 inclusion complex. *J. Chem. Soc. Chem. Commun.* **1986**, 690–691. [CrossRef]

25. Caira, M.R. On the isostructurality of cyclodextrin inclusion complexes and its practical utility. *Rev. Roum. Chim.* **2001**, *46*, 371–386.

26. de Sousa, E.G.R.; de Carvalho, E.M.; Gil, R.A.S.S.; dos Santos, T.C.; Borré, L.B.; Santos-Filho, O.A.; Ellena, J. Solution and solid state nuclear magnetic resonance spectroscopic characterization of efavirenz. *J. Pharm. Sci.* **2016**, *105*, 2656–2664. [CrossRef] [PubMed]

27. Heyes, S.J.; Clayden, N.J.; Dobson, C.M. ^{13}C-CP/MAS NMR studies of the cyclomalto-oligosaccharide (cyclodextrin) hydrates. *Carbohydr. Res.* **1992**, *233*, 1–14. [CrossRef]

28. Gidley, M.J.; Bociek, S.M. Carbon-13 CP/MAS NMR studies of amylose inclusion complexes, cyclodextrins, and the amorphous phase of starch granules: Relationships between glycosidic linkage conformation and solid-state carbon-13 chemical shifts. *J. Am. Chem. Soc.* **1988**, *110*, 3820–3829. [CrossRef]

29. Pereira, A.B.; Silva, A.M.; Barroca, M.J.; Marques, M.P.M.; Braga, S.S. Physicochemical properties, antioxidant action and practical application in fresh cheese of the solid inclusion compound γ-cyclodextrin·quercetin, in comparison with β-cyclodextrin·quercetin. *Arab. J. Chem.* **2020**, *13*, 205–215. [CrossRef]

30. Pais, J.M.; Barroca, M.J.; Marques, M.P.M.; Paz, F.A.A.; Braga, S.S. Solid-state studies and antioxidant properties of the γ-cyclodextrin·fisetin inclusion compound. *Beilstein J. Org. Chem.* **2017**, *13*, 2138–2145. [CrossRef] [PubMed]

31. Catenacci, L.; Sorrenti, M.; Bonferoni, M.C.; Hunt, L.; Caira, M.R. Inclusion of the phytoalexin trans-resveratrol in native cyclodextrins: A thermal, spectroscopic, and X-ray structural study. *Molecules* **2020**, *25*, 998. [CrossRef]

32. Kommavarapu, P.; Maruthapillai, A.; Palanisamy, K. Preparation and characterization of efavirenz nanosuspension with the application of enhanced solubility and dissolution rate. *HIV AIDS Rev.* **2016**, *15*, 170–176. [CrossRef]

33. Cristofoletti, R.; Nair, A.; Abrahamsson, B.; Groot, D.W.; Kopp, S.; Langguth, P.; Polli, J.E.; Shah, V.P.; Dressman, J.B. Biowaiver monographs for immediate release solid oral dosage forms: Efavirenz. *J. Pharm. Sci.* **2013**, *102*, 318–329. [CrossRef]

34. Job, P. Formation and stability of inorganic complexes in solution. *Anal. Chim. Appl.* **1928**, *9*, 113–203.

35. Uekama, K.; Hirayama, F.; Irie, T. Cyclodextrin drug carrier systems. *Chem. Rev.* **1998**, *98*, 2045–2076. [CrossRef] [PubMed]

Article

Thermal Degradation of Linalool-Chemotype *Cinnamomum osmophloeum* Leaf Essential Oil and Its Stabilization by Microencapsulation with β-Cyclodextrin

Hui-Ting Chang [1,*], Chun-Ya Lin [2], Li-Sheng Hsu [1] and Shang-Tzen Chang [1]

[1] School of Forestry and Resource Conservation, National Taiwan University, Taipei 10617, Taiwan; r05625034@ntu.edu.tw (L.-S.H.); peter@ntu.edu.tw (S.-T.C.)

[2] Department of Wood Based Materials and Design, National Chiayi University, Chiayi 600355, Taiwan; keocylin@mail.ncyu.edu.tw

* Correspondence: chtchang@ntu.edu.tw; Tel.: +886-2-3366-5880

Abstract: The thermal degradation of linalool-chemotype *Cinnamomum osmophloeum* leaf essential oil and the stability effect of microencapsulation of leaf essential oil with β-cyclodextrin were studied. After thermal degradation of linalool-chemotype leaf essential oil, degraded compounds including β-myrcene, *cis*-ocimene and *trans*-ocimene, were formed through the dehydroxylation of linalool; and ene cyclization also occurs to linalool and its dehydroxylated products to form the compounds such as limonene, terpinolene and α-terpinene. The optimal microencapsulation conditions of leaf essential oil microcapsules were at a leaf essential oil to the β-cyclodextrin ratio of 15:85 and with a solvent ratio (ethanol to water) of 1:5. The maximum yield of leaf essential oil microencapsulated with β-cyclodextrin was 96.5%. According to results from the accelerated dry-heat aging test, β-cyclodextrin was fairly stable at 105 °C, and microencapsulation with β-cyclodextrin can efficiently slow down the emission of linalool-chemotype *C. osmophloeum* leaf essential oil.

Keywords: *Cinnamomum osmophloeum*; linalool; β-cyclodextrin; microencapsulation

Citation: Chang, H.-T.; Lin, C.-Y.; Hsu, L.-S.; Chang, S.-T. Thermal Degradation of Linalool-Chemotype *Cinnamomum osmophloeum* Leaf Essential Oil and Its Stabilization by Microencapsulation with β-Cyclodextrin. *Molecules* 2021, 26, 409. https://doi.org/10.3390/molecules26020409

Academic Editors: Marina Isidori, Margherita Lavorgna and Rosa Iacovino

Received: 24 December 2020
Accepted: 13 January 2021
Published: 14 January 2021

Publisher's Note: MDPI stays neutral with regard to jurisdictional claims in published maps and institutional affiliations.

1. Introduction

Natural products from cinnamon plants (*Cinnamomum* spp., Lauraceae) exhibit various bioactivities, such as antimicrobial, insecticidal, anti-inflammatory, antidiabetic activities and, etc. [1–5]. *Cinnamomum osmophloeum* Kanehira is commonly used as folk medicines and food flavors. It is scientifically reported that extracts and essential oils of *C. osmophloeum* exhibit the antioxidant, antibacterial, anxiolytic, xanthine oxidase inhibitory effects, and so forth [6–9].

Parts of natural plant products, especially essential oils, are highly volatile in the ambient environment and, therefore, may result in thermal oxidation/degradation. Microencapsulation or nanoencapsulation of the essential oils and extracts could provide protection and enhance the stabilization [10–12]. Encapsulation materials used include gelatin, cyclodextrins, gum arabic, caseinates, alginates, cellulose derivatives, chitosans, etc. [13–16]. Properties of natural plant products after microencapsulation would be influenced by the kinds of core-shell structures and coating materials [17–20].

Cyclodextrins are amphiphilic hollow cyclic oligosaccharides and form the inclusion complexes with versatile molecules [17,21,22]. β-Cyclodextrin, composed of seven α-D-glucopyranose units, is the most common cyclodextrin product and of good durability; it is a multifunctional encapsulation material to keep the bioactive ingredients from volatilization, oxidization, and etc. [23–25]. Many researchers have reported the microencapsulation of the bioactive constituents, essentials oil, or supercritical fluid extracts with β-cyclodextrin [9,10,26,27]. Ramos et al. proved that the inclusion of isopulegol, an alcoholic monoterpene, in β-cyclodextrin is an effective method to improve its antiedematogenic and anti-inflammatory activities [28]. Cyclodextrins are appropriate carriers for delivering or releasing natural products in the pharmaceutical, food and cosmetic industries [16,29–33].

The aims of this study were to investigate the thermal degradation of linalool-chemotype *C. osmophloeum* leaf essential oil, find the optimal reaction conditions of microencapsulation of leaf essential oil with β-cyclodextrin, and evaluate the stabilization of leaf essential oil microcapsules. Through our research, it is expected to reveal the changes in the chemical structure of linalool during thermal decomposition and properly preserve the linalool-chemotype *C. osmophloeum* leaf essential oil by microencapsulating with β-cyclodextrin.

2. Materials and Methods

2.1. Hydrodistillation

Fresh *C. osmophloeum* leaves were collected from Lienhuachih Research Center of Taiwan Forestry Research Institute in Nantou County, Taiwan. Leaves were hydrodistilled in a Clevenger apparatus for 6 h to obtain the leaf essential oil. The yield of leaf essential oil was 3.46 ± 0.06% (*w/w*). Leaf essential oil was stored in the dark glass bottle and kept in the refrigerator at 4 °C.

2.2. GC–MS Analysis

The constituents of the leaf essential oil were analyzed by Thermo Trace GC Ultra gas chromatograph equipped with a Polaris Q MSD mass spectrometer (Thermo Fisher Scientific, Austin, TX, USA). Each 1 μL analyte was injected into the DB-5MS capillary column (Crossbond 5% phenyl methyl polysiloxane, 30 m length × 0.25 mm i.d. × 0.25 μm film thickness). The temperature program was: 60 °C initial temperature for 1 min; 4 °C/min up to 220 °C and hold for 2 min; and 20 °C/min up to 250 °C and hold for 3 min. The flow rate of carrier gas helium was 1 mL/min, and the split ratio was 1:10. Constituents were identified by comparing the mass spectra (*m/z* 50–600 amu) with NIST and Wiley library data, Kovats index (KI) [34] and authentic standards. Quantification of constituents was analyzed by integrating the peak area of the chromatogram using GC equipped with the flame ionization detector (FID).

2.3. Microencapsulation

Leaf essential oil microencapsulated with β-cyclodextrin was using the co-precipitation method with slight revisions [35–37]. β-Cyclodextrin (5 g) was first dissolved in 300 mL different ratio of ethanol/water solution at 50 °C for 5 min; the solution was cooled down to 25 °C. Linalool/leaf essential oil (0.88 g) was dissolved in 10 mL ethanol; and then added to the β-cyclodextrin solution, stirred at 600 rpm for 1 h. The solution was kept in the refrigerator at 4 °C overnight; the co-precipitated microcapsules were filtered and then washed with 50 mL distilled water. Microcapsules were dried at 50 °C in the oven for 24 h. The yield of leaf oil microcapsules was calculated by the following Formula (1).

$$\text{Yield (\%)} = \text{microcapsules (g)}/[\text{β-cyclodextrin (g)} + \text{leaf essential oil (g)}] \times 100 \quad (1)$$

2.4. Accelerated Dry-Heat Aging Test

An accelerated dry-heat aging test (ISO 5630–1; CNS 12,887–1) was used to evaluate the thermostability of leaf oil microcapsules. Microcapsules were heated at 105 °C in a ventilated oven. Weights of leaf oil microcapsules were measured during the accelerated dry-heat aging test (1, 2, 4, and 8 days). After the aging test, weight losses of leaf oil microcapsules were calculated.

2.5. Statistical Analysis

Results data were statistically analyzed using Scheffé's test of the SAS system (version 9.2 Cary, NC, USA) with a 95% confidence interval. Scheffé's test is a post hoc multiple comparison method with stringent error control.

3. Results and Discussion

3.1. Changes in Composition of C. osmophloeum Leaf Oil after Thermal Degradation

Constituents of *C. osmophloeum* leaf essential oil were analyzed by GC–MS; a gas chromatogram of leaf essential oil is shown in Figure 1A. The main constituent of leaf essential oil was linalool (93.30%), the other minor constituents were 2-methyl benzofuran (1.99%), α-pinene (0.66%), cinnamyl acetate (0.63%), limonene (0.61%), β-caryophyllene (0.59%), methyl chavicol (0.57%), and *trans*-cinnamaldehyde (0.52%), as listed in Table 1. Due to the high content of linalool, *C. osmophloeum* leaf essential oil was classified into the linalool-chemotype.

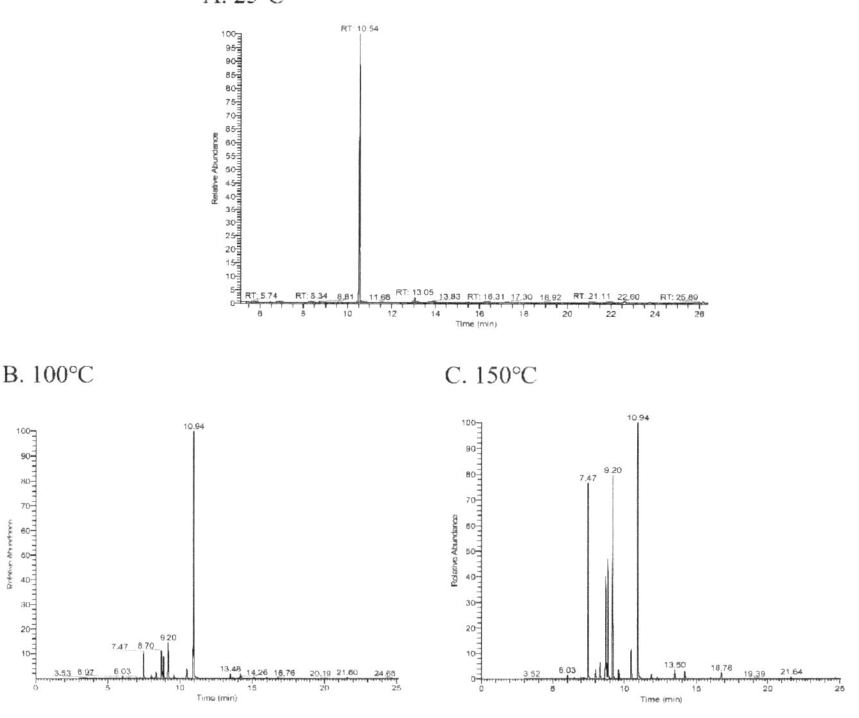

Figure 1. Gas chromatogram of linalool-chemotype *C. osmophloeum* leaf oil after thermal degradation. (**A**) 25 °C; (**B**) 100 °C; (**C**) 150 °C.

As presented in Figure 1B, several significant peaks occurred in the gas chromatogram of leaf essential oil after the heat treatment at 100 °C for 30 min. The content of the main constitute, linalool, was reduced from 93.30% to 64.01% (Table 1). New constituents were observed for β-myrcene (5.56%), α-phellandrene (0.91%), α-terpinene (1.49%), *cis*-ocimene (4.70%), γ-terpinene, and terpinolene (2.53%) in the thermally degraded leaf essential oil. Major variations were found for the increasing contents of limonene and *trans*-ocimene, which were 0.61% and 0.32% in the raw leaf essential oil and obviously increased to 7.77% and 7.94% in the thermally degraded specimen.

Similar degradation was observed from the gas chromatogram of leaf essential oil after the heat treatment at 150 °C for 30 min (Figure 1C). Linalool had a remarkable decrease in content from 93.30% in the original leaf essential oil to 27.54% in the 150 °C-degraded specimens. The peaks of the new compounds generated under more severe heat treatment became more obvious; *trans*-ocimene was present in the degraded leaf essential oil of 20.08%, β-myrcene 17.89%, *cis*-ocimene 11.72%, limonene 11.40%, terpinolene 3.37%, and α-terpinene 1.69%, in comparison with the original leaf essential oil, where these contents were much smaller. The increased amount (65.22%) of these compounds was close to the decrease in linalool (65.76%). Figure 2 illustrates the degradation mechanism of linalool and chemical structures of degradation products. After heat treatments at 100 °C and 150 °C, compounds β-myrcene, *cis*-ocimene and *trans*-ocimene were formed through the dehydroxylation of linalool. Moreover, ene cyclization occurred to linalool and its dehydroxylated products, further formed the compounds limonene, terpinolene and α-terpinene.

Table 1. Compositions of linalool-chemotype *Cinnamomum osmophloeum* leaf essential oil after thermal degradation.

Rt (min)	KI	rKI	Constituent	Content (%)		
				Original	100 °C	150 °C
6.03	938	939	α-Pinene	0.66	0.36	0.29
7.21	982	979	β-Pinene	0.28	0.07	0.06
7.47	991	990	β-Myrcene	-	5.56	17.89
8.01	1007	1002	α-Phellandrene	-	0.91	0.98
8.32	1017	1017	α-Terpinene	-	1.49	1.69
8.70	1027	1029	Limonene	0.61	7.77	11.40
8.86	1032	1037	*cis*-Ocimene	-	4.70	11.72
9.20	1049	1050	*trans*-Ocimene	0.32	7.94	20.08
9.60	1055	1059	γ-Terpinene	-	0.86	0.97
10.48	1082	1088	Terpinolene	-	2.53	3.37
10.94	1100	1096	Linalool	93.30	64.01	27.54
13.48	1180	-	2-Methylbenzofuran	1.99	1.23	1.07
14.17	1186	1188	α-Terpineol	0.28	1.50	1.12
14.28	1198	1196	Methyl chavicol	0.57	-	-
16.80	1273	1270	*trans*-Cinnamaldehyde	0.52	0.15	0.74
21.64	1420	1424	β-Caryophyllene	0.59	0.22	0.17
22.42	1445	1446	Cinnamyl acetate	0.63	0.02	-

RT: retention time (min); KI: Kovats index relative to *n*-alkanes (C9 – C24) on a DB-5MS column; rKI: Kovats index on a DB-5MS column in the reference [34].

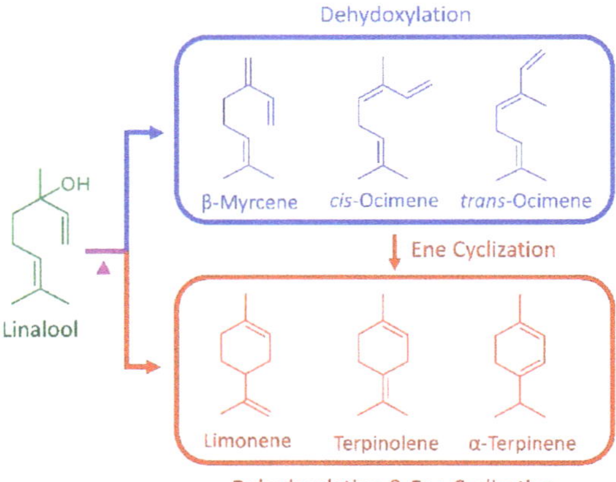

Figure 2. Schematic illustration of the thermal degradation of linalool.

Leiner et al. (2013) investigated the pyrolysis behavior of linalool. Linalool was pyrolyzed in a temperature range of 350–600 °C under nitrogen and underwent ene-type cyclization reactions leading to plinols, four cyclopentanol compounds [38]. The result varied from this study may be due to the different heating temperatures and the environment (under N2 or under air).

3.2. Optimization of Microencapsulation of Leaf Essential Oil with β-Cyclodextrin

The preparation method of microencapsulation can influence the property of β-cyclodextrin microcapsules. Kfoury et al. (2016) studied the aroma release effect from the solid inclusion complexes of β-cyclodextrin with trans-anethole by two preparation methods, freeze-drying (FD) and co-precipitation coupled to FD (Cop-FD). Cop-FD microcapsules retained more efficiently trans-anethole than that of FD microcapsules; it revealed co-precipitation method provide superior inclusion characteristics [12].

The specimen to β-cyclodextrin ratio and the solvent ratio is the important factors that influence the yield of microcapsules. Yields of linalool and leaf essential oil microencapsulated with β-cyclodextrin by different reaction conditions are presented in Table 2. β-Cyclodextrin completely dissolved in the solution (ethanol/water, 1:2 *v/v*) by heating at 50 °C for 5 min, then the solution was cooled down to 25 °C without adding the core material, and no powders/crystals formed or precipitated even at 4 °C. The highest yield of microcapsule was 94.2% at the linalool to the β-cyclodextrin ratio of 15:85 (*w/w*), which was quite close to the molar ratio (linalool:β-cyclodextrin) of 1:1.

Table 2. Yields of linalool and leaf essential oil microencapsulated with β-cyclodextrin.

Specimen	Specimen: β-CD (*w/w*)	EtOH: H_2O (*v/v*)	Yield (%)
	0:100	1:2	0.0 ± 0.0 [d*]
	5:95	1:2	54.3 ± 3.1 [a*]
Linalool	10:90	1:2	86.9 ± 0.2 [b]
	15:85	1:2	94.2 ± 0.4 [c]
	20:80	1:2	91.0 ± 0.5 [b,c]
	15:85	1:7	93.3 ± 0.6 [B]
	15:85	1:5	98.1 ± 0.1 [C]
Linalool	15:85	1:3	97.0 ± 0.4 [C]
	15:85	1:2	94.2 ± 0.4 [B]
	15:85	1:1	83.4 ± 1.8 [A]
Leaf essential oil	15:85	1:5	96.5 ± 0.2

*: Different letters ([a–d] and [A–C]) in Table refer to statistically significant difference at the level of $p < 0.05$ according to Scheffé's test.

As for the ethanol/water ratio, the highest yield of microcapsule was 98.1% under the 1:5 ratio of ethanol to water. There was no statistically significant difference in the microcapsule yields between the solvent ratio of 1:3 and 1:5 ($p < 0.05$). Using the optimal reaction conditions, the yield of linalool-chemotype leaf essential oil microencapsulated with β-cyclodextrin was 96.5%.

3.3. Stabilization and Release of Leaf Essential Oil Microcapsules

The constituents of common essential oils from aroma plants are small molecular weight and highly volatile. The encapsulation of limonene would influence its properties, such as evaporation, stabilization and controlled release, by different encapsulation methods and selected materials. The retention of limonene in extruded starch (non-encapsulation) was quite low (8.0%) compared with that of limonene microencapsulated with β-cyclodextrin (92.2%) [20].

Using the accelerated dry-heat aging test to evaluate the stability and release of leaf essential oil microcapsules. Figure 3 shows the weight losses of β-cyclodextrin (β-CD), linalool-chemotype leaf essential oil (LL oil), and leaf oil microcapsules (β-CD/LL oil) at 105 °C during the accelerated aging period. The weight loss of β-CD was very slight (less than 0.5%) after 8 days of the accelerated aging test; it indicated that β-CD was thermostable at 105 °C. Trotta et al. (2000) investigated the thermal behavior of β-CD; the decomposition temperature of β-cyclodextrin was 250 °C by using the thermogravimetric analysis (TGA) [39]. Weight losses of linalool-chemotype leaf essential oil was 99.1% after

30 min at 105 °C (data not shown in Figure 2); leaf essential oil exhibited highly volatile in the high-temperature environment. Weight losses of leaf essential oil microcapsules were 6.73%, 9.33%, 12.14%, and 13.40% after 1, 2, 4, and 8 days of the aging test, respectively. Results revealed that microencapsulation with β-cyclodextrin slowed down the release/emission of leaf essential oil in the dry-heat aging test and improved the thermal stabilization of linalool-chemotype leaf essential oil.

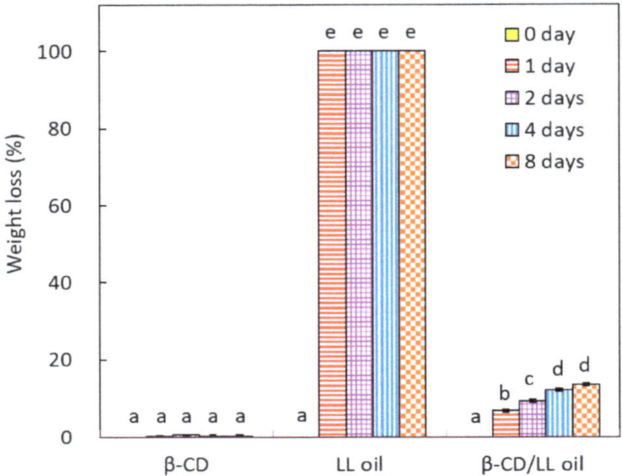

Figure 3. Changes in weight loss of leaf essential oil microcapsules during the dry-heat aging period β-CD: β-cyclodextrin; LL oil: *C. osmophloeum* leaf oil; β-CD/LL oil: leaf oil microcapsules; different letters (a–e) in the figure refer to the statistically significant difference at the level of $p < 0.05$ according to the Scheffé's test.

4. Conclusions

The thermal degradation of linalool-chemotype *C. osmophloeum* leaf oil is investigated by GC–MS. After the heat treatment, compounds β-myrcene, cis-ocimene and trans-ocimene form through the dehydroxylation of linalool and compounds limonene, terpinolene and α-terpinene by a further ene cyclization. The significantly high microcapsule yield of 96.5% is obtained from the optimal reaction conditions with the leaf essential oil to the β-cyclodextrin ratio of 15:85 and ethanol to water ratio of 1:5. Based on the accelerated dry-heat aging assay, β-cyclodextrin is stable under the environment at 105 °C, and microencapsulation with β-cyclodextrin effectively slows down the release/emission of leaf essential oil.

Author Contributions: Conceptualization, S.-T.C. and H.-T.C.; methodology, C.-Y.L., L.-S.H. and H.-T.C.; formal analysis and Investigation, C.-Y.L., L.-S.H. and H.-T.C.; writing and Editing, S.-T.C. and H.-T.C. All authors have read and agreed to the published version of the manuscript.

Funding: This research was funded by Forestry Bureau, Council of Agriculture, Executive Yuan, Taipei, Taiwan.

Data Availability Statement: The data are available are available from the corresponding author on reasonable request.

Acknowledgments: The authors would like to thank the Taiwan Forestry Bureau for the financial support and the Taiwan Forestry Research Institute for the collection of plant material.

Conflicts of Interest: The authors declare no conflict of interest.

Sample Availability: Samples are available the corresponding author.

References

1. Bandara, T.; Uluwaduge, I.; Jansz, E.R. Bioactivity of cinnamon with special emphasis on diabetes mellitus: A review. *Int. J. Food Sci. Nutr.* **2012**, *63*, 380–386. [CrossRef] [PubMed]
2. Melgarejo-Flores, B.G.; Ortega-Ramirez, L.A.; Silva-Espinoza, B.A.; Gonzalez-Aguilar, G.A.; Miranda, M.R.A.; Ayala-Zavala, J.F. Antifungal protection and antioxidant enhancement of table grapes treated with emulsions, vapors, and coatings of cinnamon leaf oil. *Postharvest Biol. Technol.* **2013**, *86*, 321–328. [CrossRef]
3. Rao, P.V.; Gan, S.H. Cinnamon: A multifaceted medicinal plant. *Evid. Based Complement. Altern. Med.* **2014**, *2014*, 642942. [CrossRef] [PubMed]
4. Davaatseren, M.; Jo, Y.J.; Hong, G.P.; Hur, H.J.; Park, S.; Choi, M.J. Studies on the Anti-oxidative function of *trans*-cinnamaldehyde-included β-cyclodextrin complex. *Molecules* **2017**, *21*, 1644. [CrossRef] [PubMed]
5. Kumar, S.; Kumari, R.; Mishra, S. Pharmacological properties and their medicinal uses of Cinnamomum: A review. *J. Pharm. Pharmacol.* **2019**, *71*, 1735–1761. [CrossRef]
6. Wu, C.L.; Chang, H.T.; Hsui, Y.R.; Hsu, Y.W.; Liu, J.Y.; Wang, S.Y.; Chang, S.T. Antioxidant-enriched leaf water extracts of *Cinnamomum osmophloeum* from eleven provenances and their bioactive flavonoid glycosides. *BioResources* **2013**, *5*, 571–580. [CrossRef]
7. Chang, S.T.; Chen, P.F.; Chang, S.C. Antibacterial activity of leaf essential oils and their constituents from *Cinnamomum osmophloeum*. *J. Ethnopharmacol.* **2001**, *77*, 123–127. [CrossRef]
8. Cheng, B.H.; Sheen, L.Y.; Chang, S.T. Evaluation of anxiolytic potency of essential oil and S-(+)-linalool from *Cinnamomum osmophloeum* ct. linalool leaves in mice. *J. Tradit. Complement. Med.* **2015**, *5*, 27–34. [CrossRef]
9. Huang, C.Y.; Yeh, T.F.; Hsu, F.L.; Lin, C.Y.; Chang, S.T.; Chang, H.T. Xanthine oxidase inhibitory activity and thermostability of cinnamaldehyde-chemotype leaf oil of *Cinnamomum osmophloeum* microencapsulated with β-cyclodextrin. *Molecules* **2018**, *23*, 1107. [CrossRef]
10. Kaiser, C.S.; Rompp, H.; Schmidt, P.C. Supercritical carbon dioxide extraction of chamomile flowers: Extraction efficiency, stability, and in-line inclusion of chamomile-carbon dioxide extract in β-cyclodextrin. *Phytochem. Anal.* **2004**, *15*, 249–256. [CrossRef]
11. Wang, J.; Cao, Y.; Sun, B.; Wang, C. Physicochemical and release characterisation of garlic oil-β-cyclodextrin inclusion complexes. *Food Chem.* **2011**, *127*, 1680–1685. [CrossRef]
12. Kfoury, M.; Auezova, L.; Greige-Gerges, H.; Larsen, K.L.; Fourmentin, S. Release studies of *trans*-anethole from β-cyclodextrin solid inclusion complexes by multiple headspace extraction. *Carbohydr. Polym.* **2016**, *151*, 1245–1250. [CrossRef] [PubMed]
13. Pires, M.A.S.; Santos, R.A.S.; Sinisterra, R.D. Pharmaceutical composition of hydrochlorothiazide: β-cyclodextrin: Preparation by three different methods, physico-chemical characterization and in vivo diuretic activity evaluation. *Molecules* **2011**, *16*, 4482–4499. [CrossRef] [PubMed]
14. Moya-Ortega, M.D.; Alvarez-Lorenzo, C.; Concheiro, A.; Loftsson, T. Cyclodextrin- based nanogels for pharmaceutical and biomedical applications. *Int. J. Pharm.* **2012**, *428*, 152–163. [CrossRef] [PubMed]
15. Szente, L.; Szeman, J.; Sohajda, T. Analytical characterization of cyclodextrins: History, official methods and recommended new techniques. *J. Pharm. Biomed. Anal.* **2016**, *130*, 347–365. [CrossRef]
16. Braga, S.S. Cyclodextrins: Emerging medicines of the new millennium. *Biomolecules* **2019**, *9*, 801. [CrossRef]
17. Guimaraes, A.G.; Oliveira, M.A.; Alves, R.S.; Menezes, P.P.; Serafini, M.R.; Araujo, A.A.S.; Bezerra, D.P.; Quintans, L.J. Encapsulation of carvacrol, a monoterpene present in the essential oil of oregano, with β-cyclodextrin, improves the pharmacological response on cancer pain experimental protocols. *Chem. Biol. Interact.* **2015**, *227*, 69–76. [CrossRef]
18. Nascimento, S.S.; Araujo, A.A.S.; Brito, R.G.; Serafini, M.R.; Menezes, P.P.; DeSantana, J.M.; Lucca, W.; Alves, P.B.; Blank, A.F.; Oliveira, R.C.M.; et al. Cyclodextrin-complexed *Ocimum basilicum* leaves essential oil increases fos protein expression in the central nervous system and produce an antihyperalgesic effect in animal models for fibromyalgia. *Int. J. Mol. Sci.* **2015**, *16*, 547–563. [CrossRef]
19. Bomfim, L.M.; Menezes, L.R.; Rodrigues, A.C.; Dias, R.B.; Rocha, C.A.; Soares, M.B.; Neto, A.F.; Nascimento, M.P.; Campos, A.F.; Silva, L.C.; et al. Antitumour activity of the microencapsulation of *Annona vepretorum* essential oil. *Basic Clin. Pharmacol. Toxicol.* **2016**, *118*, 208–213. [CrossRef]
20. Ibanez, M.D.; Sanchez-Ballester, N.M.; Blazquez, M.A. Encapsulated limonene: A pleasant lemon-like aroma with promising application in the agri-food industry. A review. *Molecules* **2020**, *25*, 2598. [CrossRef]
21. Szejtli, J. Introduction and general overview of cyclodextrin chemistry. *Chem. Rev.* **1998**, *98*, 1743–1753. [CrossRef] [PubMed]
22. Spada, G.; Gavini, E.; Cossu, M.; Rassu, G.; Carta, A.; Giunchedi, P. Evaluation of the effect of hydroxypropyl β-cyclodextrin on topical administration of milk thistle extract. *Carbohydr. Polym.* **2013**, *92*, 40–47. [CrossRef] [PubMed]
23. Hill, L.E.; Gomes, C.; Taylor, T.M. Characterization of beta-cyclodextrin inclusion complexes containing essential oils (*trans*-cinnamaldehyde, eugenol, cinnamon bark, and clove bud extracts) for antimicrobial delivery applications. *LWT-Food Sci. Technol.* **2013**, *51*, 86–93. [CrossRef]
24. Maes, C.; Bouquillon, S.; Fauconnier, M.L. Encapsulation of essential oils for the development of biosourced pesticides with controlled release: A review. *Molecules* **2019**, *24*, 2539. [CrossRef]
25. Becerril, R.; Nerin, C.; Silva, F. Encapsulation systems for antimicrobial food packaging components: An update. *Molecules* **2020**, *25*, 1134. [CrossRef] [PubMed]

26. Quintans, J.S.S.; Menezes, P.P.; Santos, M.R.V.; Bonjardim, L.R.; Almeida, J.R.G.S.; Gelain, D.P.; Araujo, A.A.S.; Quintans-Junior, L.J. Improvement of *p*-cymene antinociceptive and anti-inflammatory effects by inclusion in β-cyclodextrin. *Phytomedicine* **2013**, *20*, 436–440. [CrossRef] [PubMed]
27. Jiang, S.; Li, J.N.; Jiang, Z.T. Inclusion reactions of β-cyclodextrin and its derivatives with cinnamaldehyde in *Cinnamomum loureirii* essential oil. *Eur. Food. Res. Technol.* **2010**, *230*, 543–550. [CrossRef]
28. Ramos, A.G.B.; de Menezes, I.R.A.; da Silva, M.S.A.; Pessoa, R.T.; de Lacerda Neto, L.J.; Passos, F.R.S.; Coutinho, H.D.M.; Iriti, M.; Quintans-Junior, L.J. Antiedematogenic and anti-inflammatory activity of the monoterpene isopulegol and its β-cyclodextrin (β-CD) inclusion complex in animal inflammation models. *Foods* **2020**, *9*, 630. [CrossRef]
29. Waleczek, K.J.; Marques, H.M.C.; Hempel, B.; Schmidt, P.C. Phase solubility studies of pure (2)-α-bisabolol and camomile essential oil with β-cyclodextrin. *Eur. J. Pharm. Biopharm.* **2003**, *55*, 247–251. [CrossRef]
30. Iacovino, R.; Caso, J.V.; Rapuano, F.; Russo, A.; Isidori, M.; Lavorgna, M.; Gaetano Malgieri, G.; Isernia, C. Physicochemical characterization and cytotoxic activity evaluation of hydroxymethylferrocene: β-cyclodextrin inclusion complex. *Molecules* **2012**, *17*, 6056–6070. [CrossRef]
31. Iacovino, R.; Rapuano, F.; Caso, J.V.; Russo, A.; Lavorgna, M.; Russo, C.; Isidori, M.; Russo, L.; Malgieri, G.; Isernia, C. β-Cyclodextrin inclusion complex to improve physicochemical properties of pipemidic acid: Characterization and bioactivity evaluation. *Int. J. Mol. Sci.* **2013**, *14*, 13022–13041. [CrossRef] [PubMed]
32. Ciobanu, A.; Mallard, I.; Landy, D.; Brabie, G.; Nistor, D.; Fourmentin, S. Retention of aroma compounds from *Mentha piperita* essential oil by cyclodextrins and crosslinked cyclodextrin polymers. *Food Chem.* **2013**, *138*, 291–297. [CrossRef] [PubMed]
33. Donato, C.D.; Lavorgna, M.; Fattorusso, R.; Isernia, C.; Isidori, M.; Malgieri, G.; Piscitelli, C.; Russo, C.; Russo, L.; Iacovino, R. Alpha- and beta-cyclodextrin inclusion complexes with 5-fluorouracil: Characterization and cytotoxic activity evaluation. *Molecules* **2016**, *22*, 1868. [CrossRef] [PubMed]
34. Adams, R.P. *Identification of Essential Oil Components by Gas Chromatography/Mass Spectrometry*; Allured Publishing Corporation: Carol Stream, IL, USA, 2007; pp. 6–398. ISBN 978-1932633214.
35. Ayala-Zavala, J.F.; Soto-Valdez, H.; Gonzalez-Leon, A.; Alvarez-Parrilla, E.; Martin-Belloso, O.; Gonzalez-Aguilar, G.A. Microencapsulation of cinnamon leaf (*Cinnamomum zeylanicum*) and garlic (*Allium sativum*) oils in β-cyclodextrin. *J. Incl. Phenom. Macrocycl. Chem.* **2008**, *60*, 359–368. [CrossRef]
36. Liu, H.; Yang, G.; Tang, Y.; Cao, D.; Qi, T.; Qi, Y.; Fan, G. Physicochemical characterization and pharmacokinetics evaluation of β-caryophyllene/β-cyclodextrin inclusion complex. *Int. J. Pharm.* **2013**, *450*, 304–310. [CrossRef]
37. Abarca, R.L.; Rodriguez, F.J.; Guarda, A.; Galotto, M.J.; Bruna, J.E. Characterization of beta-cyclodextrin inclusion complexes containing an essential oil component. *Food Chem.* **2016**, *196*, 968–975. [CrossRef]
38. Leiner, J.; Stolle, A.; Ondruschka, B.; Netscher, T.; Bonrath, W. Thermal behavior of pinan-2-ol and linalool. *Molecules* **2013**, *18*, 8358–8375. [CrossRef]
39. Trotta, F.; Zanetti, M.; Camino, G. Thermal degradation of cyclodextrins. *Polym. Degrad. Stab.* **2000**, *69*, 373–379. [CrossRef]

Article

The Budesonide-Hydroxypropyl-β-Cyclodextrin Complex Attenuates ROS Generation, IL-8 Release and Cell Death Induced by Oxidant and Inflammatory Stress. Study on A549 and A-THP-1 Cells

Jules César Bayiha [1], Brigitte Evrard [2], Didier Cataldo [3], Pascal De Tullio [4] and Marie-Paule Mingeot-Leclercq [1,*]

[1] Cellular and Molecular Pharmacology Unit, Louvain Drug Research Institute, Université catholique de Louvain, Avenue E. Mounier 73, B1.73.05, 1200 Brussels, Belgium; jules.bayiha@gmail.com

[2] Laboratoire de Technologie Pharmaceutique et Biopharmacie, CIRM, Université de Liège, 4000 Liège, Belgium; B.Evrard@uliege.be

[3] Laboratory of Tumor & Development Biology, GIGA-Cancer, Université de Liège and CHU, 4000 Liège, Belgium; Didier.Cataldo@uliege.be

[4] Laboratoire de Chimie Pharmaceutique, CIRM, Université de Liège, 4000 Liège, Belgium; P.DeTullio@uliege.be

* Correspondence: marie-paule.mingeot@uclouvain.be

Academic Editors: Marina Isidori, Margherita Lavorgna, Rosa Iacovino and Georgia N. Valsami

Received: 11 August 2020; Accepted: 15 October 2020; Published: 22 October 2020

Abstract: Synthetic glucocorticoids such as budesonide (BUD) are potent anti-inflammatory drugs commonly used to treat patients suffering from chronic inflammatory diseases. A previous animal study reported a higher anti-inflammatory activity with a 2-hydroxypropyl-β-cyclodextrin (HPβCD)-based formulation of BUD (BUD:HPβCD). This study investigated, on cellular models (A549 and A-THP-1), the effect of BUD:HPβD in comparison with BUD and HPβCD on the effects induced by oxidative and inflammatory stress as well as the role of cholesterol. We demonstrated the protective effect afforded by BUD:HPβCD against cytotoxicity and ROS generation induced by oxidative and inflammatory stress. The effect observed for BUD:HPβCD was comparable to that observed with HPβCD with no major effect of cholesterol content. We also demonstrated (i) the involvement of the canonical molecular pathway including ROS generation, a decrease in PI3K/Akt activation, and decrease in phosphorylated/unphosphorylated HDAC2 in the effect induced by BUD:HPβCD, (ii) the maintenance of IL-8 decrease with BUD:HPβCD, and (iii) the absence of improvement in glucocorticoid insensitivity with BUD:HPβCD in comparison with BUD, in conditions where HDAC2 was inhibited. Resulting from HPβCD antioxidant and anticytotoxic potential and protective capacity against ROS-induced PI3K/Akt signaling and HDAC2 inhibition, BUD:HPβCD might be more beneficial than BUD alone in a context of concomitant oxidative and inflammatory stress.

Keywords: cyclodextrins; HPβCD; budesonide; inflammation; ROS; Akt; HDAC; cholesterol

---

1. Introduction

Inhaled corticosteroids were first discovered 50 years ago and are used as anti-inflammatory drugs. They are very effective controllers of asthma and largely used in chronic obstructive pulmonary disease (COPD) to prevent exacerbations and improve quality of life in COPD patients [1,2] despite the appearance of corticosteroid insensitivity [3]. Several alternatives to glucocorticoids have been developed in the past few years [4–6] but efforts are still essential to address the lack of treatment options in COPD smoking patients for whom a loss of sensitivity to glucocorticoids is observed [7]. The cellular and molecular mechanisms underlying steroid insensitivity in severe asthma and COPD are still not fully understood [7]. Oxidative stress, an increase in phosphoinositide-3-kinase/Akt (PI3K/Akt) signaling leading to the phosphorylation of HDAC2, associated with a loss of HDAC2 activity, could be critical [3,8].

Budesonide (BUD) is one of the most extensively used inhaled glucocorticoids including in the prophylactic management of asthma [9] and smoking-induced COPD [10,11]. However, frequent dosing remains a major concern in the use of budesonide. Moreover, the therapeutic potential of budesonide might be limited by its low solubility at a physiological pH. The development of budesonide formulations that can enhance drug solubility and the dissolution rate in biological fluids will likely achieve higher tissue concentrations and effectiveness.

With the aim to improve the use of inhaled corticoids with sustained release, Dufour et al. [12] evaluated in a mouse model of asthma a new formulation where budesonide was complexed with cyclodextrin (2-hydroxypropyl-β-cyclodextrin; HPβCD). In a model of smoking-induced COPD in mammals, Cataldo et al. (Cataldo et al., Patent, 2014) suggested a potential interest in a pharmaceutical preparation resulting from the complexation of budesonide with HPβCD.

Cyclodextrins are typically cone-shaped cyclic oligosaccharides of six (α-CD), seven (β-CD) or eight (γ-CD) glucose units. They possess a hydrophobic cavity allowing them to host hydrophobic molecules. They are widely used as complexing agents for low water-soluble drugs to improve their physicochemical properties including solubility, bioavailability and stability, but they also have many other applications in food, cosmetics, or textiles, for example [13–16]. β-CD and its derivatives can form a soluble inclusion complex with cholesterol and are often used to extract it from biological material [15]. Among β-CDs, methyl-β-CD (MβCD) is the most effective and the most used method to extract cholesterol but has limited clinical application, unlike HPβCD whose clinical application is broader [17–20].

The main anti-inflammatory mechanism of glucocorticoids involves the activation of glucocorticoid receptors in the cytosol after glucocorticoid binding, leading to their translocation to the nucleus, where they recruit histone deacetylase 2 (HDAC2) to the activated inflammatory gene complex. HDAC2 then reduces the acetylation of histones and glucocorticoid receptors, allowing chromatin condensation and the trans-repression of inflammatory transcription factors, respectively [21,22]. Through a decrease in activity and expression of HDAC2 in lung airways and alveolar macrophages, corticosteroid treatment is poorly effective for patients suffering from COPD [21,22]. Based on the characterized interaction between cyclodextrins and cholesterol, we hypothesized that this interaction could be involved in the effects of the BUD:HPβCD complex. Cholesterol is largely known for its effect on biophysical membrane properties and cholesterol-enriched domains are linked to membrane signaling [23–26] including pathways involved in PI3K/Akt signaling and inflammation processes. On giant unilamellar vesicles (GUVs) and lipid monolayers, BUD:HPβCD induced the disruption of cholesterol-enriched raft-like liquid ordered domains—an increase in membrane permeability and fluidity [27]. Except for membrane fluidity, all these effects were enhanced when HPβCD was complexed with budesonide as compared with HPβCD [27]. On cellular models, this could involve signal transduction pathways such as ROS generation, inflammatory cytokines expression and cell death.

The current study aimed to characterize the effect of the BUD:HPβCD complex in comparison with BUD and HPβCD on the response of human alveolar epithelial cells (A549) or human monocytes (A-THP1) to a mix of hydrogen peroxide and lipopolysaccharide (H_2O_2 + LPS) mimicking stressful effects including those from cigarette smoke [28–30] or from environmental toxicants. In detail, we pursued four objectives: first, to establish the potential interest of BUD:HPβCD on the cytotoxicity induced by oxidative and inflammatory stressors; second, to investigate the cellular effect of BUD:HPβCD on the signaling pathway involved in corticosteroid effects including ROS generation, PI3K/Akt activation, HDAC2 activity and the release of pro-inflammatory cytokines such as IL-8;

third, to question the role of cholesterol in the effect induced by BUD:HPBCD on ROS generation and PI3K/Akt activation, with the two first membranous events leading to inflammation, and fourth, to determine the effect of BUD:HPβCD as compared to BUD in glucocortioid resistance and the role of HDAC2 in mediating the loss of the glucocorticoid anti-inflammatory effect.

This study is a part of the continuing efforts to develop novel drug delivery systems, as the complex between budesonide and cyclodextrins, with the aim to improve the treatment of patients suffering from smoking-induced COPD.

2. Results

2.1. BUD:HPβCD Complex and HPβCD Attenuate H_2O_2 + LPS-Induced Cytotoxicity

Since alveolar cell death is one feature observed in the lung of patients suffering from smoking-induced COPD [31], the potential effect of BUD:HPβCD on A549 human alveolar epithelial cells submitted to oxidant and inflammatory stressors was investigated. Cells were incubated with H_2O_2 – LPS for 2 h. Cytotoxicity, as reflected by lactate dehydrogenase (LDH) release, was observed with a 1.7-fold increase as compared to untreated cells (Figure 1A). The increase in cytotoxicity between 2 and 6 h is low (1.8-fold at 6 h) suggesting the cytotoxicity almost reached its maximum at 2 h. H_2O_2 + LPS-induced cytotoxicity seemed to result from the addition of H_2O_2 and LPS.

The effect of the BUD:HPβCD complex on cytotoxicity induced by H_2O_2 + LPS was followed. Incubation of A549 cells with the BUD:HPβCD complex together with H_2O_2 + LPS induced a decrease in cytotoxicity (Figure 1D). This protective effect appeared to not evolve further after 2 h of incubation. A similar effect was recorded with HPβCD (Figure 1C), whereas BUD showed no effect whatever the dosage (Figure 1B). These results suggest that the BUD:HPβCD complex and HPβCD would have anticytotoxic potential against H_2O_2 + LPS-induced cytotoxicity in A549 cells.

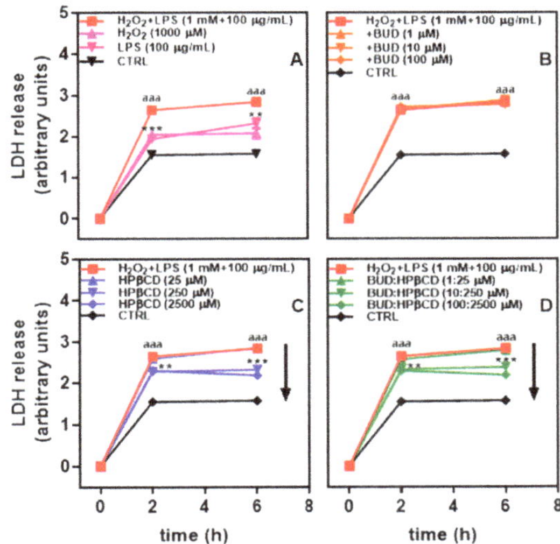

Figure 1. Lactate dehydrogenase (LDH) release after A549 cells incubation with H_2O_2, Lipopolysaccharides (LPSs), and H_2O_2 + LPS for up to 6 h (**A**) and effect of budesonide (BUD) (**B**), HPβCD (**C**), and BUD:HPβCD complex (**D**) on H_2O_2 + LPS-induced cytotoxicity. Each point represents the mean ± SEM of at least 4 independent means of triplicated measures; where not visible, error bars are included in the symbol. The difference was considered significant for a *p*-value < 0.05. (aaa) indicates $p < 0.001$ versus untreated group; (**) and (***) corresponds to $p < 0.01$ and 0.001 versus H_2O_2 +LPS-treated group, respectively.

To determine if apoptosis is involved in the cell death process for which BUD:HPβCD could protect, apoptosis was monitored by counting condensed/fragmented nuclei using HOECHST dye on H_2O_2 + LPS-treated cells. We also determined if the BUD:HPβCD complex as well as BUD, and HPβCD could attenuate apoptosis.

A549 cells were incubated for 2 h with H_2O_2 + LPS with/without BUD:HPβCD in comparison with BUD or HPβCD. H_2O_2 + LPS induced significant apoptosis, which appeared to result from the addition of the individual effects of H_2O_2 and LPS (Figure 2A). Concomitant incubation with each of the selected compounds induced a concentration-dependent decrease in H_2O_2 + LPS-induced apoptosis. These results suggest that the BUD:HPβCD complex (Figure 2D) and HPβCD (Figure 2C), at the highest selected dosages could protect cells against H_2O_2 + LPS-induced apoptosis in A549 cells. A non-significant decrease was observed with BUD (Figure 2B).

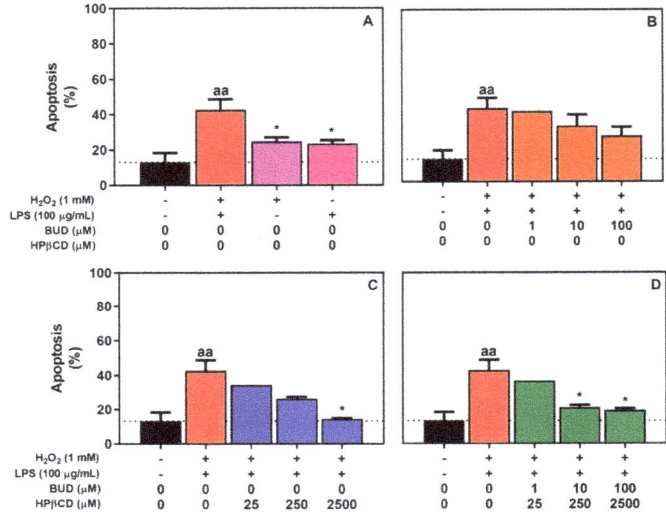

Figure 2. A549 cells apoptosis after treatment with H_2O_2, LPS, and H_2O_2 + LPS for 2 h (**A**) and effect of BUD (**B**), HPβCD (**C**), and BUD:HPβCD complex (**D**) on H_2O_2 + LPS-induced apoptosis. Apoptosis was quantified by counting condensed/fragmented nuclei after HOECHST staining. Each bar represents the mean of 3 ± SEM or 2 independent measures. A one-way ANOVA with Dunett post-test was used to compare the mean of a test group with the mean of the untreated group or H_2O_2 +LPS-treated group. The difference was considered significant for a *p*-value < 0.05. (aa) indicates *p* < 0.01 versus untreated group, (*) indicates *p* < 0.05 versus H_2O_2 +LPS-treated group.

2.2. BUD:HPβCD Complex and HPβCD Protect against H_2O_2 + LPS-Induced Oxidative Stress in A549 Cells: Dose and Time-Dependent Effects

Because oxidative stress is critical for numerous pathologies including smoking-induced COPD [32], and with the aim to understand the mechanism of action behind the effects observed with BUD:HPβCD, the potential antioxidant effect of the BUD:HPβCD complex in A549 human alveolar epithelial cells was monitored. A549 cells were incubated for 2 h with a H_2O_2 + LPS mix to model a concomitant oxidative and inflammatory environment. ROS generation induced by the BUD:HPβCD complex was monitored in comparison with the effect induced by budesonide or HPβCD.

In comparison with control cells, we observed a 1.6-fold significant increase in intracellular ROS production when cells were incubated with H_2O_2 + LPS (Figure 3A) This effect was similar to the effect induced by treatment with H_2O_2 alone (1.5-fold increase), unlike the treatment with LPS alone, which did not show any significant effects, suggesting that H_2O_2 + LPS-induced oxidative stress would be mainly driven by H_2O_2 in A549 cells.

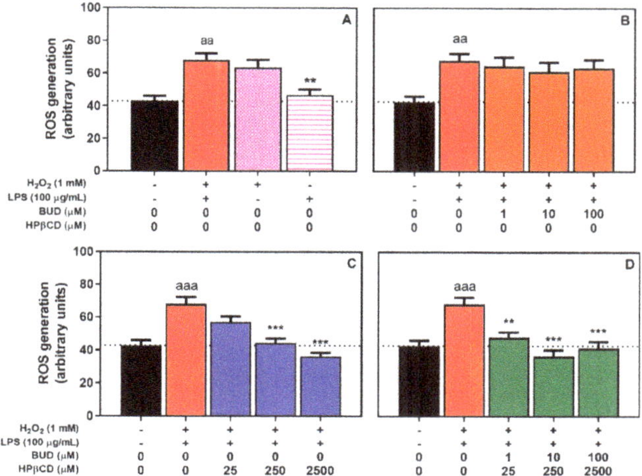

Figure 3. ROS generation in A549 cells after treatment with H_2O_2, LPS, or H_2O_2 + LPS for 2 h (**A**) and effect of BUD (**B**), HPβCD (**C**), and BUD:HPβCD complex (**D**) on H_2O_2 + LPS-induced ROS generation. ROS generation was evaluated by measuring the fluorescence of dichlorofluorescein (DCF). Each bar represents the mean ± SEM of 4 independent means of triplicated measures. A one-way ANOVA with Dunett post-test was used to compare the mean of a test group with the mean of untreated group or H_2O_2 + LPS-treated group. The difference was considered significant for a *p*-value < 0.05. (aa), and (aaa), indicate *p* < 0.01, and 0.001 versus untreated group, respectively); (**), and (***), correspond to *p* < 0.01 and 0.001 versus H_2O_2 + LPS-treated group, respectively.

A concomitant incubation of A549 cells with H_2O_2 + LPS and increasing concentrations of the BUD:HPβCD complex (1:25, 10:250, 100:2500 µM; Figure 3D) was associated with a decrease in ROS production as compared with the experimental conditions in which the complex was not present. This suggests a protective effect of the BUD:HPβCD complex against the oxidative stress induced by H_2O_2 + LPS. A similar effect was observed with increasing concentrations of HPβCD (25–2500 µM; Figure 3C). In contrast no effect of BUD (1–100 µM; Figure 3B) was observed.

Regarding the effect of time (Figure 4), ROS production in the presence of H_2O_2 + LPS was already marked after 30 min compared to untreated cells. This effect was maintained throughout the entire time period investigated (6 h) and seemed to evolve in parallel with untreated cells after 2 h. When the BUD:HPβCD complex was added together with H_2O_2 + LPS, a lowering effect on ROS production was observed throughout the entire time period investigated (Figure 4B). The extent of the effect depended upon the dose and was largely similar to the effect observed in the presence of HPβCD (Figure 4C). No significant change was observed after 2 h of incubation with H_2O_2 + LPS. Again, during the entire period investigated, no significant effect was observed in the presence of BUD (Figure 4A). Altogether, these results suggest that the BUD:HPβCD complex and HPβCD have a similar antioxidant potential against H_2O_2 + LPS in A549 cells.

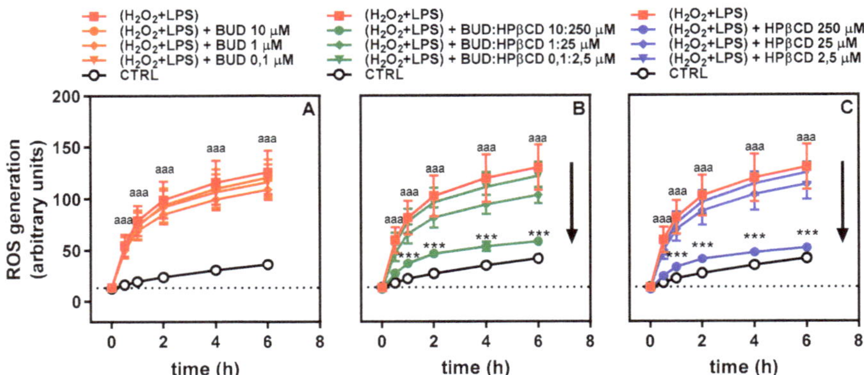

Figure 4. Effect of BUD (**A**), BUD:HPβCD complex (**B**) and HPβCD (**C**) on H_2O_2 + LPS-induced ROS generation for 0 to 6h of incubation. ROS generation was evaluated by measuring the fluorescence of dichlorofluorescein (DCF). Each bar represents the mean ± SEM of 3 independent means of triplicated measures. These results and the results illustrated in Figure 3 are independent. When deviations are not visible they are too small to be seen. The difference was considered significant for a *p*-value < 0.05. (aaa) corresponds to $p < 0.001$ versus untreated group; (***) indicates $p < 0.001$ versus H_2O_2 + LPS-treated group.

2.3. BUD:HPβCD Complex and HPβCD Attenuate H_2O_2 + LPS-Induced Phosphoinositide-3-Kinase/Akt Signaling in A549 Cells

Oxidative stress-induced glucocorticoid insensitivity involves an increase in PI3K/Akt signaling [8,33,34] as reflected by Akt phosphorylation. To validate in in vitro model the relationship between oxidative stress and increase in PI3K/Akt signaling, the phosphorylation of Akt, in the absence or in the presence of antioxidants (N-acetyl-L-cysteine (NAC), vitamin C (Vit C)) and of a PI3K inhibitor (LY294002) (Figure 5A,C) was measured. An incubation of A549 cells with H2O2 + LPS for 2 h increased Akt phosphorylation. Concomitant incubation with NAC or vitamin C or LY294002 was associated with a lower phosphorylation of Akt (of approximately 36% (NAC) and 65% (Vit C)) or a complete suppression of phosphorylation (LY294002) (Figure 5A,C). Regarding the effect of the BUD:HPβCD complex or of HPβCD at a concentration at which a significant antioxidant effect was observed, we demonstrated a decrease in Akt phosphorylation (Figure 5B,D). The decrease was approximately 32% both for the BUD:HPβCD complex and for HPβCD (Figure 5B,D). Thus, the BUD:HPβCD complex and HPβCD inhibited H2O2 + LPS-induced PI3K/Akt signaling increases in a similar way in A549 cells.

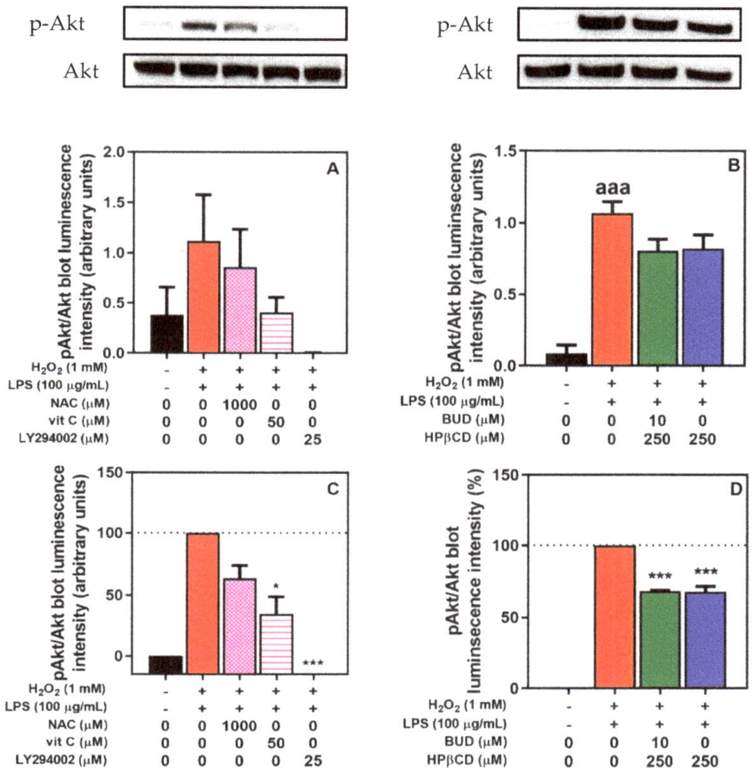

Figure 5. Akt phosphorylation induced by H_2O_2 + LPS in A549 cells. Effect of *N*-acetyl-L-cysteine (NAC), vitamin C (VitC) and LY294002 (**A**,**C**) and BUD:HPβCD complex and HPβCD (**B**,**D**) after 2 h of incubation. Data are expressed in absolute values (A/B; with representative blots) or in relative values (in comparison with the pAkt/Akt ration of cells incubated with H_2O_2 + LPS; C/D). Akt phosphorylation was quantified after a Western blot by measuring the proportion of phosphorylated-Akt (p-Akt) blot luminescence intensity/total Akt (Akt) blot luminescence intensity. Each bar represents the mean of 3 independent measures. A one-way ANOVA with Dunett post-test was used to compare the mean of each test group with the mean of untreated group or H_2O_2 + LPS-treated group. The difference was considered significant for a p-value < 0.05. (aaa) correspond to $p < 0.001$ versus untreated group, (*) and (***) indicate $p < 0.05$, and 0.001 versus H_2O_2 + LPS-treated group, respectively.

2.4. Cholesterol Might Limit the Effects of BUD:HPβCD Complex and HPβCD in ROS Generation and PI3K/Akt Signaling Induced by H_2O_2 + LPS

To give insight on the molecular mechanisms involved in the protective effect of BUD:HPBCD and HPBCD, the potential role of cholesterol on ROS generation and PI3/Akt phosphorylation induced by H_2O_2 + LPS as well as the protective effects of the BUD:HPβCD complex and HPβCD were investigated. The rationale was derived from the ability of cyclodextrins to interact with cholesterol [35], the effects of BUD:HPβCD and HPβCD on the biophysical membrane properties of cholesterol-enriched domains [27], the importance of lipid-ordered domains enriched in cholesterol in membrane called rafts for ROS generation [36,37] and PI3K/Akt signaling [23,26].

2.4.1. Cholesterol Content Might Influence the Effects of the BUD:HPβCD Complex and HPβCD in ROS Generation Induced by H_2O_2 + LPS

In conditions where cholesterol was partly depleted (see Figure S1) we observed (Figure 6A) an increase in basal intracellular ROS levels, although non-significant. A greater and significant increase in H_2O_2- and H_2O_2 + LPS-induced intracellular oxidant generation was also observed, suggesting that cholesterol content plays a role in H_2O_2 + LPS-related oxidative signaling in A549 cells.

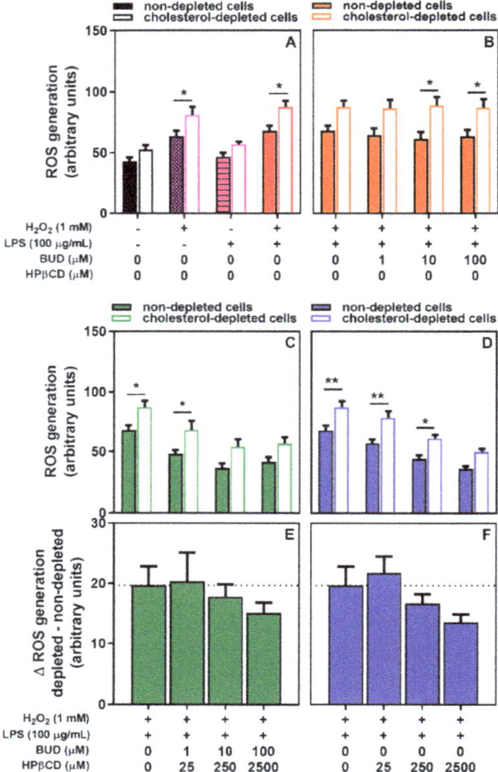

Figure 6. ROS generation in cholesterol-non-depleted or cholesterol-depleted A549 cells after treatment with H_2O_2, LPS, or H_2O_2 + LPS for 2 h (**A**) and effect of BUD (**B**), BUD:HPβCD complex (**C**) and HPβCD (**D**) on H_2O_2 + LPS-induced ROS generation. Panels E and F show the difference (Δ) between the effect observed in cholesterol-depleted and -non-depleted cells of BUD:HPβCD complex (**E**) and HPβCD (**F**) on H_2O_2 + LPS-induced oxidant generation after treatment for 2 h. Each bar represents the mean ± SEM of 4 independent means of triplicated measures. H_2O_2 + LPS-treated bar is the same for each panel. A one-way ANOVA with Dunett post-test was used to compare the mean of each test group in panels (**E,F**) with the mean of the H_2O_2 + LPS-treated group. A two-way ANOVA with Tukey multiple comparison post-test was used to compare the mean of non-depleted group with the mean of cholesterol-depleted group in the same concentration (panels **A–D**). The difference was considered significant for a p-value < 0.05. (*) and (**) indicate, respectively, $p < 0.05$ and 0.01 between non-depleted and cholesterol-depleted group (panels **A–D**).

Compared to non-depleted cells, the ability of the BUD:HPβCD complex (Figure 6C) and HPβCD (Figure 6D) to protect against H_2O_2 + LPS-induced ROS production was preserved. Moreover, when the difference between H_2O_2 + LPS-induced ROS production in cholesterol-depleted and non-depleted cells was considered, a concentration-dependent increase in this protective effect was observed (Figure 6E,F), although this was non-significant. Again, no matter the cholesterol status of the cells, budesonide did not show any protective effects (Figure 6B). Thus, cholesterol content might influence the antioxidant effect of the BUD:HPβCD complex and HPβCD.

In contrast with cholesterol, sphingomyelin, another major component from raft and also interacting with HPβCD did not play a critical role on neither the oxidant generation induced by H_2C_2 + LPS nor on the ability of the BUD:HPβCD complex and HPβCD to protect against H_2O_2 + LPS-induced oxidant generation (Figure S2).

2.4.2. Cholesterol Limits the Effects of the BUD:HPβCD Complex and HPβCD in PI3K/Akt Signaling Induced by H_2O_2 + LPS

Since cholesterol-enriched plasma membrane domains may also play a critical role in the activation of PI3K/Akt signaling [23,25,26], which could be modulated by oxidative stress, the effect of cholesterol depletion on the protective effects of the BUD:HPβCD complex and HPβCD on the phosphorylation of Akt was studied. A decrease in Akt phosphorylation was preserved (Figure 7), without difference in cholesterol-depleted or not depleted cells.

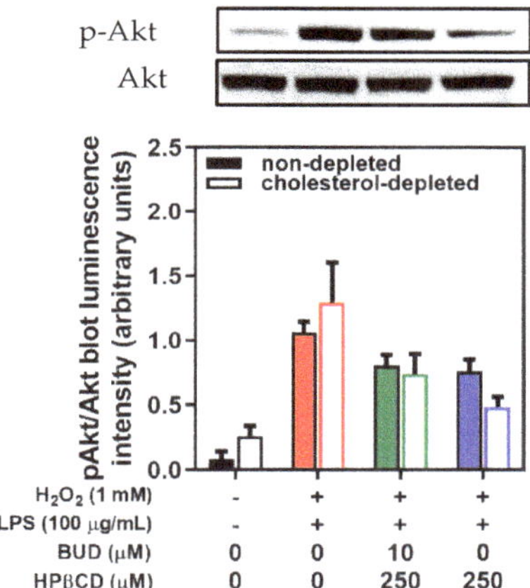

Figure 7. Effect of the BUD:HPβCD complex versus HPβCD on H_2O_2 + LPS-induced Akt phosphorylation (p-Akt) in cholesterol-depleted and non-depleted A549 cells after 2 h of incubation with representative blot (cholesterol-depleted cells); each bar represents the mean of 3 independent measures.

2.5. BUD:HPβCD Complex and HPβCD Protect against H_2O_2 + LPS-Induced Increase in HDAC2 Phosphorylation in A549 Cells

The increase in PI3K/Akt signaling induced by oxidative stress results in the phosphorylation of HDAC2, a critical step in oxidative stress-related glucocorticoid insensitivity [38,39]. The relationship between oxidative stress and HDAC2 phosphorylation as well as the relationship between the increase in PI3K/Akt signaling and HDAC2 phosphorylation was investigated by using NAC and LY294002, respectively (Figure 8A,C). The treatment of A549 cells with H_2O_2 + LPS for 2 h was associated with an increase in phosphorylated HDAC2. Concomitant incubation with NAC or LY294002 was associated with a lower phosphorylation of HDAC2 of approximately 62% (NAC) and 40% (LY294002) (Figure 8A,C). This confirms that H_2O_2 + LPS increases HDAC2 phosphorylation through a mechanism involving oxidative stress and PI3K/Akt signaling in A549 cells.

A concomitant incubation with the BUD:HPβCD complex or HPβCD with H_2O_2 + LPS was associated with a lower phosphorylation of HDAC2 of approximately 53% (BUD:HPβCD) and 74%

(HPβCD) (Figure 8B,D). Thus, the BUD:HPβCD complex and HPβCD inhibited the decrease in HDAC2 activity induced by H_2O_2 + LPS treatment in A549 cells with a higher effect induced by HPβCD.

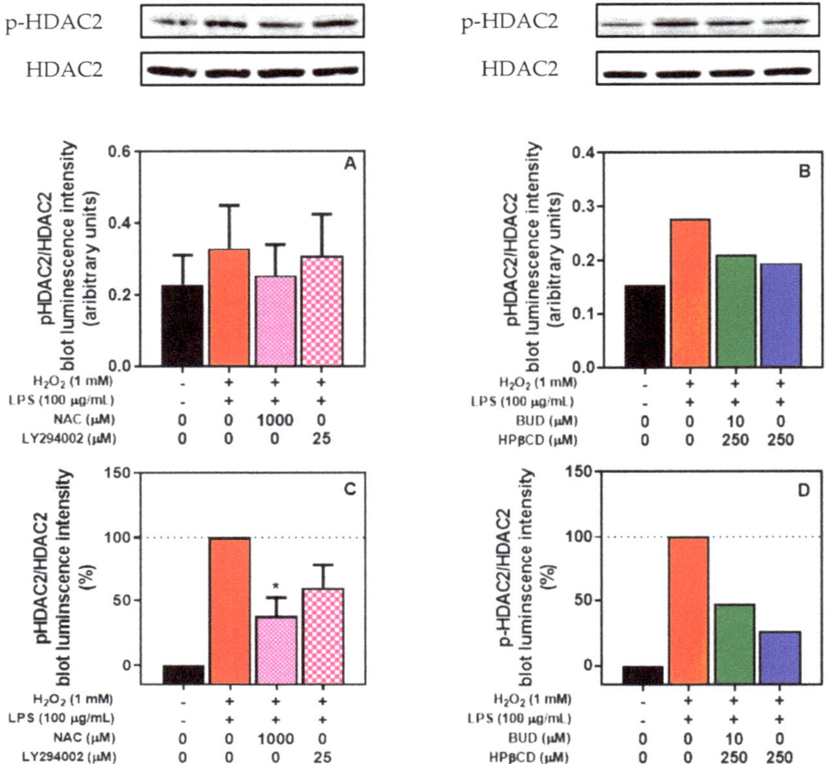

Figure 8. HDAC2 phosphorylation induced by H_2O_2 + LPS in A549 cells. Effect of NAC and LY294002 (**A**) and the BUD:HPβCD complex versus HPβCD (**B**) after 2 h of incubation. Data are expressed in absolute values (**A/B**; with representative blots) or in relative values (in comparison with the pHDAC2/HDAC2 ration of cells incubated with H_2O_2 + LPS; **C/D**). HDAC2 phosphorylation was quantified after a Western blot by measuring the proportion of phosphorylated-HDAC2 (p-HDAC2) blot luminescence intensity/total HDAC2 (HDAC2) blot luminescence intensity. Each bar represents the mean of 3 ± SEM or 2 independent measures. A one-way ANOVA with Dunett post-test was used to compare the mean of each test group with the mean of the control group (H_2O_2 + LPS-treated group). The difference was considered significant for a p-value < 0.05. (*) indicate $p < 0.05$ versus control group.

2.6. BUD:HPβCD Complex and HPβCD Attenuate H_2O_2 + LPS-Induced Inflammatory Response in THP-1 Cells

Since persistent inflammatory response in the lung is a major feature of smoking-induced COPD [31], the anti-inflammatory potential of the BUD:HPβCD complex in comparison with BUD or HPβCD was evaluated. Thus, the effect of the BUD:HPβCD complex, BUD or HPβCD on H_2O_2 + LPS-induced IL-8 release in A549 cells and TH-P1 cells was determined. Because A549 cells appeared not sensitive to LPS [40], phorbol myristate acetate-activated THP-1 (A-THP-1) cells [41], a widely used model for human monocytes, which are highly sensitive to LPS treatment, were used. For the sake of comparison, we also treated A549 cells with TNF-α for inflammatory stress.

First, the incubation of A-THP-1 cells with H_2O_2 + LPS for 2 h was significantly associated with an 8.6-fold increase in IL-8 release (Figure 9A). Treatment with LPS alone induced a significant 6.6-fold increase, whereas H_2O_2 alone did not induce any effects on IL-8 expression [41].

Concomitant incubation of the BUD:HPβCD complex with H_2O_2 + LPS was associated with a lower release of IL-8 of approximately 45% no matter the concentration of the BUD:HPβCD complex used (Figure 9D). In the presence of BUD, a decrease in of IL-8 release was also observed. The effect was similar with that induced by the BUD:HPβCD complex (budesonide 1 and 10 μM), but higher at higher budesonide concentration (100 μM) (67%) (Figure 9B). The presence of HPβCD was also associated with a non-dependent dose-type decrease in IL-8 release. The effect was slightly lower as compared to BUD and the BUD:HPβCD complex (approximately 34%) (Figure 9C). These results suggest that the BUD:HPβCD complex and BUD have similar anti-inflammatory properties, except at high concentrations at which budesonide was more efficient.

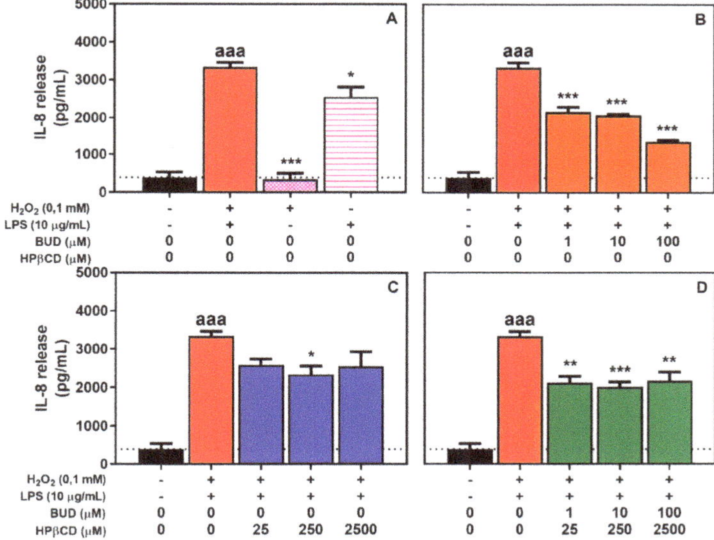

Figure 9. IL-8 release by A-THP-1 cells after treatment with H_2O_2, LPS or H_2O_2 + LPS for 2 h (**A**) and effect of BUD (**B**), HPβCD (**C**), and BUD:HPβCD complex (**D**) on H_2O_2 + LPS-induced IL-8 release. IL-8 release was measured in the extracellular medium by sandwich ELISA. Each bar represents the mean ± SEM of 3 independent means of triplicated measures. A one-way ANOVA with Dunett post-test was used to compare the mean of each test group with the mean of untreated group or H_2O_2 + LPS-treated group. The difference was considered significant for a p-value < 0.05; (aaa), indicate $p < 0.001$ versus untreated group; (*), (**), and (***) correspond to $p < 0.05$, 0.01 and 0.001 versus H_2O_2 + LPS-treated group, respectively.

A similar effect on IL-8 release was reported when comparing the effect of TNFα on A549 cells with the effect LPS on A-THP1. An anti-inflammatory potential of BUD (1 μM) and the BUD:HPβCD complex (1:25 μM) was observed with a lower potential of HPβCD to decrease IL-8 release (Figure 10).

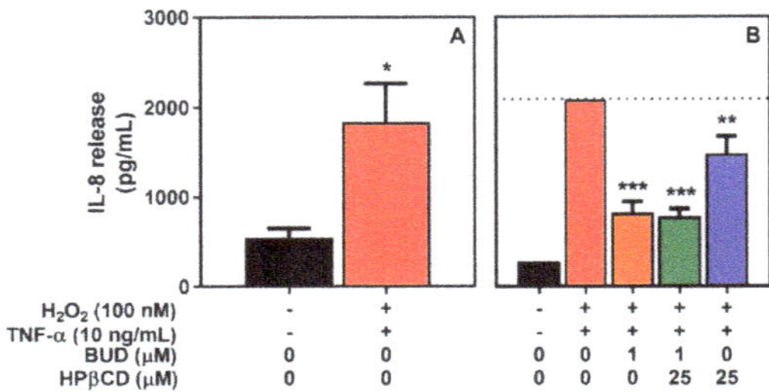

Figure 10. IL-8 release by A549 cells after treatment with H_2O_2 + TNF-α for 2 h (**A**) and effect of the BUD:HPβCD complex versus BUD and HPβCD on H_2O_2 + TNF-α-induced IL-8 release (**B**). IL-8 release was measured in the extracellular medium by sandwich ELISA. Results on panel B were normalized relative to untreated cells (0%) and H_2O_2 + TNF-α-treated cells (100%). Each bar represents the mean ± SEM of 3 independent means of triplicated measures. (*), (**), and (***), indicate, respectively, $p < 0.05$, 0.01, and 0.001 versus non-treated cells (**A**) or H_2O_2+LPS-treated cells (**B**).

As HDAC2 is recruited by the activated glucocorticoid receptor to repress the transcription of proinflammatory genes [42] and to study the potential role of HDAC2 in the protection afforded by the BUD:HPβCD complex in comparison with BUD or HPβCD on IL-8 release, IL-8 release induced by TNF-α in conditions where cells were preincubated with or without trichostatin, a pharmacological HDAC2 inhibitor [43], was measured. We pretreated for 30 min A549 cells with increasing concentrations of trichostatin (0–250 nM) and determined IL-8 release after incubation for 2 h of cells with TNF-α (20 ng/mL) and BUD:HPβCD complex or BUD or HPβCD (Figure 11).

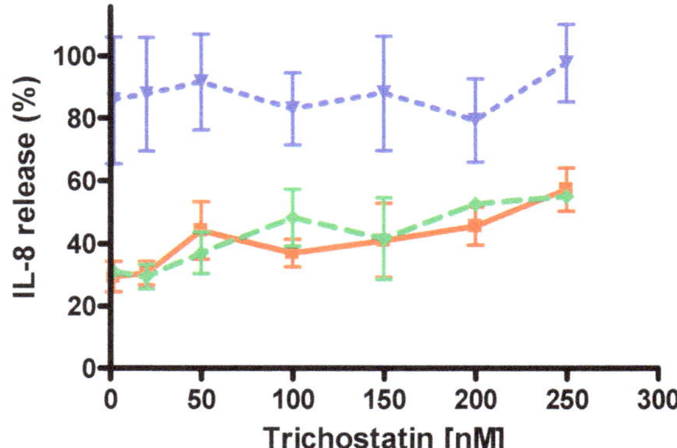

Figure 11. Percentage of IL-8 released after A549 cells pretreatment for 30 min with trichostatin (TSA) and incubation for 2 h with TNF-α in presence of BUD:HPβCD complex, BUD or HPβCD. Results are expressed in percentage of IL-8 released. 100% corresponds to cells preincubated for 30 min with TSA and incubated for 2 h with TNF-α only. IL-8 release was measured in the extracellular medium by sandwich ELISA. Data are from 3 independent experiments in triplicates. ▼HPβCD; ♦ BUD:HPβCD; ■ BUD.

IL-8 release induced by TNFα was markedly reduced (around 70%) by BUD and BUD:HPβCD whereas HPβCD alone did not shown any effect or a very slight effect. When trichostatin was used in preincubation to inhibit the HDAC2 activity, the IL-8 release was increased, in a dose-dependent fashion in the presence of BUD or BUD:HPβCD. BUD:HPβCD failed to improve the response of glucocorticoids in the condition of HDAC2 inhibition. Again, no or a very slight effect was observed with HPβCD (Figure 11).

3. Discussion

In animal models, Dufour et al. [12] suggested that budesonide (BUD) complexed with 2-hydroxypropyl-β-cyclodextrin (HPβCD) might be an alternative to BUD alone in the treatment of smoking-induced COPD. The current study was designed to characterize the effect of the BUD:HPβCD complex on the response of human alveolar epithelial cells (A549) or human monocytes (A-THP1) to a mix of hydrogen peroxide and lipopolysaccharides (H_2O_2 + LPSs) mimicking stressful effects including those from cigarette smoke [28,44,45] or from environmental toxicants. We characterized the effect of the BUD:HPβCD complex on (i) ROS generation (oxidative stress), (ii) Akt phosphorylation (PI3K/Akt signaling activation), (iii) HDAC2 phosphorylation (HDAC2 inhibition of activity), and (iv) IL-8 release (inflammatory response) in comparison with the effects induced by BUD or HPβCD.

We demonstrated the protective effect afforded by BUD:HPβCD against cytotoxicity and ROS generation induced by oxidative and inflammatory stress in comparison with BUD. The effect observed for BUD:HPBCD was comparable to that observed with HPBCD and might be limited by cholesterol. We also demonstrated (i) the involvement of the canonical molecular pathway including ROS generation, decrease in PI3K/Akt activation, decrease in HDAC2 activity and insensitivity to glucocorticoid in the effect induced by BUD:HPβCD, (ii) maintenance of IL-8 decrease with BUD:HPβCD—even BUD at a high concentration (100 μM) induced a slightly higher effect—and (iii) the absence of improvement in glucocorticoid insensitivity with BUD:HPβCD in comparison with BUD, in conditions where HDAC2 was inhibited.

Improvement of cell viability after oxidative and inflammatory stress induced by BUD:HPβCD is likely due to HPβCD and linked to a decrease in ROS generation. The literature has reported that cyclodextrins, including HPβCD, may improve the toxicological profile of drugs by complexing them [46,47]. Additionally, the antioxidant potential of HPβCD has been reported. Anraku et al. [48] showed HPβCD remove pro-oxidants such as uremic toxins from the blood in a rat model of chronic renal failure. Zimmer et al. [49] showed that HPβCD decreases aortic ROS generation in a mouse model of atherosclerosis. Other reports reviewed by López-Nicolás et al. [50] described HPβCD as a protective agent of lipophilic nutrients and antioxidants against oxidation in foods. The demonstration of HPβCD's antioxidant potential is interesting given the major role played by oxidative stress in numerous pathologies including COPD [51]. Here, the molecular mechanism leading to a decrease in ROS is still unclear but a direct effect through the interaction of HPβCD with H_2O_2 (Figure S3) is unlikely.

An indirect effect through changes in biophysical membrane properties could be suggested as an alternative explanation. We initially suggested that membrane cholesterol would play a major role in the occurrence of BUD:HPβCD-related cytoprotective effects. It has been extensively demonstrated that βCDs, including HPβCD, can interact with lipid membranes, and change membrane biophysical properties [17,35] closely related to signal transduction. This agrees with our previous experiments on giant unilamellar vesicles (GUVs), since we demonstrated BUD:HPβCD and HPβCD disrupted the liquid-disordered/liquid-ordered (Ld/Lo) phase separation observed in the presence of cholesterol for the benefit of the Ld phase, a process hindered in the presence of cholesterol [27]. Here, we observed an increase in BUD:HPβCD-related antioxidant effects in cholesterol-depleted cells suggesting that cholesterol might hinder ROS generation. The BUD:HPβCD-related antioxidant effect was preserved and even increased in cells partially depleted in cholesterol (50% cholesterol depletion after 30 min exposition to MβCD at 5 mM; no or very small cholesterol depletion induced by HPβCD for 2 h at the highest concentrations used in this work; Figure S4). Extracellular mechanisms are unlikely since we observed (i) no cellular uptake of HPβCD over the entire incubation period (Figure S5), and (ii) no neutralization of extracellular signals potentially responsible for oxidative stress, namely H_2O_2 and free radicals (Figure S3). BUD:HPβCD and HPβCD-related cytoprotective effects could be seen as a membrane-mediated mechanism involving membrane lipid disorganization with limited lipid extraction after 2 h (cholesterol extraction induced by BUD:HPβCD or HPβCD reached 18% and 12%, respectively, while no cholesterol extraction in cholesterol-depleted cells was observed (Figure S4)), agreeing with the work of Lopez et al. [52].

One remaining question is the cross-talk between the antioxidant effect and inhibiting effect on oxidant-induced PI3K/Akt signaling. NAC and Vit C concentrations that inhibit more than 75% of ROS generation after 2 h of incubation (Figure S6) showed a protective effect against H_2O_2 + LPS-induced increase in PI3K/Akt signaling of about 36% (NAC) and ~65% (Vit C). The effect was not related to the effect of LPS which might induce an increase in PI3K/Akt signaling [53,54] since we observed that H_2O_2 + LPS-induced increase in PI3K/Akt signaling was almost exclusively associated with the presence of H_2O_2 (Figure S7). The activity of endogenous antioxidant enzymes GSH peroxidase against H_2O_2 [55], difference in the location within the bilayer between the effect induced by BUD:HPBCD and the location of enzymes involved in ROS generation or PI3K/Akt activation could be also involved.

Focusing on the final attempt for BUD:HPBCD, meaning its ability to decrease the release of inflammatory cytokines after oxidant and inflammatory stress, we could have expected a higher anti-inflammatory effect of the BUD:HPβCD complex compared to the BUD alone. Dufour et al. [12] in a murine asthma model showed that similar anti-inflammatory effects could be obtained with a 2.5-fold lower BUD concentration when given as a complex with HPβCD. Zimmer et al. [49] reported anti-inflammatory effects of HPβCD in vivo in a mice model of atherosclerosis. At the cellular level, George et al. [56] assumed an anti-inflammatory property of HPβCD after showing that its presence along with plasticized poly(vinyl chloride) (PVC) reduced LPS-induced TNF-α expression in human monocyte-like U937 cells while PVC alone had no effect. Matassoli et al. [57] showed that HPβCD can inhibit LPS-induced TNF-α secretion in primary human monocytes. The higher effect on reduction of IL-8 release induced by BUD at a high concentration (100 μM) as compared to the effect of BUD:HPβCD could be linked to an inflammatory effect observed at high doses of HPβCD [58]. Cell-type cellular components of inflammation [59] and changes in the release of BUD from HPβCD hydrophobic cavity, depending upon the concentrations, could also play a role.

Lastly, in a potential translational perspective, the design of studies and concentrations have to be questioned. First, cells were exposed with BUD and H_2O_2 + LPS at the same time, meaning that BUD had time to prevent the inflammatory response before the decrease in glucocorticoid sensitivity induced by oxidative stress could take place. We reproduced experiments by changing the time course and by preincubating cells for 30 min with the oxidant and inflammatory stress before the incubation of cells with BUD:HPβCD for 2 h. No differences were observed. Another critical parameter would be the equilibrium between the free and bound forms of BUD or HPBCD [60]. In the presence of a lipophilic membrane, drug partitioning from the complex into the membranes can occur, promoting drug release from the CD hydrophobic cavity. The latter point agrees with our work. Indeed, we showed that in pure phospholipid monolayers there is an increase in membrane surface pressure with the BUD:HPβCD complex but not with HPβCD. This increase is usually associated with the insertion of a molecule within the monolayer. Since the only difference between HPβCD and the BUD:HPβCD complex is the presence of BUD, we could assume that BUD was inserted within the membrane. The critical importance of the equilibrium between free and complexed budesonide was also evidenced when we determined the effect of a mix of BUD and HPβCD on IL-8 release for cells treated with increasing concentrations of trichostatin, a pharmacological inhibitor of HDAC2. The protective effect against IL-8 release of the mixture was higher than that afforded by the complex (Figure S8).

Second, BUD concentrations and/or amount of oxidant and inflammatory stressors used are relevant for patho-physio-logical conditions. BUD dry powder for inhalation (Pulmicort®), was recommended for COPD patient administration—up to 1000 μg/day on average. If we assume that about 30% of the nominal dose inhaled with a dry powder inhaler might reach the lungs [61,62], therefore 300 μg of BUD dry powder in Pulmicort® administered in patients might reach the lungs. If the 300 μg of BUD will disperse in the lung lining fluid (20–40 mL in a human of 70 kg) [63], then the pulmonary BUD concentration could be approximately 17–35 μM, which is in the range of concentrations used in our study (1–100 μM). However, we must remain cautious since the amount of

BUD deposited in the lung is difficult to predict. The question of the relevance of the quantity of H_2O_2 + LPS is also raised. Here again, it appears difficult to properly assess the exposition of alveolar cells to H_2O_2 and LPS during smoking—e.g., many factors should be considered, such as the frequency of smoking, the number and type of cigarettes smoked per day, the duration of smoking, the distribution of smoke in the lungs, the half-life of each molecular species generated in cigarette smoke, their own biodisponibility, and so on. Nakayama et al. [28] and Hasday et al. [45], respectively, reported that extract amounts of H_2O_2 ranging from 500 nmol to 4 µmol of H_2O_2 per cigarette and 6 to 9 µg of active LPS per gram of a cigarette can be extracted. The amount of H_2O_2 used in this work appears less important (up to 200 nmol in 200 µL), whereas the amount of LPS appears in the same range (up to 10 µg in 100 µL).

In conclusion, we demonstrated the anticytotoxic, antioxidant, anti-inflammatory properties of the BUD:HPβCD complex with protective activity against PI3K/Akt signaling activation and HDAC2 inhibition induced by oxidative stress. The antioxidant and anticytotoxic properties appeared essentially due to HPβCD while the anti-inflammatory properties appeared mainly to be due to BUD. Further investigations are clearly needed for a more complete view of the potential of the BUD:HPβCD complex in other relevant models of oxidative stress-induced glucocorticoid insensitivity in vitro or in vivo.

4. Material and Methods

4.1. Material

A549 (ATCC® CCL185™) and THP-1 (ATCC® TIB-202™) cells were purchased from the American Type Culture Collection (Manassas, VA, USA). HPβCD was obtained from Roquette, Lestrem, France. H_2O_2, Lipopolysaccharides (LPSs), Budesonide (BUD), Phorbol Myristate Acetate (PMA), MβCD, sphingomyelinase from Bacillus cereus, *N*-acetyl-ʟ-cysteine (NAC), 2,2-Diphenyl-1-picrylhydrazyl (DPPH•), and ʟ-ascorbic-acid (vitamin C) were ordered from Sigma-Aldrich (Saint Louis, MO, USA). A Cytotoxicity Detection KitPLUS (LDH) was ordered from Roche (Mannheim, Germany) and HOECHST® 33,342 staining solution from Life technologies (Eugene, OR, USA). LY294002 was ordered from Gibco (Camarillo, CA, USA). Phospho-Akt (Ser473) (D9E) XP® and Akt rabbit monoclonal antibodies were obtained from Cell Signaling Technology® (Beverly, MA, USA). β-actin (C4), mouse IgGκ light chain binding protein (m-IgGκ BP) conjugated to horseradish peroxidase (HRP) and mouse antirabbit IgG-HRP monoclonal antibodies were obtained from Santa Cruz Technology (Dallas, TX, USA). Horseradish Peroxidase (HRP) was ordered from Thermo Scientific™ (Rockford, IL, USA). Anti-HDAC2 (Ab-394) and Anti-phospho-HDAC2 (pSer394) antibodies produced in rabbit were ordered from Sigma-Aldrich (Saint Louis, MO, USA). Phenolsulfonphtalein (phenol red) was obtained from Merck (Darmstadt, Germany). A Human IL-8/CXCL8 DuoSet ELISA kit was obtained from R&D systems (Minneapolis, MN, USA). Trimethylsilyl-3-propionide acid-*d4* (TMSP) and deuterium oxide (99.96% D) were purchased from Eurisotop (Gif-sur-Yvette, France). Certified maleic acid and phosphate buffer powder were provided by Sigma-Aldrich (Karlsruhe, Germany). NMR measurements were recorded on a Bruker Avance spectrometer operating at 500.13 MHz for the proton signal acquisition and equipped with a 5-mm TCI cryoprobe with a Z-gradient.

4.2. BUD:HPβCD Complex Stock Solution Preparation and Characterization

We adapted the method of Dufour et al. [64]. The BUD:HPβCD complex stock solutions were prepared by adding 200 mM HPβCD (molar substitution = 0.64) in deionized water to BUD powder at a final concentration of 8.13 mM (BUD:HPβCD 1:25 molar ratio). The solution was then thoroughly mixed for 1 h 30 min (13,500 rpm) with a T25 basic Ultra-Turrax® homogenizer from IKA (Staufen, Germany) and filtered (0.22-µm filter unit). BUD and HPβCD were quantified in the solution obtained by HPLC-UV and ^{1}H-NMR, respectively, and checked for complexation as described by Dufour et al. [64]. Solutions were stored at 4 °C and renewed every 2 months.

4.3. Cell Handling

A549 [40] and THP-1 [41] cells were grown in DMEM (1X) and RPMI-1640 medium (1X) (Gibco, Paisley, UK), respectively, and were both supplemented with FBS (10%) and penicillin-streptomycin (1%) (Gibco, Grand Island, NY, USA) at 37 °C in a 5% CO_2 humidified atmosphere. Sub-cultures were performed according to the manufacturer's instructions. To activate THP-1 in macrophage-like cells (A-THP-1), cells were resuspended in fresh media, and phorbol 12-myristate 13-acetate (PMA) was added (final concentration 200 μg/L) to the THP-1-containing medium [65]. A549 and THP-1 cells were then seeded in culture microplates, dishes or flasks depending on the experiment and incubated until sub-confluent (A549, ~80%) or 24 h (THP-1) at 37 °C in a 5% CO_2 humidified atmosphere. For experiments, test molecules were dissolved in 1% FBS-supplemented medium, unless otherwise mentioned. When prior cell cholesterol or sphingomyelin depletion was required, cells were preincubated for 30 min with 5 mM methyl-β-cyclodextrin (MβCD) or 50 mU/mL of sphingomyelinase from *Bacillus cereus* [66] in 1% FBS-supplemented medium, respectively.

4.4. Cytotoxicity Studies

4.4.1. Lactate Dehydrogenase Assay

A549 cells in 96-well plates were incubated with increasing concentrations of BUD, HPβCD, or BUD:HPβCD complex, or with H_2O_2 + LPS with/without increasing concentrations of BUD, HPβCD, or BUD:HPβCD complex. The activity of lactate dehydrogenase (LDH) released by non-viable cells in the supernatant was quantified using the Cytotoxicity Detection KitPLUS (LDH) from Roche (Mannheim, Germany) according to the manufacturer's instructions.

4.4.2. HOECHST Nuclear Staining

A549 cells in ibiTreat μ-slides 2 wells from ibidi (Martinsried, Germany) were incubated with H_2O_2 + LPS with/without BUD, HPβCD, or BUD:HPβCD complex. Cells were then washed with PBS, covered with a 2000-fold dilution of HOECHST® 33,342 staining solution (Life technologies, Eugene, OR, USA) in PBS, and incubated 5 min at room temperature protected from light. Cells were then washed with PBS and imaged with a fluorescence microscope ($\lambda_{ex}/_{em}$ = 350/461, DAPI filter set). Cells with bright and/or fragmented nuclei were considered apoptotic. The proportion of apoptotic cells was calculated from a total cell count of 400/well. H_2O_2 + LPS concentrations were those preselected for LDH assay.

4.5. DCF Assay for Determining ROS Generation

We adapted the method of Wang and Joseph [67]. Briefly, A549 cells in 96-well plates were incubated for 30 min with 10 or 50 μM membrane-permeant and non-fluorescent 2′,7′-dichlorofluorescein diacetate (DCFDA) (Sigma-Aldrich, Saint Louis, MO, USA), which was deacetylated by non-specific intracellular esterases into the membrane-impermeant and non-fluorescent $DCFH_2$. Cells were then washed with Hank's balanced salt solution (HBSS) and incubated with H_2O_2 + LPS with/without BUD, HPβCD, or BUD:HPβCD complex in HBSS. Oxidative stress was evaluated through the measure of the fluorescence of DCF resulting from the oxidation of $DCFH_2$ by intracellular oxidants (λ_{ex} = 490 nm; λ_{em} = 523 nm).

4.6. Evaluation of Protein Quantity by Western Blotting

A549 cells in 6-well plates or 60 × 15 mm culture dishes were incubated with H_2O_2 + LPS with/without BUD, HPβCD, or BUD:HPβCD complex. After incubation, cells were washed with ice-cold PBS and scraped off with a cold scraper in the presence of ice-cold RIPA or Biovision's cell lysis buffer supplemented with protease and phosphatase inhibitor cocktails. Detached cells in lysis buffer were then incubated for 30 min at 4 °C with agitation in a 2-mL microcentrifuge tube and centrifuged for 10 min (10,000× g, 4 °C). The supernatant (whole-cell lysate) was stored at −80 °C at least overnight. A quantity of 30 µg of proteins per sample was mixed with 1X NuPAGE LDS sample buffer and 1X NuPAGE sample reducing agent (Thermo Scientific[TM], Carlsbad, CA, USA) and heated for 10 min at 70 °C. Samples were then electrophoresed on precasted NuPAGE Bis-Tris gels in the presence of MOPS (3-(N-morpholino)propanesulfonic acid) running buffer 1X, transferred to PVDF (Polyvinylidene difluoride) transfer membranes (Thermo Scientific[TM], Rockford, IL, USA) in the presence of NuPAGE transfer buffer 1X (Thermo Scientific[TM], Carlsbad, CA, USA) and blocked for 1 h in 5% non-fat dry milk in 20 mL of tris-buffered saline 1X containing 0.05% Tween 20 (TBS-T). Membranes were incubated overnight at 4 °C with primary antibodies with gentle agitation, washed 3 times with TBS-T, then incubated for 1 h at room temperature with the appropriated HRP-conjugated secondary antibodies. The manufacturer's recommendations were followed for antibody dilutions. After washing 3 times with TBS-T, blots were revealed using the SuperSignal West Pico Chemiluminescent Substrate (Thermo Scientific[TM], Rockford, IL, USA), the Fusion Pulse 7 apparatus and Fusion Capt Advance Pulse 7 software. To reveal proteins with similar migration profiles, membranes were washed in TBS-T after the first reveal, and antibodies were stripped with a 10-min bath in Restore[TM] Western Blot Stripping Buffer (Thermo Scientific[TM], Rockford, IL, USA) and washed again with TBS-T. Then, the Western blot protocol was repeated from the block for 1 h in 5% non-fat dry milk in 20 mL of TBS-T.

4.7. Evaluation of Inflammatory Cytokine (IL-8) Expression by Sandwich ELISA

A-THP-1 cells in 96-well plates were incubated with H_2O_2 + LPS with/without BUD, HPβCD, or BUD:HPβCD complex. A 4-fold dilution of the supernatant in RPMI-1640 medium was stored overnight at −80 °C. IL-8 cytokine levels in diluted supernatant were quantified using the Human IL-8/CXCL8 DuoSet ELISA kit (R&D systems, Minneapolis, MN, USA) according to the manufacturer's instructions.

4.8. Data Analysis

GraphPad Prism® (version 4.03 for Windows, GraphPad Prism Software, San Diego, CA, USA) was used for graphic illustrations and statistical analysis. The statistical tests used to study the significance of the results are described in the captions of the corresponding figures.

Supplementary Materials: The following are available online. Figure S1: Cholesterol content in A549 cells untreated (control), incubated with methyl-β-cyclodextrin (MβCD) for 30 min, and incubated with methyl-β-cyclodextrin (MβCD) for 30 min and thereafter in medium for 2 h. Figure S2: Effect of the BUD: HPβCD complex (green) and HPβCD (blue) on oxidant generation induced by H_2O_2 + LPS after treatment for 2 h in non-depleted and sphingomyelin-depleted A549 cells. Figure S3: Effect of the BUD:HPβCD complex versus HPβCD on H_2O_2 (**A**) and 2,2-Diphenyl-1-picrylhydrazyl (DPPH•) (**B**) after 25 min (**A**) and 1 h of incubation (**B**). Figure S4: Cholesterol content in A549 cells incubated with the BUD:HPβCD complex, HPβCD or BUD for 2 h. Figure S5: HPβCD relative quantity in A549 extracellular medium after 2 h of incubation with the BUD:HPβCD complex (green bars) and HPβCD (blue bars). HPβCD was quantified using proton nuclear magnetic resonance spectroscopy. Figure S6: Oxidant generation kinetic in A549 cells after treatment with H_2O_2 + LPS (1 mM + 100 µg/mL) for 6 h and effect of N-acetyl-L-cysteine (NAC) and Vit C on H_2O_2 + LPS-induced oxidant generation. Figure S7: Akt phosphorylation induced by H_2O_2 + LPS, H_2O_2 and LPS in A549 cells after 2 h of incubation. Figure S8: Percentage of IL-8 released after A 549 cells pretreatment for 30 min with trichostatin (TSA) and incubation for 2 h with TNF-α in presence of BUD + HPβCD complex, BUD or HPβCD [68–71].

Author Contributions: Conceptualization, D.C., J.C.B. and M.-P.M.-L.; Methodology, J.C.B.; P.D.T.; B.E. and M.-P.M.-L. Investigation, J.C.B.; P.D.T.; B.E.; Resources, P.D.T.; B.E. and M.-P.M.-L.; Writing—Original Draft Preparation, J.C.B.; Writing—Review and Editing, J.C.B. and M.-P.M.-L.; Supervision, M.-P.M.-L.; Project Administration, D.C. and M.-P.M.-L.; Funding Acquisition, D.C., B.E. and M.-P.M.-L. All authors have read and agreed to the published version of the manuscript.

Funding: This research received no external funding.

Acknowledgments: JCB thanks WB and OJ for the BUD:HPβCD complex used in this study and V. Mohymont who provided dedicated technical assistance. This work was supported by Walloon Region (AEROGAL) and FRIA.

Conflicts of Interest: The authors declare no conflict of interest.

References

1. Cazzola, M.; Rogliani, P.; Stolz, D.; Matera, M.G. Pharmacological treatment and current controversies in COPD. *F1000Research* **2019**, *8*. [CrossRef] [PubMed]
2. Nici, L.; Mammen, M.J.; Charbek, E.; Alexander, P.E.; Au, D.H.; Boyd, C.M.; Criner, G.J.; Donaldson, G.C.; Dreher, M.; Fan, V.S.; et al. Pharmacologic Management of Chronic Obstructive Pulmonary Disease. An Official American Thoracic Society Clinical Practice Guideline. *Am. J. Respir. Crit Care Med.* **2020**, *201*, e56–e69. [CrossRef] [PubMed]
3. Marwick, J.A.; Adcock, I.M.; Chung, K.F. Overcoming reduced glucocorticoid sensitivity in airway disease: Molecular mechanisms and therapeutic approaches. *Drugs* **2010**, *70*, 929–948. [CrossRef] [PubMed]
4. Barnes, P.J. Corticosteroid resistance in patients with asthma and chronic obstructive pulmonary disease. *J. Allergy Clin. Immunol.* **2013**, *131*, 636–645. [CrossRef] [PubMed]
5. Barnes, P.J. Inflammatory mechanisms in patients with chronic obstructive pulmonary disease. *J. Allergy Clin. Immunol.* **2016**, *138*, 16–27. [CrossRef] [PubMed]
6. Barnes, P.J. Glucocorticosteroids. *Handb. Exp. Pharmacol.* **2017**, *237*, 93–115.
7. Mei, D.; Tan, W.S.D.; Wong, W.S.F. Pharmacological strategies to regain steroid sensitivity in severe asthma and COPD. *Curr. Opin. Pharmacol.* **2019**, *46*, 73–81. [CrossRef]
8. Bi, J.; Min, Z.; Yuan, H.; Jiang, Z.; Mao, R.; Zhu, T.; Liu, C.; Zeng, Y.; Song, J.; Du, C.; et al. PI3K inhibitor treatment ameliorates the glucocorticoid insensitivity of PBMCs in severe asthma. *Clin. Transl. Med.* **2020**, *9*, 22. [CrossRef]
9. Pelaia, G.; Vatrella, A.; Busceti, M.T.; Fabiano, F.; Terracciano, R.; Matera, M.G.; Maselli, R. Molecular and cellular mechanisms underlying the therapeutic effects of budesonide in asthma. *Pulm. Pharmacol. Ther.* **2016**, *40*, 15–21. [CrossRef]
10. Tashkin, D.P.; Lipworth, B.; Brattsand, R. Benefit: Risk Profile of Budesonide in Obstructive Airways Disease. *Drugs* **2019**, *79*, 1757–1775. [CrossRef]
11. Janson, C. Treatment with inhaled corticosteroids in chronic obstructive pulmonary disease. *J. Thorac. Dis.* **2020**, *12*, 1561–1569. [CrossRef]
12. Dufour, G.; Bigazzi, W.; Wong, N.; Boschini, F.; de Tullio, P.; Piel, G.; Cataldo, D.; Evrard, B. Interest of cyclodextrins in spray-dried microparticles formulation for sustained pulmonary delivery of budesonide. *Int. J. Pharm.* **2015**, *495*, 869–878. [CrossRef] [PubMed]
13. Loftsson, T.; Saokham, P.; Sa Couto, A.R. Self-association of cyclodextrins and cyclodextrin complexes in aqueous solutions. *Int. J. Pharm.* **2019**, *560*, 228–234. [CrossRef] [PubMed]
14. Jansook, P.; Ogawa, N.; Loftsson, T. Cyclodextrins: Structure, physicochemical properties and pharmaceutical applications. *Int. J. Pharm.* **2018**, *535*, 272–284. [CrossRef] [PubMed]
15. Crini, G. Review: A history of cyclodextrins. *Chem. Rev.* **2014**, *114*, 10940–10975. [CrossRef]
16. Braga, S.S. Cyclodextrins: Emerging Medicines of the New Millennium. *Biomolecules* **2019**, *9*, 801. [CrossRef] [PubMed]
17. Zidovetzki, R.; Levitan, I. Use of cyclodextrins to manipulate plasma membrane cholesterol content: Evidence, misconceptions and control strategies. *Biochim. Biophys. Acta* **2007**, *1768*, 1311–1324. [CrossRef]
18. di Cagno, M.P. The Potential of Cyclodextrins as Novel Active Pharmaceutical Ingredients: A Short Overview. *Molecules* **2016**, *22*, 1. [CrossRef]
19. Gould, S.; Scott, R.C. 2-Hydroxypropyl-beta-cyclodextrin (HP-beta-CD): A toxicology review. *Food Chem. Toxicol.* **2005**, *43*, 1451–1459. [CrossRef]
20. Malanga, M.; Szeman, J.; Fenyvesi, E.; Puskas, I.; Csabai, K.; Gyemant, G.; Fenyvesi, F.; Szente, L. "Back to the Future": A New Look at Hydroxypropyl Beta-Cyclodextrins. *J. Pharm. Sci.* **2016**, *105*, 2921–2931. [CrossRef]
21. Barnes, P.J. Role of HDAC2 in the pathophysiology of COPD. *Annu. Rev. Physiol.* **2009**, *71*, 451–464. [CrossRef] [PubMed]
22. Ito, K.; Ito, M.; Elliott, W.M.; Cosio, B.; Caramori, G.; Kon, O.M.; Barczyk, A.; Hayashi, S.; Adcock, I.M.; Hogg, J.C.; et al. Decreased histone deacetylase activity in chronic obstructive pulmonary disease. *N. Engl. J. Med.* **2005**, *352*, 1967–1976. [CrossRef] [PubMed]
23. Lasserre, R.; Guo, X.J.; Conchonaud, F.; Hamon, Y.; Hawchar, O.; Bernard, A.M.; Soudja, S.M.; Lenne, P.F.; Rigneault, H.; Olive, D.; et al. Raft nanodomains contribute to Akt/PKB plasma membrane recruitment and activation. *Nat. Chem. Biol.* **2008**, *4*, 538–547. [CrossRef] [PubMed]

24. Calay, D.; Vind-Kezunovic, D.; Frankart, A.; Lambert, S.; Poumay, Y.; Gniadecki, R. Inhibition of Akt signaling by exclusion from lipid rafts in normal and transformed epidermal keratinocytes. *J. Investig. Dermatol.* **2010**, *130*, 1136–1145. [CrossRef]

25. Mollinedo, F.; Gajate, C. Lipid rafts as major platforms for signaling regulation in cancer. *Adv. Biol. Regul.* **2015**, *57*, 130–146. [CrossRef] [PubMed]

26. Gao, X.; Zhang, J. Spatiotemporal analysis of differential Akt regulation in plasma membrane microdomains. *Mol. Biol. Cell.* **2008**, *19*, 4366–4373. [CrossRef] [PubMed]

27. Dos Santos, A.G.; Bayiha, J.C.; Dufour, G.; Cataldo, D.; Evrard, B.; Silva, L.C.; Deleu, M.; Mingeot-Leclercq, M.P. Changes in membrane biophysical properties induced by the Budesonide/Hydroxypropyl-beta-cyclodextrin complex. *Biochim. Biophys. Acta Biomembr.* **2017**, *1859*, 1930–1940. [CrossRef]

28. Nakayama, T.; Church, D.F.; Pryor, W.A. Quantitative analysis of the hydrogen peroxide formed in aqueous cigarette tar extracts. *Free Radic. Biol. Med.* **1989**, *7*, 9–15. [CrossRef]

29. Yamaguchi, Y.; Kagota, S.; Haginaka, J.; Kunitomo, M. Peroxynitrite-generating species: Good candidate oxidants in aqueous extracts of cigarette smoke. *Jpn. J. Pharmacol.* **2000**, *82*, 78–81. [CrossRef]

30. Valenca, S.S.; Silva, B.F.; Lopes, A.A.; Romana-Souza, B.; Marinho Cavalcante, M.C.; Lima, A.B.; Goncalves Koatz, V.L.; Porto, L.C. Oxidative stress in mouse plasma and lungs induced by cigarette smoke and lipopolysaccharide. *Environ. Res.* **2008**, *108*, 199–204. [CrossRef]

31. Tuder, R.M.; Petrache, I. Pathogenesis of chronic obstructive pulmonary disease. *J. Clin. Investig.* **2012**, *122*, 2749–2755. [CrossRef] [PubMed]

32. Wiegman, C.H.; Michaeloudes, C.; Haji, G.; Narang, P.; Clarke, C.J.; Russell, K.E.; Bao, W.; Pavlidis, S.; Barnes, P.J.; Kanerva, J.; et al. Oxidative stress-induced mitochondrial dysfunction drives inflammation and airway smooth muscle remodeling in patients with chronic obstructive pulmonary disease. *J. Allergy Clin. Immunol.* **2015**, *136*, 769–780. [CrossRef] [PubMed]

33. Marwick, J.A.; Caramori, G.; Stevenson, C.S.; Casolari, P.; Jazrawi, E.; Barnes, P.J.; Ito, K.; Adcock, I.M.; Kirkham, P.A.; Papi, A. Inhibition of PI3Kdelta restores glucocorticoid function in smoking-induced airway inflammation in mice. *Am. J. Respir. Crit Care Med.* **2009**, *179*, 542–548. [CrossRef]

34. To, Y.; Ito, K.; Kizawa, Y.; Failla, M.; Ito, M.; Kusama, T.; Elliott, W.M.; Hogg, J.C.; Adcock, I.M.; Barnes, P.J. Targeting phosphoinositide-3-kinase-delta with theophylline reverses corticosteroid insensitivity in chronic obstructive pulmonary disease. *Am. J. Respir. Crit Care Med.* **2010**, *182*, 897–904. [CrossRef]

35. Hammoud, Z.; Khreich, N.; Auezova, L.; Fourmentin, S.; Elaissari, A.; Greige-Gerges, H. Cyclodextrin-membrane interaction in drug delivery and membrane structure maintenance. *Int. J. Pharm.* **2019**, *564*, 59–76. [CrossRef] [PubMed]

36. Lopez-Revuelta, A.; Sanchez-Gallego, J.I.; Hernandez-Hernandez, A.; Sanchez-Yague, J.; Llanillo, M. Membrane cholesterol contents influence the protective effects of quercetin and rutin in erythrocytes damaged by oxidative stress. *Chem. Biol. Interact.* **2006**, *161*, 79–91. [CrossRef] [PubMed]

37. Zhang, Y.; Chen, F.; Chen, J.; Huang, S.; Chen, J.; Huang, J.; Li, N.; Sun, S.; Chu, X.; Zha, L. Soyasaponin Bb inhibits the recruitment of toll-like receptor 4 (TLR4) into lipid rafts and its signaling pathway by suppressing the nicotinamide adenine dinucleotide phosphate (NADPH) oxidase-dependent generation of reactive oxygen species. *Mol. Nutr. Food Res.* **2016**, *60*, 1532–1543. [CrossRef]

38. Wu, J.; Liu, C.; Zhang, L.; Qu, C.H.; Sui, X.L.; Zhu, H.; Huang, L.; Xu, Y.F.; Han, Y.L.; Qin, C. Histone deacetylase-2 is involved in stress-induced cognitive impairment via histone deacetylation and PI3K/AKT signaling pathway modification. *Mol. Med. Rep.* **2017**, *16*, 1846–1854. [CrossRef] [PubMed]

39. Sun, X.J.; Li, Z.H.; Zhang, Y.; Zhong, X.N.; He, Z.Y.; Zhou, J.H.; Chen, S.N.; Feng, Y. Theophylline and dexamethasone in combination reduce inflammation and prevent the decrease in HDAC2 expression seen in monocytes exposed to cigarette smoke extract. *Exp. Ther. Med.* **2020**, *19*, 3425–3431. [CrossRef]

40. Schulz, C.; Farkas, L.; Wolf, K.; Kratzel, K.; Eissner, G.; Pfeifer, M. Differences in LPS-induced activation of bronchial epithelial cells (BEAS-2B) and type II-like pneumocytes (A-549). *Scand. J. Immunol.* **2002**, *56*, 294–302. [CrossRef]

41. Bosshart, H.; Heinzelmann, M. THP-1 cells as a model for human monocytes. *Ann. Transl. Med.* **2016**, *4*, 438. [CrossRef]

42. Ito, K.; Yamamura, S.; Essilfie-Quaye, S.; Cosio, B.; Ito, M.; Barnes, P.J.; Adcock, I.M. Histone deacetylase 2-mediated deacetylation of the glucocorticoid receptor enables NF-kappaB suppression. *J. Exp. Med.* **2006**, *203*, 7–13. [CrossRef]

43. Piao, J.; Chen, L.; Quan, T.; Li, L.; Quan, C.; Piao, Y.; Jin, T.; Lin, Z. Superior efficacy of co-treatment with the dual PI3K/mTOR inhibitor BEZ235 and histone deacetylase inhibitor Trichostatin A against NSCLC. *Oncotarget* **2016**, *7*, 60169–60180. [CrossRef] [PubMed]

44. Yamaguchi, M.S.; McCartney, M.M.; Falcon, A.K.; Linderholm, A.L.; Ebeler, S.E.; Kenyon, N.J.; Harper, R.H.; Schivo, M.; Davis, C.E. Modeling cellular metabolomic effects of oxidative stress impacts from hydrogen peroxide and cigarette smoke on human lung epithelial cells. *J. Breath Res.* **2019**, *13*, 036014. [CrossRef]

45. Hasday, J.D.; Bascom, R.; Costa, J.J.; Fitzgerald, T.; Dubin, W. Bacterial endotoxin is an active component of cigarette smoke. *Chest* **1999**, *115*, 829–835. [CrossRef]

46. Volobuef, C.; Moraes, C.M.; Nunes, L.A.; Cereda, C.M.; Yokaichiya, F.; Franco, M.K.; Braga, A.F.; De, P.E.; Tofoli, G.R.; Fraceto, L.F.; et al. Sufentanil-2-hydroxypropyl-beta-cyclodextrin inclusion complex for pain treatment: Physicochemical, cytotoxicity, and pharmacological evaluation. *J. Pharm. Sci.* **2012**, *101*, 3698–3707. [CrossRef]

47. Lachowicz, M.; Stanczak, A.; Kolodziejczyk, M. Characteristic of Cyclodextrins: Their role and use in the pharmaceutical technology. *Curr. Drug Targets* **2020**. [CrossRef] [PubMed]

48. Anraku, M.; Iohara, D.; Wada, K.; Taguchi, K.; Maruyama, T.; Otagiri, M.; Uekama, K.; Hirayama, F. Antioxidant and renoprotective activity of 2-hydroxypropyl-beta-cyclodextrin in nephrectomized rats. *J. Pharm. Pharmacol.* **2016**, *68*, 608–614. [CrossRef] [PubMed]

49. Zimmer, S.; Grebe, A.; Bakke, S.S.; Bode, N.; Halvorsen, B.; Ulas, T.; Skjelland, M.; De, N.D.; Labzin, L.I.; Kerksiek, A.; et al. Cyclodextrin promotes atherosclerosis regression via macrophage reprogramming. *Sci. Transl. Med.* **2016**, *8*, 333ra50. [CrossRef]

50. Lopez-Nicolas, J.M.; Rodriguez-Bonilla, P.; Garcia-Carmona, F. Cyclodextrins and antioxidants. *Crit Rev. Food Sci. Nutr.* **2014**, *54*, 251–276. [CrossRef]

51. Gross, N.J.; Barnes, P.J. New Therapies for Asthma and Chronic Obstructive Pulmonary Disease. *Am. J. Respir. Crit Care Med.* **2017**, *195*, 159–166. [CrossRef] [PubMed]

52. Lopez, C.A.; de Vries, A.H.; Marrink, S.J. Computational microscopy of cyclodextrin mediated cholesterol extraction from lipid model membranes. *Sci. Rep.* **2013**, *3*, 2071. [CrossRef] [PubMed]

53. Ueda, K.; Nishimoto, Y.; Kimura, G.; Masuko, T.; Barnes, P.J.; Ito, K.; Kizawa, Y. Repeated lipopolysaccharide exposure causes corticosteroid insensitive airway inflammation via activation of phosphoinositide-3-kinase delta pathway. *Biochem. Biophys. Rep.* **2016**, *7*, 367–373. [PubMed]

54. Zheng, X.; Zhang, W.; Hu, X. Different concentrations of lipopolysaccharide regulate barrier function through the PI3K/Akt signalling pathway in human pulmonary microvascular endothelial cells. *Sci. Rep.* **2018**, *8*, 9963. [CrossRef]

55. Aldini, G.; Altomare, A.; Baron, G.; Vistoli, G.; Carini, M.; Borsani, L.; Sergio, F. N-Acetylcysteine as an antioxidant and disulphide breaking agent: The reasons why. *Free Radic. Res.* **2018**, *52*, 751–762. [CrossRef]

56. George, S.M.; Gaylor, J.D.; Leadbitter, J.; Grant, M.H. The effect of betacyclodextrin and hydroxypropyl betacyclodextrin incorporation into plasticized poly(vinyl chloride) on its compatibility with human U937 cells. *J. Biomed. Mater. Res. B Appl. Biomater.* **2011**, *96*, 310–315. [CrossRef]

57. Matassoli, F.L.; Leao, I.C.; Bezerra, B.B.; Pollard, R.B.; Lutjohann, D.; Hildreth, J.E.K.; Arruda, L.B. Hydroxypropyl-Beta-Cyclodextrin Reduces Inflammatory Signaling from Monocytes: Possible Implications for Suppression of HIV Chronic Immune Activation. *mSphere* **2018**, *3*. [CrossRef] [PubMed]

58. Onishi, M.; Ozasa, K.; Kobiyama, K.; Ohata, K.; Kitano, M.; Taniguchi, K.; Homma, T.; Kobayashi, M.; Sato, A.; Katakai, Y.; et al. Hydroxypropyl-beta-cyclodextrin spikes local inflammation that induces Th2 cell and T follicular helper cell responses to the coadministered antigen. *J. Immunol.* **2015**, *194*, 2673–2682. [CrossRef] [PubMed]

59. Higham, A.; Karur, P.; Jackson, N.; Cunoosamy, D.M.; Jansson, P.; Singh, D. Differential anti-inflammatory effects of budesonide and a p38 MAPK inhibitor AZD7624 on COPD pulmonary cells. *Int. J. Chronic Obstruct. Pulmon. Dis.* **2018**, *13*, 1279–1288. [CrossRef] [PubMed]

60. Stella, V.J.; Rao, V.M.; Zannou, E.A.; Zia, V. Mechanisms of drug release from cyclodextrin complexes. *Adv. Drug Deliv. Rev.* **1999**, *36*, 3–16. [CrossRef]

61. Dahlstrom, K.; Thorsson, L.; Larsson, P.; Nikander, K. Systemic availability and lung deposition of budesonide via three different nebulizers in adults. *Ann. Allergy Asthma Immunol.* **2003**, *90*, 226–232. [CrossRef]

62. Thorsson, L.; Edsbacker, S.; Conradson, T.B. Lung deposition of budesonide from Turbuhaler is twice that from a pressurized metered-dose inhaler P-MDI. *Eur. Respir. J.* **1994**, *7*, 1839–1844. [CrossRef] [PubMed]

63. Fernandes, C.A.; Vanbever, R. Preclinical models for pulmonary drug delivery. *Expert Opin. Drug Deliv.* **2009**, *6*, 1231–1245. [CrossRef] [PubMed]

64. Dufour, G.; Evrard, B.; de Tullio, P. 2D-Cosy NMR Spectroscopy as a Quantitative Tool in Biological Matrix: Application to Cyclodextrins. *AAPS J.* **2015**, *17*, 1501–1510. [CrossRef]

65. Lemaire, S.; Mingeot-Leclercq, M.P.; Tulkens, P.M.; Van Bambeke, F. Study of macrophage functions in murine J774 cells and human activated THP-1 cells exposed to oritavancin, a lipoglycopeptide with high cellular accumulation. *Antimicrob. Agents Chemother.* **2014**, *58*, 2059–2066. [CrossRef]

66. Verstraeten, S.L.; Albert, M.; Paquot, A.; Muccioli, G.G.; Tyteca, D.; Mingeot-Leclercq, M.P. Membrane cholesterol delays cellular apoptosis induced by ginsenoside Rh2, a steroid saponin. *Toxicol. Appl. Pharmacol.* **2018**, *352*, 59–67. [CrossRef]

67. Wang, H.; Joseph, J.A. Quantifying cellular oxidative stress by dichlorofluorescein assay using microplate reader. *Free Radic. Biol. Med.* **1999**, *27*, 612–616. [CrossRef]

68. Pick, E.; Keisari, Y. A simple colorimetric method for the measurement of hydrogen peroxide produced by cells in culture. *J. Immunol. Methods* **1980**, *38*, 161–170. [CrossRef]

69. Bahorun, T.; Gressier, B.; Trotin, F.; Brunet, C.; Dine, T.; Luyckx, M.; Vasseur, J.; Cazin, M.; Cazin, J.C.; Pinkas, M. Oxygen species scavenging activity of phenolic extracts from hawthorn fresh plant organs and pharmaceutical preparations. *Arzneimittelforschung* **1996**, *46*, 1086–1089.

70. Brand-Williams, W.; Cuvelier, M.E.; Berset, C. Use of a free radical method to evaluate antioxidant activity. *LWT-Food Sci. Technol.* **1995**, *28*, 25–30. [CrossRef]

71. Matheus, N.; Hansen, S.; Rozet, E.; Peixoto, P.; Maquoi, E.; Lambert, V.; Noel, A.; Frederich, M.; Mottet, D.; de Tullio, P. An easy, convenient cell and tissue extraction protocol for nuclear magnetic resonance metabolomics. *Phytochem. Anal.* **2014**, *25*, 342–349. [CrossRef] [PubMed]

Sample Availability: Samples of the compounds are available from the authors. The complex has to be prepared each three months for sake of stability.

Publisher's Note: MDPI stays neutral with regard to jurisdictional claims in published maps and institutional affiliations.

 *molecules*

Article

Understanding Surface Interaction and Inclusion Complexes between Piroxicam and Native or Crosslinked β-Cyclodextrins: The Role of Drug Concentration [†]

Giuseppina Raffaini [1,2,*] and **Fabio Ganazzoli [1,2]**

[1] Department of Chemistry, Materials, and Chemical Engineering "Giulio Natta", Politecnico di Milano, Piazza L. Da Vinci 32, 20131 Milano, Italy; fabio.ganazzoli@polimi.it
[2] INSTM, National Consortium of Materials Science and Technology, Local Unit Politecnico di Milano, 20133 Milano, Italy
[*] Correspondence: giuseppina.raffaini@polimi.it; Tel.: +39-02-23993068
[†] Dedicated to the memory of the late Professor Giuseppe Allegra.

Academic Editor: Rosa Iacovino
Received: 26 May 2020; Accepted: 17 June 2020; Published: 19 June 2020

Abstract: Drug concentration plays an important role in the interaction with drug carriers affecting the kinetics of release process and toxicology effects. Cyclodextrins (CDs) can solubilize hydrophobic drugs in water enhancing their bioavailability. In this theoretical study based on molecular mechanics and molecular dynamics methods, the interactions between β-cyclodextrin and piroxicam, an important nonsteroidal anti-inflammatory drug, were investigated. At first, both host–guest complexes with native β-CD in the 1:1 and in 2:1 stoichiometry were considered without assuming any initial a priori inclusion: the resulting inclusion complexes were in good agreement with literature NMR data. The interaction between piroxicam and a β-CD nanosponge (NS) was then modeled at different concentrations. Two inclusion mechanisms were found. Moreover, piroxicam can interact with the external NS surface or with its crosslinkers, also forming one nanopore. At larger concentration, a nucleation process of drug aggregation induced by the first layer of adsorbed piroxicam molecules is observed. The flexibility of crosslinked β-CDs, which may be swollen or quite compact, changing the surface area accessible to drug molecules, and the dimension of the aggregate nucleated on the NS surface are important factors possibly affecting the kinetics of release, which shall be theoretically studied in more detail at specific concentrations.

Keywords: β-cyclodextrin; inclusion complexes; solubilization; nanosponge; nanocarriers; pharmaceutical applications; molecular recognition phenomena; molecular dynamics simulations; drug delivery; drug concentration

---

1. Introduction

Cyclodextrins (CDs) are cyclic oligosaccharides known for more than 100 years, recognized as pharmaceutical adjuvants for the past 20 years [1–3]. Thanks to their capability to form noncovalent water-soluble complexes, they are useful as functional excipients for solubilization, delivery, and greater bioavailability of drugs in many different applications [4,5]. Cyclodextrins have an approximatively truncated cone shape and may be described as a bucket with a hydrophilic outer surface and a hydrophobic central cavity. CDs may thus form noncovalent host–guest complexes hosting hydrophobic drugs and some hydrophilic guests. The intermolecular interactions are due to electrostatic interactions, weak van der Waals contributions, hydrogen bonds with secondary and/or primary rims, and hydrophobic interactions within the cavity and the nonpolar groups or π electrons of the guest. In an aqueous solution, natural CDs can interact. The more water-soluble cyclodextrin derivatives have a low tendency to form aggregates [6]. Cyclodextrin derivatives can be hydrophilic or relatively lipophilic based on their substitution and these properties can give insights into their ability to act as permeability enhancers. Natural CDs are α-CD, β-CD, and γ-CD composed of six, seven, and eight units of D-glucopyranose, respectively, with α-1,4 linkages. Low cost, easy synthetic accessibility, and suitable cavity size (0.60–0.65 nm) for the inclusion of small- and medium-sized drugs result in the wide use of β-cyclodextrin (β-CD) in pharmaceutical and food industries, on the early stages of pharmaceutical applications. Drugs often have various problems: they are hydrophobic, with low solubility in water and low stability either in vivo or in vitro. All these factors reduce the therapeutic effect of the drug. In order to overcome these problems, β-CD has been widely used. The drug can interact with the β-CD hydrophobic cavity and form host–guest inclusion complexes, possibly with different stoichiometries. Therefore, CDs enhance the bioavailability of insoluble drugs by increasing their solubility, dissolution, drug permeability, by making the drug available at the surface of biological barriers so that the in vivo and in vitro stability increases.

Over the past 20 years, it has been shown that CDs and CD complexes self-associate to form an aggregate or micelle-like structures [7,8]; sometimes, it is very difficult to detect them. Furthermore, the formation of drug/CD complex nanoparticles appears to increase the ability of CDs to enhance drug delivery through some mucosal membranes. Recently, chemically modified β-CD derivatives have been synthesized [9–14] and also theoretically studied [15] in order to improve cyclodextrin interactions with hydrophobic drugs and to enhance drug release through cell membranes. One of the possible modifications is to chemically bind aliphatic chains of different lengths on the primary or secondary CD rim in order to obtain amphiphilic cyclodextrins (aCD) [9]. This modification allows increasing the cyclodextrin interactions with biological membranes, improving their interaction with hydrophobic drugs, and inducing a higher self-assembly capacity in aqueous solutions compared to native β-CD [16]. This last property has been used to obtain β-CD-based nanosized carriers. Again, in the last years, new CD derivatives enhancing solubility and bioavailability have been synthetized considering both linear polymerized β-CDs [10] or crosslinked systems, such as β-CD nanosponges (β-CD NS) and chemically modified β-cyclodextrin derivatives [11–14].

Piroxicam (PX) is one of the most efficient nonsteroidal anti-inflammatory agents widely used for the treatment of rheumatoid arthritis, osteoarthritis, and acute pain in musculoskeletal disorders (see Scheme 1).

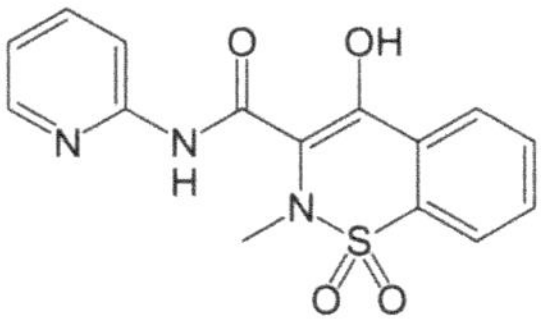

Scheme 1. Chemical structure of the piroxicam drug molecule.

Because of its very low solubility in gastrointestinal fluids, it has poor bioavailability after oral administration. The formation of inclusion complexes with β-CD may be a useful strategy in order to overcome solubility and bioavailability problems that are very important, in particular, for pediatric and geriatric patients. In 1992, Fronza et al. [17] reported an NMR study of a 1:1 β-CD/PX inclusion complex. Significant nuclear Overhauser effects were observed between inner protons of CD and the protons of both aromatic rings of the piroxicam molecule. The data indicated the possibility of having two different inclusion complexes with two different equilibrium constants within the investigated range of concentrations; furthermore, at smaller concentrations, the complex was found to be completely dissociated. Therefore, the possibility of identifying two types of host–guest inclusion compounds and the influence of the concentration for the stability and formation process are important factors that should be better investigated. In 2003, Guo et al. [18] investigated β-CDs and PX host–guest complexes using steady state fluorescence and NMR techniques, again indicating a strong interaction between the hydrophobic drug and the hydrophobic cavity of β-CDs, in particular forming stable complexes in a 1:1 stoichiometry and probably in a 1:2 stoichiometry. In order to enhance both the solubility and the fast release of PX, in 2019, Dharmasthala et al. [19] proposed a very interesting formulation for an oral film containing the β-CD/PX inclusion complex, studying the dissolution process in vitro, indicating fast drug diffusion, and permitting to obtain a better therapeutic efficiency.

Over the past 20 years, molecular mechanics (MM) and molecular dynamics (MD) simulations have been demonstrated to be a useful tool in order to atomistically investigate and describe non-covalent interactions in different systems ranging from protein adsorption on surfaces of biomaterials to formation and self-aggregation of host–guest complexes, including their stoichiometry and self-aggregation in an nonpolar solvent or in water [20–26]. Molecular simulation studies offer great insights into these phenomena in very good agreement with experimental data [20,27–29].

In this paper, a theoretical study of the host–guest inclusion complexes between β-CD and PX is reported with no a priori assumption about the inclusion stoichiometry and geometry in a 1:1 and in a 2:1 stoichiometry as suggested by NMR data. Then, the interaction with a nanosponge (NS) model containing β-CDs connected with a pyromellitic dianhydride (PMA) crosslinker [30] is studied (PMA NS). The NS was generated by linking 8 β-CDs (Model 2 in [29]) through PMA moieties (see Scheme 1 in [30]). Accordingly, each CD carries two PMA linking agents bound to primary hydroxyls at diametrically opposite sides of the macrocycle. Increasing the number of drug host molecules, a ratio between the number of β-CDs in the NS model and the number of PX molecules (8:4, 8:8, 8:16, 8:40), i.e., a β-CD/PX ratio of 2:1, 1:1, 1:2, 1:5 was investigated. The results for the interaction with the NS at different drug concentrations are then reported. The conformational changes during the MD runs are also investigated. The types of inclusion complexes also formed in a crosslinked system and the surface interaction will be highlighted. The drug concentration influencing formation of the host–guest complexes and the surface β-CD/drug interactions related to the NS flexibility as well as the drug self-association will also be described.

2. Results and Discussion

2.1. β-CD/PX Interaction

In this section, the β-CD/PX interaction forming host–guest inclusion complexes in 1:1 and 2:1 stoichiometries in implicit water is studied. The theoretical results are reported and discussed in comparison with NMR experiments in the literature.

2.1.1. β-CD/PX Interaction: Complex Formation in a 1:1 Host–Guest Stoichiometry

At first, the PX conformation was studied, and then the β-CD reported by Raffaini [22] was considered for the interaction between piroxicam and β-CD in a 1:1 stoichiometry considering possible host–guest inclusion arrangements without assuming any a priori inclusion complexes. Using a simulation protocol adopted in the previous work [20,21], four different starting geometries between piroxicam and the primary and secondary rim of β-CD are considered in the 1:1 stoichiometry as reported in Figure 1.

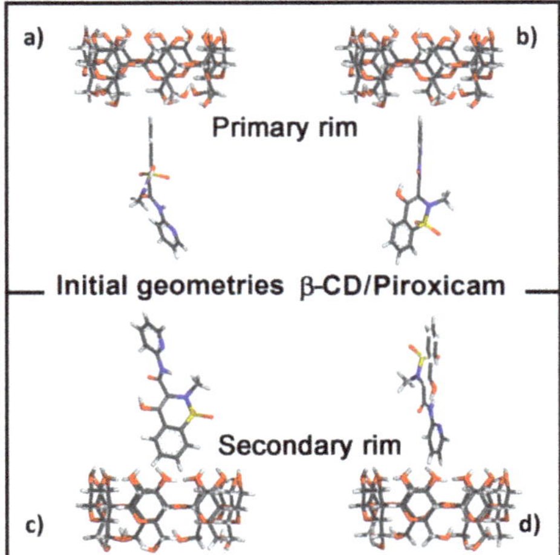

Figure 1. Side view of the four initial non-optimized geometries studied for the interaction between PX and β-CD with the primary rim and with the secondary rim at the top and the bottom, indicated respectively as (**a–d**). The two molecules are colored by atoms. Color codes: C atoms are grey, O atoms are red, N atoms are blue, S atoms are yellow, and H atoms are white.

These four initial geometries were optimized and after the initial energy minimizations, favorable host–guest complexes were formed. The final optimized arrangements found after these MM calculations are shown in Figure 2.

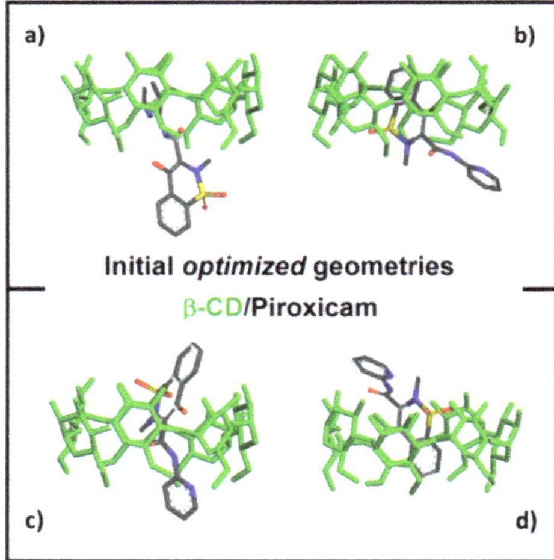

Figure 2. Side view of the 1:1 host–guest complexes formed by PX with β-CD, in the initial optimized geometries found after the MM calculations, starting from the four geometries reported in Figure 1 indicated respectively as (**a–d**). The β-CDs are green, the PX molecule is colored by atoms. Color codes: C atoms are grey, O atoms are red, N atoms are blue, S atoms are yellow, and H atoms are omitted for clarity.

After these initial energy minimizations, starting from all the four different geometries, MD runs were performed in implicit water and the final conformations assumed by the system at the end of each MD run were analyzed. Two different inclusion complexes were found, the most and the less stable host–guest inclusion complexes in the 1:1 stoichiometry being reported in Figure 3, while Table 1 reports the potential energy, the calculated interaction energies, and the information about the intermolecular and β-CD intramolecular H-bonds for all four optimized geometries. In Figure S1, the two metastable geometries reported in Table 1 indicated as B and C D are shown. The animation of the MD runs is shown in Supplemtary Information SI.

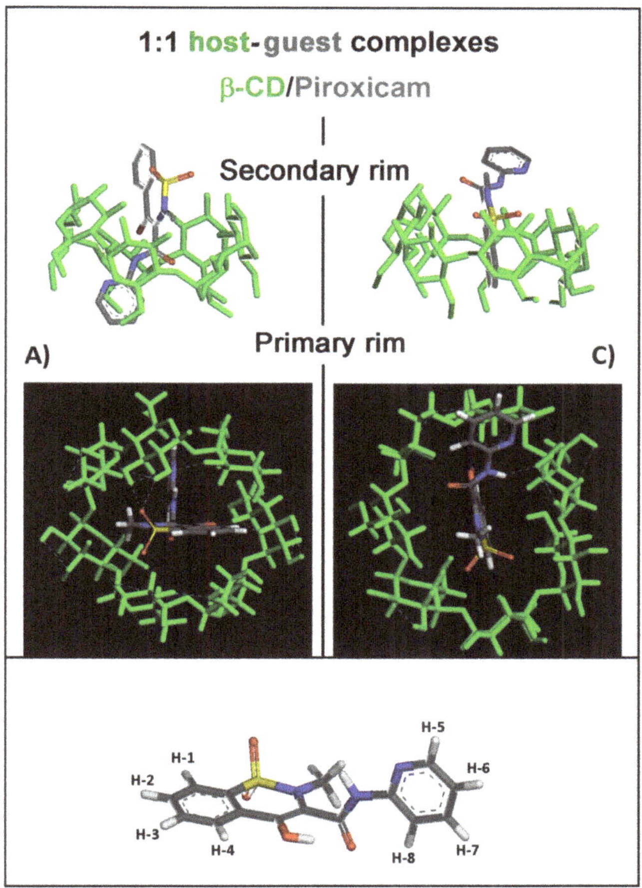

Figure 3. The side view and the top view with intermolecular H-bonds (light blue dotted line) of the 1:1 host–guest complexes formed by PX with β-CD (**A**) in the most stable and (**C**) in the less stable optimized geometries found after the MM, MD calculations. The β-CD is green, the PX molecule is colored by atoms (color codes: C atoms are grey, O atoms are red, N atoms are blue, S atoms are yellow, and in the side view, H atoms are omitted for clarity). Below, the conformation of the optimized isolated PX is reported with the diagnostic hydrogen highlighted.

Table 1. Potential energy and interaction energy in kJ/mol calculated in the optimized geometries assumed at the end of the MD runs for the most and the less stable final geometries. Intermolecular β-CD/PX H-bonds and intramolecular β-CD H-bonds are reported.

Geometry	E_{pot} (kJ/mol)	E_{int} (kJ/mol)	Intermolecular H-Bonds	Intramolecular β-CD H-Bonds
A) Most stable	766.5	−173.2	1 bidentate (with II rim)	16
B) Metastable	769.4	−170.3	1 (with I rim)	17
C) Less stable	776.6	−163.1	1 (with II rim)	8
D) Metastable	771.5	−168.1	1 bidentate (with II rim)	12

As reported by Fronza et al. [17] using NMR experiments, two different inclusion complexes having different stability were found. The most stable geometry reported in Figure 3 (geometry A) displays the pyridyl aromatic part of PX included in the hydrophobic cavity of β-CD near the primary rim interacting with its hydrogens, while the second geometry of Figure 3 (geometry C) is less stable, and displays the pyridyl moiety included in the hydrophobic β-CD cavity facing the secondary rim interacting with the latter hydrogens. The most stable geometry has an interaction energy equal to −173.2 kJ/mol, while the other one is less stable by 10 kJ/mol; similar favorable interactions stabilize these host–guest complexes thanks to the hydrophobic interaction in the β-CD cavity and to the H-bonds at the primary and especially at the secondary rim. It is important to underline that these theoretical results are in good agreement with the NMR data reported by Fronza et al. [16] who proposed two different inclusion compounds. It is interesting to note that these two inclusion complexes are also found when β-CD are crosslinked in the PMA NS model as it will be reported later in Section 2.2. Guo et al. [18] proposed a partial inclusion due to the absence of a Nuclear Overhauser Effect NOE of H-2, H-3 guest hydrogens (see Figure 3, H-2 and H-3 are the hydrogens of the PX phenyl ring) and diagnostic hydrogen β-CD, suggesting that the dimension of β-CD is too small to host the whole PX. In fact, in the most stable geometry with a 1:1 stoichiometry reported in Figure 3 (geometry A), the hydrogens of the phenyl ring are far from the β-CD cavity; only in the less stable and less populated inclusion complex, the phenyl ring is in the midst of the hydrophobic cavity, relatively closer to H-5′ protons, while the pyridyl moiety is exposed to the solvent far from the secondary rim (geometry C in Figure 3). Guo et al. reported the 2D-ROESY (Rotating-Frame Overhauser Effect Spectroscopy) spectrum. In this spectrum, strong sizable contacts between the guest H-6, H-7, and H-8 protons (the hydrogens of the pyridyl moiety) and the host H-3′ and H-5′ are shown, probably due to deeper insertion, indicating that both phenyl and pyridyl rings are in the β-CD cavity. For this reason, Guo also suggested the presence of a β-CD and PX complex in a 2:1 stoichiometry with CDs facing their secondary rims. It will be important to carry out new reliable NMR experiments for the determination of the structure of cyclodextrin inclusion complexes in a solution [31], in particular, off-resonance ROESY or T-ROESY (Transverse Rotating-frame Overhauser Enhancement Spectroscopy) experiments must be carried out in order to determine the structure of complexes in a solution. In this work, increasing the CD concentration, a theoretical study of host–guest complexes in a 2:1 stoichiometry was carried out as reported in the next Section.

2.1.2. β-CD/PX Interaction: Complex Formation in a 2:1 Host–Guest Stoichiometry

The β-CD/PX in a 2:1 host–guest geometry was investigated considering both the most and the less stable inclusion complexes in a 1:1 stoichiometry reported in Figure 3: four different interaction geometries between these complexes and another β-CD facing both the secondary and the primary rim were considered (Figure S2). After the MD runs and final energy minimizations, the optimized geometries for the most stable and the less stable arrangements are shown in Figure 4. In Figure S3, the two metastable geometries are reported.

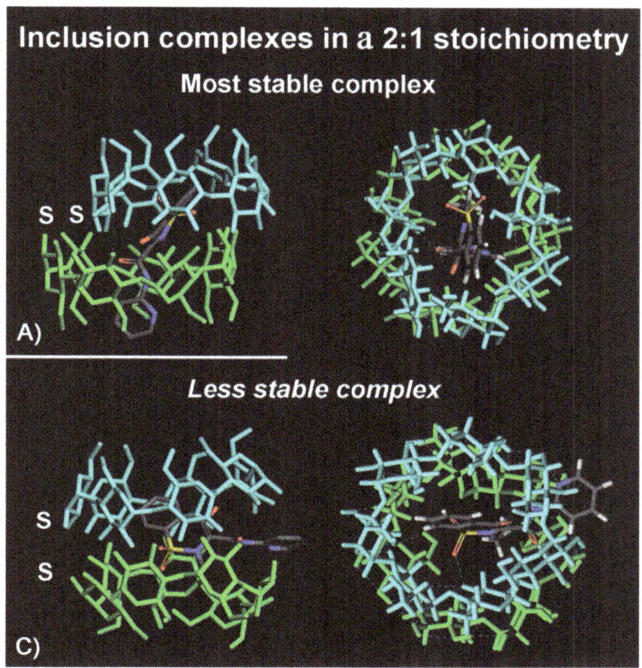

Figure 4. The side view (left) and the top view with H-bonds (light blue dotted lines) (right) of the 2:1 host–guest complexes formed by PX with two β-CDs, (**A**) in the most stable and (**C**) in the less stable optimized geometries found after the MD runs. The β-CD of the first 1:1 complex is in green, the second one facing the secondary rim is in light blue. The piroxicam molecule is colored by atoms. Color codes: C atoms are grey, O atoms are red, N atoms are blue, S atoms are yellow. In the side view, the H atoms are omitted for clarity.

In both cases, the dimers formed facing the secondary rims of two CDs display a more favorable interaction energy than the 1:1 inclusion complex (see Table 2), indicating a stabilization more than twice as large. Only few intermolecular β-CD/drug H-bonds were found in these dimers, one in the most stable geometry and two in the less stable one involving not only PX oxygen atoms, but also the nitrogen of the pyridyl moiety. As proposed by Guo et al. from the 2D-ROESY spectrum, PX can effectively interact with two CDs facing their secondary rims in the most stable complexes in a 2:1 stoichiometry. The concentration of β-CD in a solution as indicated by Fronza et al. and by Guo et al. influences the type of the inclusion complex, so that it becomes important to study the variable concentration of host molecules [6–8]. It will be necessary to confirm the presence of the 2:1 complex or others complexes by Isothermal Titration Calorimetry (ITC) experiments [32].

Table 2. Interaction energy of the optimized geometries assumed at the end of the MD runs for the most stable and the less stable final geometries in a 2:1 stoichiometry. Intermolecular β-CD/PX and β-CD H-bonds and intramolecular β-CD H-bonds are also reported.

Geometry	E_{pot} (kJ/mol)	E_{int} (kJ/mol)	Intermolecular β-CD/PX H-bonds	Intermolecular β-CD H-bonds	Intramol.first β-CD H-bonds	Intramol.second β-CD H-bonds
A) Most stable	233.4	−381.7	1 (with II rim)	5	9	8
B) Less stable	247.6	−322.8	2 (with II rim)	4	18	15
C) Metastable	241.4	−348.8	1 (with II rim)	3	13	15
D) Metastable	242.8	−342.5	2 (1 with I and II rims)	5	16	18

2.2. PMA β-CD NS Model/PX Interaction at Different Concentrations

In this Section, the PX interaction with a nanosponge (NS) model containing β-CDs connected by a pyromellitic dianhydride PMA crosslinker [30] is investigated. The NS model was generated by linking 8 β-CDs (Model 2 in [30]) through PMA moieties (see Scheme 1 in [30]). Accordingly, each CD carries two PMA linking agents bound to a primary hydroxyl at diametrically opposite sides of the macrocycle. An increasing number of host molecules was considered, with β-CD:PX equal to 8:4, 8:8, 8:16, 8:40, therefore, with β-CD:PX in a 2:1, 1:1, 1:2, 1:5 ratio.

Since it will be relevant in the following, it is important to report that the solvent-accessible surface area calculated for the β-CD NS model reported in Figure 5 is equal to 4,315 Å^2 and the radius of gyration is equal to 12.5 Å. In the following, the interaction between the PMA β-CD NS model [30] and PX molecules considered at smaller and larger concentrations were studied and the results are reported and discussed.

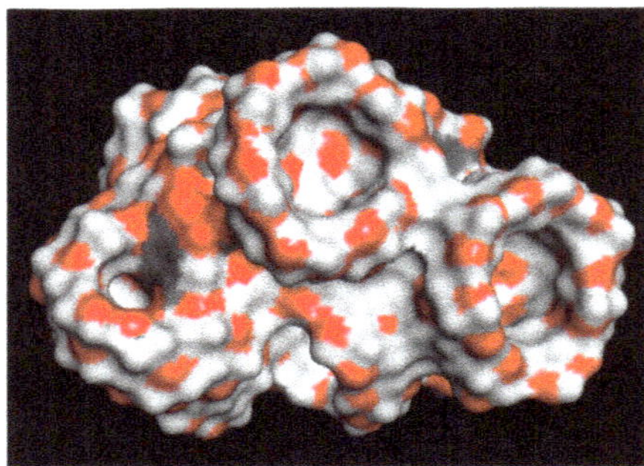

Figure 5. The solvent-accessible surface area of a β-CD NS (β-cyclodextrin Nanosponge) model with a pyromellitic dianhydride (PMA) crosslinker as reported in Model 2 in [29] colored by atoms. Color codes: C atoms are grey, oxygen atoms re red, and hydrogen atoms are white.

2.2.1. NS/PX Interaction: 8 β-CDs in the NS Model and 4 PX Molecules (β-CD/Drug in a 2:1 Stoichiometry)

Using the simulation protocol proposed in the previous work [20–22], as the first step, energy minimization of the NS model and 4 PX molecules, considering the β-CD/drug in a 2:1 stoichiometry, was carried out starting from a random disposition. An MD simulation run was then performed in order to understand possible conformational changes and interactions, in particular, the possible inclusion complexes or the surface interaction with the exposed PMA β-CD NS atoms. In Figure 6, the potential energy and the van der Waals contribution calculated during the MD run are reported. After an initial energy decrease, some conformational changes of the nanosponge take place in order to maximize its interactions with the four PX molecules, increasing the surface area accessible to the drug molecules. It should be noted that hydrophobic drug/drug interactions are absent. The final optimized conformation assumed by the system is reported in Figure 7. See the animation file of the MD run in SI.

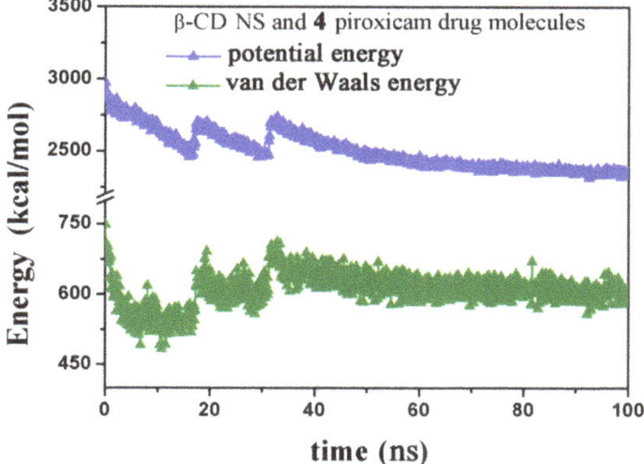

Figure 6. The potential energy and the van der Waals contribution calculated during the MD run for the NS model interacting with 4 piroxicam molecules.

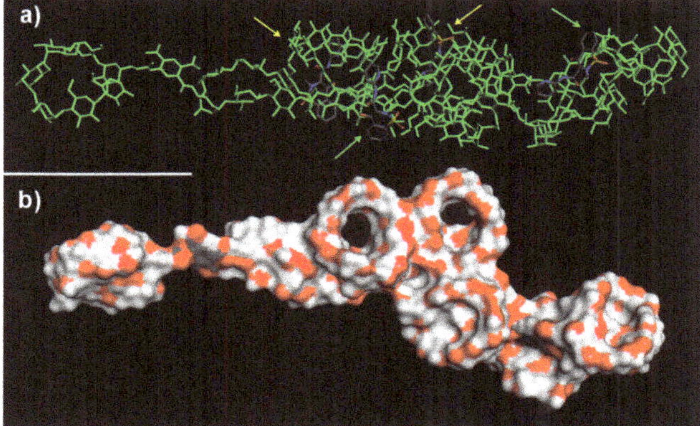

Figure 7. Panel (**a**) shows the optimized geometry assumed by the NS/4 PX molecules system at the end of the MD run displaying two included molecules in a 1:1 stoichiometry (see the yellow arrows) and two PX molecules interacting on the external surface of the NS model. The NS is in green for clarity and the PX molecules are colored by atoms, see the color codes in Figure 3. Panel (**b**) shows the solvent-accessible surface area without the PX molecules for clarity (see the color codes in Figure 5).

It is interesting to note that upon removal of the PX molecules as shown in panel b) of Figure 7, the solvent-accessible surface area of the NS in the optimized final geometry is equal to 5,263 Å^2, with an increase of about 20% compared to the starting geometry, while the radius of gyration increases by more than 90% to a value of 23.3 Å. It is well-known that β-CD nanosponges are able to swell in a water solution [14,30]. They are very flexible systems that can have significant conformational changes in order to maximize the favorable interactions with drugs. Significant flexibility was displayed by this β-CD crosslinked system during the MD run. The final optimized geometry reported in panel a) of Figure 7 displays an evident elongation of the NS structure interacting with PX molecules forming two different host–guest complexes (see the yellow arrows) in a 1:1 stoichiometry studied and reported in Section 2.1; one complex displays the hydrogens of the pyridyl part preferentially in the β-CD cavity and the phenyl part far from the hydrophobic cavity and the second one displays the pyridyl part far from the β-CD cavity. Interestingly, these geometries are very similar to the two inclusion complexes considering the native β-CD discussed in Section 2.1.1 and reported in Figure 3 (see geometries A and

C in Figure 3). Again, it is important to note that the other two piroxicam molecules interact with the external part of the NS, in particular, with the exposed crosslinker (see the green arrows in Figure 7).

2.2.2. NS/PX Interaction: 8 β-CDs in the NS Model and 8 PX Molecules (β-CD/Drug in a 1:1 Stoichiometry)

In this Section, the interaction between PMA β-CD NS with PX was studied considering 8 drug molecules. The NS contains 8 β-CD [30], so that the β-CD/drug in a 1:1 stoichiometry is investigated. Using the same strategy as before, after the MD run in the final optimized geometry reported in Figure 8, PX was found to interact both with the NS in the accessible hydrophobic cavities and with the NS surface. The NS shows large conformational changes in order to adsorb the PX molecules well. Interestingly, at this concentration, possible drug/drug hydrophobic interactions take place. In the final optimized geometry, again, two types of inclusion complexes are found together with the surface interaction on the exposed atoms of this flexible crosslinked system. As already found for the smaller concentration, the nanosponge maximizes the possible interaction with a drug.

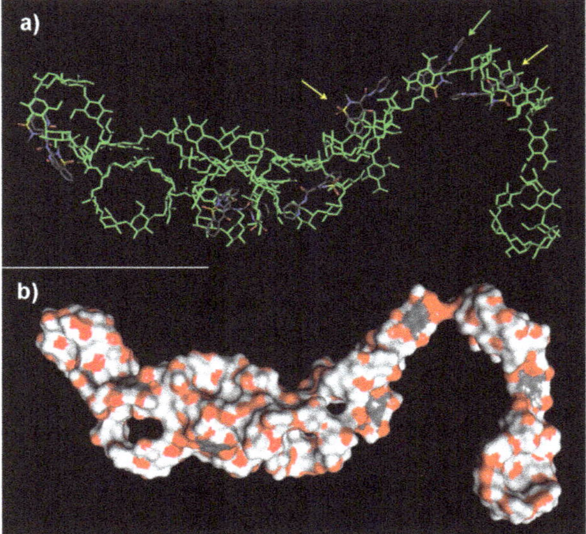

Figure 8. In the Figure, panel (**a**) shows the optimized geometry assumed by the NS/8 PX drug molecules system at the end of the MD run displaying two types of interactions with hydrophobic cavities (see the yellow arrows) and six PX molecules interacting on the external surface of the NS. The NS is in green for clarity and the PX molecules are colored by atoms according to the color codes in Figure 3. Panel (**b**) shows the solvent-accessible surface area without the PX molecules for clarity (see the color codes in Figure 5).

The NS becomes well-elongated due to the favorable interactions with the PX drug forming two inclusion complexes (see the yellow arrows) and interacting with the crosslinkers (an example is highlighted by the green arrow). At this larger concentration, two PX molecules interact owing to the hydrophobic drug/drug interactions (see Figure 8, top left). Interestingly, the solvent-accessible surface area of the NS (see panel b) in Figure 8) is equal to 6,112 Å^2 with the increase of 40% with respect to the initial geometry, and the radius of gyration increases by more than 90% to a value of 25.5 Å.

2.2.3. NS/PX Interaction: 8 β-CDs in the NS Model and 16 PX Molecules (β-CD/Drug in a 1:2 Stoichiometry)

In this Section, the results concerning the β-CD NS/drug in a 1:2 stoichiometry will be presented and discussed. After the initial energy minimization of the NS and 16 PX molecules in a random disposition, an MD simulation run was performed as before in order to understand the possible inclusion complexes or surface interactions between piroxicam and the NS. Figure 9 reports the potential energy and the van der Waals components calculated during the MD run.

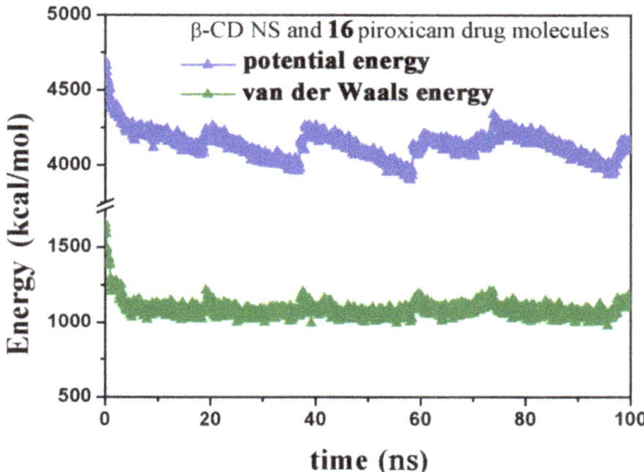

Figure 9. The Figure shows the potential energy and the van der Waals components calculated during the MD run for the NS interacting with 16 PX molecules.

After the initial energy decrease, some conformational changes of the NS take place in order to maximize the interactions with the 16 PX molecules. The final optimized conformation assumed by the system at the end of the MD run is reported in Figure 10. During the MD run at this larger PX concentration, an equilibrium between a more compact NS conformation with a PX aggregate on its surface (see panel a) in Figure 10) and a more elongated NS conformation maximizing the interactions with the PX molecules (see panel b) in Figure 10) are observed.

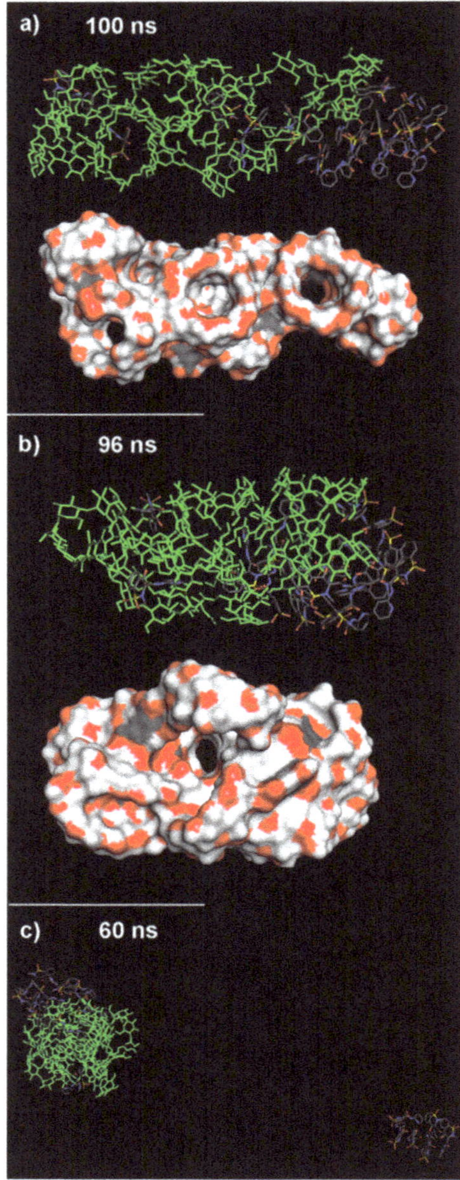

Figure 10. Panel (**a**) shows the optimized geometry assumed by the NS/16 PX drug molecules system at the end of the MD run lasting 100 ns and panel (**b**) displays the optimized geometry assumed by the system at 96 ns. In the top part of each panel the NS is in green for clarity and the PX molecules are colored according to the color codes of Figure 3, and in the bottom part, the solvent-accessible surface area without the PX molecules is reported for clarity (see the color codes in Figure 5). Panel (**c**) shows the geometry assumed by the system at 60 ns.

The solvent-accessible surface areas of the optimized system calculated at 100 ns and 96 ns are reported in Figure 10. In particular, this area is equal to 5,664 Å^2 with a 30% increase over the initial geometry for the conformation assumed at 100 ns, while the radius of gyration is equal to 17.0 Å, with the increase of 35%. The more compact conformation assumed by the system at 96 ns corresponds to a lower potential energy value, 2424 kcal/mol, compared to 2534 kcal/mol for the optimized geometry assumed by the system at the end of the MD run. The solvent-accessible surface area reported in Figure 10 in panel b) (bottom) without the PX molecules is equal to 5161 Å^2, with the increase of 20%, while the radius of gyration is equal to 13.0 Å, with a 13% increase. It is interesting to note in Figure 10 the formation of a nanopore in the central part of the β-CD NS in the optimized conformation at 96 ns. It is more evident that during the MD run, the system maximizes the interaction with PX owing to high NS flexibility. In both cases, there are some inclusion complexes and a PX drug aggregate interacting on the external NS surface at 100 ns and 96 ns formed by 10 and 11 PX molecules, respectively, therefore, with about the same number of molecules that is stable in time.

Interestingly, during the MD run at 60 ns, a spherical aggregate formed by 6 molecules of the PX drug detaches from the NS surface as shown in Figure 10 (panel c)). For the animation file of the MD run, see SI. The potential energy of the system calculated during the simulation is the lowest one, indicating also an equilibrium during the MD run between the drug molecules still attached on the NS surface for the larger part of the MD run and the detached ones diffusing as a single cluster for some ns before the new adsorption on the NS surface. The solvent-accessible surface area of the system at 60 ns is equal to 5,026 Å^2, with a 16% increase over the initial geometry, while the radius of gyration is equal to 13.0 Å, with the increase of only 4%.

2.2.4. NS/PX Interaction: 8 β-CDs in the NS and 40 PX Molecules (β-CD/Drug in a 1:5 Stoichiometry)

Finally, the results concerning the β-CD/drug in a 1:5 stoichiometry of these NS/PX systems are presented and discussed. Using the same methodology as before, the potential energy and the van der Waals components calculated during the MD run are reported in Figure 11. In this case, at the largest piroxicam concentration considered here, after the initial fast decrease of the system energy, fluctuations around average values were observed with the NS in a compact structure in contact with the PX aggregate due to drug association owing to hydrophobic π–π interactions induced by the first layer of PX molecules adsorbed on the NS surface. For the animation file of the MD run, see SI.

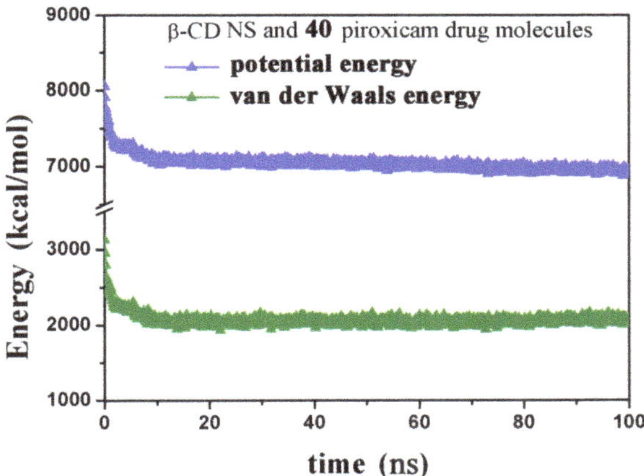

Figure 11. The figure shows the potential energy calculated during the MD run and the van der Waals components for the NS model and 40 PX molecules.

The optimized conformation assumed by the system at the end of the MD run is reported in Figure 12 (panel a)).

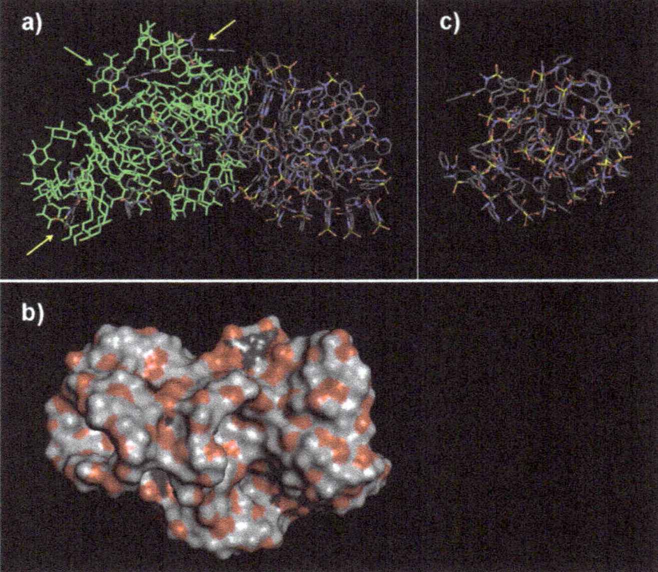

Figure 12. Panel (**a**) shows the optimized geometry assumed by the NS/40 PX drug molecules system at the end of the MD run displaying some host–guest inclusion complexes (see the arrows). All the other molecules are on the external surface of the NS. Panel (**b**) shows the solvent-accessible surface area without the PX molecules for clarity (see the color codes in Figure 5). Panel (**c**) shows the PX aggregate formed by 32 molecules adsorbed on the NS surface well-ordered in an almost spherical shape in detail.

In the optimized conformation assumed at the end of the MD run (see panel a) in Figure 12), some inclusion geometries are found, in particular, two host–guest complexes in a "type A" geometry (see the yellow arrows), the most stable one being reported in the left part of Figure 3 having the pyridyl moiety in the hydrophobic β-CD cavity, and one host–guest complex in a "type C" geometry (see the green arrow, the less stable geometry) in the right part of Figure 3 with an included phenyl. All the other drug molecules interact with the external surface of the NS. In particular, it is interesting to note the almost spherical aggregate that is nucleated on the external surface after the initial favorable interaction of the first layer of adsorbed PX molecules. β-CD acts as a surface of the nucleation process which subsequently takes place thanks to π–π interactions between the drug molecules. Considering this final optimized geometry, a self-ordered structure is clearly seen in the right part of panel a) of Figure 12, with a parallel arrangement of the aromatic rings of the drug molecules producing an almost spherical aggregate formed by 32 PX molecules represented in another orientation in panel c). The conformation of the β-CD NS having the PX aggregate adsorbed is compact and, in fact, its radius of gyration is equal to 13.5 Å, larger only by 7% with respect to the isolated NS, and the solvent-accessible surface area is equal to 4,920 Å^2 (see panel b) in Figure 12), larger by 14% than in the isolated NS. Interestingly, in this case, the calculated increase was smaller than in the simulations at relatively smaller PX concentrations.

3. Materials and Methods

The interaction of PX with the β-CD NS model was studied using molecular mechanics (MM) and molecular dynamics (MD) methods at the constant temperature of 300 K. The MM and MD simulations are performed using the consistent valence force field (CVFF) [33], and the Materials Studio and Insight II/Discover packages [34] adopting the same simulation protocol as proposed in yjr previous work [20–23]: after the initial energy minimization, MD runs were carried out until the equilibrium state was achieved followed by final geometry optimizations of the final configuration and of some conformations assumed during the MD run. The PX structure (see Scheme 1) was generated using the Insight II/Discover Module Builder and the final most stable geometry (Figure 3) was obtained after an MD run lasting for 1 ns, with the final optimization of different conformations assumed by the system during the MD run. At first, the possible β-CD/PX inclusion compounds in a 1:1 and in a 2:1 stoichiometry were studied. Then, the interaction between the β-CD NS model and PX was considered. The structure of the β-CD NS model reported by Raffaini [30] was generated by linking 8 β-CD (Model 2 in [30]) with pyromellitic dianhydride (PMA): each β-CD carries two PMA linking agents bound to a primary hydroxyl at diametrically opposite sides of the macrocycle, linking also two topologically distant β-CDs. The drug molecules were initially placed in a random distribution cell far from to the NS surface in order to study the interactions without assuming any a priori inclusion complexes. The interaction was studied considering 4, 8, 16, and 40 PX molecules randomly distributed around the NS model in a 220 Å cubic cell with a varying CDs/PX ratio. All simulations were performed in implicit water using a distance-dependent dielectric constant with periodic boundary conditions. All energy minimizations were carried out up to an energy gradient $< 4 \times 10^{-3}$ kJ mol^{-1} Å^{-1}. The MD simulations were performed in an NVT ensemble (canonical ensemble, see Abbreviations) at a constant temperature (300 K) controlled using a Berendsen thermostat. The integration of dynamical equations was carried out with the Verlet algorithm using a time step of 1 fs and the instantaneous coordinates were periodically saved for further analysis. The MD runs in implicit water lasted for 10 ns in Section 2.1 and 100 ns for the NS systems in Section 2.2. During the MD runs, the time evolution of the potential energy and of the van der Waals components was calculated in order to monitor the significant conformational changes.

4. Conclusions

Drug concentration plays an important role in the interaction with drug carriers affecting the kinetics of the release process and the toxicology effects. Hydrophobic drugs can be solubilized in water through CD in order to enhance their bioavailability. Solubility of PX, an important nonsteroidal anti-inflammatory drug, is enhanced by β-CDs. In the present work, MD simulation of their host–guest complexes in a 1:1 and a 2:1 stoichiometry are reported, with results in good agreement with the NMR experimental literature data. Different host–guest inclusion geometries were obtained considering both native β-CD and crosslinked β-CD nanosponges (NS). When PX interacts with an NS, at smaller concentration, the drug molecules form inclusion complexes and feature some interactions on the external NS surface, dynamically forming some H-bonds or hydrophobic interaction with the PMA crosslinker, while at larger concentrations, they also aggregate on the NS surface, forming spherical droplets. Therefore, the β-CD NS enhances the solubility of hydrophobic PX. At small drug concentration, during the MD runs, the NS shows equilibrium between an extended structure, when the NS maximizes the interactions with drug molecules exposing a larger surface area, and a more compact structure, also forming a temporary nanopore when drug/drug self-aggregation due to π–π interactions takes place on its external surface. The noncovalent interactions are favored by the great NS flexibility, enhancing solubilization and PX transport. The drug concentration plays an important role. At larger concentrations, a spherical PX nanoaggregate is observed on the external NS surface besides drug inclusion complexes, because the first layer of adsorbed PX molecules induces aggregation of other hydrophobic molecules, the NS acting as a nucleation surface. During an MD run, an equilibrium is also observed between the drugs adhered on the NS surface and the detached ones in a small aggregate that diffuses before subsequent adsorption on the NS surface. Finally, it is important to highlight that the β-CD/drug and the drug/drug interactions and association will likely also affect the kinetics of the release process, depending on the site of interaction and on the flexibility of the crosslinked β-CDs. It will be important to theoretically investigate this system at a specific concentration to model the release process in order to better understand the specific interactions and the geometry and stability of the aggregates, which may affect the kinetics of the release process. Theoretical and experimental studies of the β-CD NS/PX interaction in a specific stoichiometry and drug release are in progress and will be published in a separate paper.

Supplementary Materials: Supplementary data can be found in the online version.

Author Contributions: Conceptualization, G.R.; data curation, G.R.; formal analysis, G.R.; funding acquisition, G.R.; investigation, G.R.; methodology, G.R.; resources, G.R.; supervision, G.R. and F.G.; validation, G.R.; visualization, G.R.; writing—original draft preparation, G.R.; writing—review and editing, G.R. and F.G.; funding acquisition, G.R. All authors have read and agreed to the published version of the manuscript.

Funding: This research was funded by MIUR (Ministero dell'Istruzione, dell'Università e della Ricerca)—FIRB (Fondo per gli Investimenti della Ricerca di Base) Futuro in Ricerca 2008 (Surface-Associated Selective Transfection—SAST, grant RBFR08XH0H) and by INSTM (Consorzio Interuniversitario Nazionale per la Scienza e Tecnologia dei Materiali) (INSTMMIP07 project) (G.R.).

Acknowledgments: G. Raffaini gratefully acknowledges financial support from MIUR—FIRB 2008 (Surface-Associated Selective Transfection—SAST, grant RBFR08XH0H) and from INSTM (INSTMMIP07 project) (G.R.). Helpful discussions with Roberto Vittorio Pivato are gratefully acknowledged.

Conflicts of Interest: The authors declare no conflicts of interest.

Abbreviations

MM	molecular mechanics
MD	molecular dynamics
β-CD	β-cyclodextrin
PX	piroxicam
PMA	pyromellitic dianhydride
NS	nanosponge
β-CD NS	β-cyclodextrin nanosponge
NVT ensemble	Number of particles, Volume and Temperature are constant

References

1. Loftsson, T.; Brewster, M.E. Pharmaceutical applications of cyclodextrins: basic science and product development. *J. Pharm. Pharmacol.* **2010**, *62*, 1607–1621. [CrossRef] [PubMed]
2. Vyas, A.; Saraf, S.; Saraf, S. Cyclodextrin based novel drug delivery systems. *J. Incl. Phenom. Macrocycl. Chem.* **2008**, *62*, 23–42. [CrossRef]
3. Jambhekar, S.S.; Breen, P. Cyclodextrins in pharmaceutical formulations I: structure and physicochemical properties, formation of complexes, and types of complex. *Drug Discov. Today* **2016**, *21*, 356–362. [CrossRef] [PubMed]
4. Loftsson, T.; Stefansson, E. Cyclodextrins in ocular drug delivery: theoretical basis with dexamethasone as a sample drug. *J. Drug Deliv. Sci. Technol.* **2007**, *17*, 3–9. [CrossRef]
5. Loftsson, T.; Stefánsson, E. Cyclodextrins and topical drug delivery to the anterior and posterior segments of the eye. *Int. J. Pharm.* **2017**, *531*, 413–423. [CrossRef]
6. Szente, L.; Szejtli, J.; Kis, G.L. Spontaneous Opalescence of Aqueous γ-Cyclodextrin Solutions: Complex Formation or Self-Aggregation? *J. Pharm. Sci.* **1998**, *87*, 778–781. [CrossRef]
7. Mele, A.; Mendichi, R.; Selva, A.; Molnar, P.; Toth, G. Non-covalent associations of cyclomaltooligosaccharides (cyclodextrins) with carotenoids in water. A study on the α- and β-cyclodextrin/ψ,ψ-carotene (lycopene) systems by light scattering, ionspray ionization and tandem mass spectrometry. *Carbohydr. Res* **2002**, *337*, 1129–1136. [CrossRef]
8. Loftsson, T.; Saokham, P.; Couto, A.R.S. Self-association of cyclodextrins and cyclodextrin complexes in aqueous solutions. *Int. J. Pharm.* **2019**, *560*, 228–234. [CrossRef]
9. Bondi', M.L.; Scala, A.; Sortino, G.; Amore, E.; Botto, C.; Azzolina, A.; Balasus, D.; Cervello, M.; Mazzaglia, A. Nanoassemblies Based on Supramolecular Complexes of Nonionic Amphiphilic Cyclodextrin and Sorafenib as Effective Weapons to Kill Human HCC Cells. *Biomacromolecules* **2015**, *16*, 3784–3791. [CrossRef]

10. Giglio, V.; Viale, M.; Bertone, V.; Maric, I.; Vaccarone, R.; Vecchio, G. Cyclodextrin polymers as nanocarriers for sorafenib. *Investig. New Drugs* **2017**, *36*, 370–379. [CrossRef]

11. Swaminathan, S.; Vavia, P.; Trotta, F.; Torne, S. Formulation of betacyclodextrin based nanosponges of itraconazole. *J. Incl. Phenom. Macrocycl. Chem.* **2007**, *57*, 89–94. [CrossRef]

12. Swaminathan, S.; Cavalli, R.; Trotta, F.; Ferruti, P.; Ranucci, E.; Gerges, I.; Manfredi, A.; Marinotto, D.; Vavia, P.R. In vitro release modulation and conformational stabilization of a model protein using swellable polyamidoamine nanosponges of β-cyclodextrin. *J. Incl. Phenom. Macrocycl. Chem.* **2010**, *68*, 183–191. [CrossRef]

13. Mele, A.; Castiglione, F.; Malpezzi, L.; Ganazzoli, F.; Raffaini, G.; Trotta, F.; Rossi, B.; Fontana, A.; Giunchi, G. HR MAS NMR, powder XRD and Raman spectroscopy study of inclusion phenomena in βCD nanosponges. *J. Incl. Phenom. Macrocycl. Chem.* **2010**, *69*, 403–409. [CrossRef]

14. Trotta, F.; Mele, A. *Nanosponges Synthesis and Applications*; Wiley-VCH: Weinheim, Germany, 2019; ISBN 978-3-527-34099-6.

15. Khuntawee, W.; Wolschann, P.; Rungrotmongkol, T.; Wong-Ekkabut, J.; Hannongbua, S. Molecular Dynamics Simulations of the Interaction of Beta Cyclodextrin with a Lipid Bilayer. *J. Chem. Inf. Model.* **2015**, *55*, 1894–1902. [CrossRef]

16. Putaux, J.L.; Lancelon-Pi, C.; Legrand, F.-X.; Pastrello, M.; Choisnard, L.; Geze, A.; Rochas, C.; Wouessidjewe, D. Self-Assembly of Amphiphilic Biotranseterified beta-Cyclodextrins: Supramolecular Structure of Nanoparticles and Surface Properties. *Langmuir* **2017**, *33*, 7917–7928. [CrossRef]

17. Fronza, G.; Mele, A.; Redenti, E.; Ventura, P. Proton Nuclear Magnetic Resonance Spectroscopy Studies of the Inclusion Complex of Piroxicam with β-Cyclodextrin. *J. Pharm. Sci.* **1992**, *81*, 1162–1165. [CrossRef]

18. XiLiang, G.; Yu, Y.; Guoyan, Z.; Guomei, Z.; Jianbin, C.; ShaoMin, S. Study on inclusion interaction of piroxicam with beta-cyclodextrin derivatives. *Spectrochim. Acta Part. A: Mol. Biomol. Spectrosc.* **2003**, *59*, 14. [CrossRef]

19. Dharmasthala, S.; Shabaraya, A.R.; Andrade, G.S.; Shriram, R.G.; Hebbar, S.; Dubey, A. Fast Dissolving Oral Film of Piroxicam: Solubility Enhancement by forming an Inclusion Complex with β-cyclodextrin, Formulation and Evaluation. *J. Young- Pharm.* **2018**, *11*, 1–6. [CrossRef]

20. Raffaini, G.; Ganazzoli, F.; Malpezzi, L.; Fuganti, C.; Fronza, G.; Panzeri, W.; Mele, A. Validating a Strategy for Molecular Dynamics Simulations of Cyclodextrin Inclusion Complexes through Single-Crystal X-ray and NMR Experimental Data: A Case Study. *J. Phys. Chem. B* **2009**, *113*, 9110–9122. [CrossRef]

21. Raffaini, G.; Ganazzoli, F. Molecular dynamics study of host–guest interactions in cyclodextrins: methodology and data analysis for a comparison with solution data and the solid-state structure. *J. Incl. Phenom. Macrocycl. Chem.* **2007**, *57*, 683–688. [CrossRef]

22. Raffaini, G.; Ganazzoli, F. Hydration and flexibility of α-, β-, γ- and δ-cyclodextrin: a molecular dynamics study. *Chem. Phys.* **2003**, *333*, 625–635. [CrossRef]

23. Raffaini, G.; Ganazzoli, F. Surface hydration of polymeric (bio)materials: A molecular dynamics simulation study. *J. Biomed. Mater. Res. Part. A* **2009**, *9999*, 1382–1391. [CrossRef] [PubMed]

24. Raffaini, G.; Ganazzoli, F. Protein adsorption on biomaterial and nanomaterial surfaces: a molecular modeling approach to study non-covalent interactions. *J. Appl. Biomater. Biomech.* **2011**, *8*, 135–145. [CrossRef]

25. Raffaini, G.; Mazzaglia, A.; Ganazzoli, F. Aggregation behaviour of amphiphilic cyclodextrins: the nucleation stage by atomistic molecular dynamics simulations. *Beilstein J. Org. Chem.* **2015**, *11*, 2459–2473. [CrossRef]

26. Raffaini, G.; Ganazzoli, F.; Mazzaglia, A. Aggregation behavior of amphiphilic cyclodextrins in a nonpolar solvent: evidence of large-scale structures by atomistic molecular dynamics simulations and solution studies. *Beilstein J. Org. Chem.* **2016**, *12*, 73–80. [CrossRef]

27. Raffaini, G.; Ganazzoli, F. A Molecular Dynamics Study of the Inclusion Complexes of C60 with Some Cyclodextrins. *J. Phys. Chem. B* **2010**, *114*, 7133–7139. [CrossRef]

28. Raffaini, G.; Ganazzoli, F. A Molecular Dynamics Study of a Photodynamic Sensitizer for Cancer Cells: Inclusion Complexes of γ-Cyclodextrins with C70. *Int. J. Mol. Sci.* **2019**, *20*, 4831. [CrossRef]

29. Mele, A.; Ganazzoli, F.; Raffaini, G.; Juza, M.; Schurig, V. Macrocycle conformation and self-inclusion phenomena in octakis(3-O-butanoyl-2,6-di-O-n-pentyl)-γ-cyclodextrin (Lipodex E) by NMR spectroscopy and molecular dynamics. *Carbohydr. Res.* **2003**, *338*, 625–635. [CrossRef]

30. Raffaini, G.; Ganazzoli, F.; Mele, A.; Castiglione, F. A molecular dynamics study of cyclodextrin nanosponge models. *J. Incl. Phenom. Macrocycl. Chem.* **2012**, *75*, 263–268. [CrossRef]

31. Pean, C.; Djedaïni-Pilard, F.; Perly, B. Reliable NMR Experiments for the Determination of the Structure of Cyclodextrin Inclusion Complexes in Solution. In *Proceedings of the Ninth International Symposium on Cyclodextrins*; Labandeira, J.J.T., Vila-Jato, J.L., Eds.; Springer: Dordrecht, The Netherlands, 1999; pp. 659–662.
32. Landy, D. Measuring Binding Constants of Cyclodextrin Inclusion Compounds. In *Advanced Nanostructured Materials for Environmental Remediation*; Fourmentin, S., Crini, G., Lichtfouse, E., Eds.; Cyclodextrin Fundamentals, Reactivity and Analysis; Springer International Publishing: Cham, Switzerland, 2018; pp. 223–255, (Environmental Chemistry for a Sustainable World). Available online: https://doi.org/10.1007/978-3-319-76159-6_5.
33. Hwang, M.-J.; Ni, X.; Waldman, M.; Ewig, C.S.; Hagler, A.T. Derivation of class II force fields. VI. Carbohydrate compounds and anomeric effects. *Biopolym.* **1998**, *45*, 435–468. [CrossRef]
34. Materials Studio BIOVIA. *Accelrys Inc. InsightII 2000*; Accelrys Inc.: San Diego, CA, USA, 2000.

Sample Availability: Samples of the compounds are not available from the authors.

 *molecules*

Article

Biological Effect Evaluation of Glutathione-Responsive Cyclodextrin-Based Nanosponges: 2D and 3D Studies

Monica Argenziano [1,†], Federica Foglietta [1,†], Roberto Canaparo [1], Rita Spagnolo [1], Carlo Della Pepa [1], Fabrizio Caldera [2], Francesco Trotta [2], Loredana Serpe [1,‡] and Roberta Cavalli [1,*,‡]

[1] Department of Drug Science and Technology, University of Torino, Via Pietro Giuria 9, 10125 Torino, Italy; monica.argenziano@unito.it (M.A.); federica.foglietta@unito.it (F.F.); roberto.canaparo@unito.it (R.C.); rita.spagnolo@unito.it (R.S.); carlo.dellapepa@unito.it (C.D.P.); loredana.serpe@unito.it (L.S.)

[2] Department of Chemistry, University of Torino, Via Pietro Giuria 7, 10125 Torino, Italy; fabrizio.caldera@unito.it (F.C.); francesco.trotta@unito.it (F.T.)

* Correspondence: roberta.cavalli@unito.it; Tel.: +39-011-670-7190; Fax: +39-011-670-7162

† These authors shared co-first authorship.

‡ These authors shared co-last authorship.

Academic Editor: Marina Isidori

Received: 29 May 2020; Accepted: 13 June 2020; Published: 16 June 2020

Abstract: This study aims to evaluate the bioeffects of glutathione-responsive β-cyclodextrin-based nanosponges (GSH-NSs) on two- (2D) and three-dimensional (3D) cell cultures. The bioeffects of two types of GSH-NS formulations, with low (GSH-NS B) and high (GSH-NS D) disulfide-bond content, were evaluated on 2D colorectal (HCT116 and HT-29) and prostatic (DU-145 and PC3) cancer cell cultures. In particular, the cellular uptake of GSH-NS was evaluated, as their effects on cell growth, mitochondrial activity, membrane integrity, cell cycle distribution, mRNA expression, and reactive oxygen species production. The effect of GSH-NSs on cell growth was also evaluated on multicellular spheroids (MCS) and a comparison of the GSH-NS cell growth inhibitory activity, in terms of inhibition concentration $(IC)_{50}$ values, was performed between 2D and 3D cell cultures. A significant decrease in 2D cell growth was observed at high GSH-NS concentrations, with the formulation with a low disulfide-bond content, GSH-NS B, being more cytotoxic than the formulation with a high disulfide-bond content, GSH-NS D. The cell growth decrease induced by GSH-NS was owing to G_1 cell cycle arrest. Moreover, a significant down-regulation of mRNA expression of the cyclin genes *CDK1*, *CDK2*, and *CDK4* and up-regulation of mRNA expression of the cyclin inhibitor genes *CDKN1A* and *CDKN2A* were observed. On the other hand, a significant decrease in MCS growth was also observed at high GSH-NS concentrations, but not influenced by the nanosponge disulfide-bond content, with the MCS IC_{50} values being significantly higher than those obtained on 2D cell cultures. GSH-NSs are suitable nanocarries as they provoke limited cellular effects, as cell cycle arrest only occurred at concentrations significantly higher than those used for drug delivery.

Keywords: nanosponges; β-cyclodextrin; glutathione; cancer; multicellular spheroids

1. Introduction

The ideal nanoparticle-based drug delivery system assures the safe delivery and selective action of a drug to a target site. Indeed, nanomaterials can add further functionality to the conjugated/loaded drug and, taking advantage of their unique size, are able to play a crucial therapeutic role. This has triggered an increased interest in nanopharmaceuticals [1,2] and the development of a wide range of nanoparticle systems, such as liposomes, nanoparticles, micelles, dendrimers, and nanotubes [3–5]. However, only a few nanoparticle-based systems have been FDA-approved for cancer therapy to date [6,7]. Although nanoscience in drug development is in its early stages, the fusion of engineered nanomaterials and nanopharmaceutical research is paving the way for the development of stimuli-responsive drug delivery systems, especially in cancer treatment. Interestingly, several chemically modified polymers [8,9] and cross-linked cyclodextrin-based polymers have been proposed to obtain compounds responsive to the external environment [10]. In this regard, cyclodextrin-based nanosponges are of particular interest [11,12].

Nanosponges are hyper-cross-linked cyclodextrin polymers generally obtained from α, β, and γ cyclodextrins, containing suitable amounts of linear dextrin cross-linked with a proper cross-linking agent. A cage-like structure is obtained via the cross-linking of cyclodextrins, thus creating nanochannels in the polymer matrix that can be modulated employing different types of cross-linking agents and/or varying the amount used [13,14]. It is worth noting that active carbonyl compounds, like carbonyl diimidazole, diphenyl carbonate, and organic dianhydrides such as pyromellitic dianhydride, can be used as cross-linker in the preparation of nanosponges [15,16].

Nanosponges offer several features, such as sustained and controlled release, improvement of aqueous solubility, bioavailability, and stability of the hosted molecules, which could be advantageously exploited for drug delivery [15,17,18]. Indeed, previous research highlighted the capability of nanosponges to encapsulate different active molecules and magnify their activity in either in vitro or in vivo studies [15,19,20]. In particular, several anticancer drugs such as doxorubicin, paclitaxel, and camptothecin have been efficiently incorporated in cyclodextrin-based nanosponges, showing an improved antitumor effect [21–24].

Interestingly, nanosponge-based drug delivery systems can be tuned to form 'stimuli-responsive' nanocarriers that modify their structure in response to external changes, such as pH or redox potential [25–27]. Therefore, glutathione (GSH)-responsive nanocarriers have been developed for targeted intracellular anticancer drug release [28], as the GSH tripeptide has a higher intracellular than extracellular concentration [29]. Several intracellular compartments, such as cytosol, mitochondria, and the cell nucleus, contain higher GSH concentrations than do extracellular fluids and circulation. Moreover, GSH intracellular concentration is further increased in cancer cells and, above all, in chemoresistant cells [30]. Oxidative stress has long been implicated in cancer development and progression [31]. An increase in reactive oxygen species (ROS) usually induces a cell adaptive response and the compensatory up-regulation of antioxidant systems to restore redox homeostasis; GSH/GSH disulfide is the major redox combination in mammalian cells [32]. Moreover, many primary tumors have high levels of overexpression of antioxidant enzymes [33,34].

Trotta et al. have developed a next generation of nanosponges that are bioresponsive to GSH external concentration [35]. This behavior may be an ideal trigger for rapid nanocarrier destabilization inside cells, leading to efficient intracellular drug release through disulfide-bond cleavage [35]. Indeed, the disulfide bridge remains stable in extracellular fluids for long periods before being reduced upon internalization in the cytosol, having a higher GSH concentration, thus improving drug bioavailability [36]. As depleting endogenous antioxidants, like GSH, make cancer cells more chemosensitive, this reduction-sensitive nanosystem is further suited to anticancer therapy owing to its ability to enhance the anticancer activity of such drugs. Previously, the encapsulation of doxorubicin and strigolactone analogues in GSH-responsive nanosponges was in vitro and in vivo evaluated. In vitro release kinetics studies from GSH-NS revealed a GSH concentration-based drug release profile over time. Moreover, the GSH-NSs were able to release the payload as a function of in-cell GSH concentration [37,38]. This behavior might favor the selectively controlled release in target cancer cells and enhance the cytotoxic effect. Indeed, both of the compounds loaded in GSH-NS were more effective in inhibiting the cell viability than the corresponding free drugs, particularly in cancer cells presenting a higher GSH content. In addition, a greater reduction of prostate cancer growth was observed for doxorubicin incorporated in GSH-NS compared with the free drug in xenograft mice models [37].

The fact that GSH-NS may well represent an efficient stimuli-responsive drug delivery system for anticancer drugs prompted in-depth study into their biological effect per se on cell growth reported herein. As preliminary cellular evaluations of nanocarriers are usually carried out on 2D cell cultures, previously, the effects of cyclodextrin-based nanosponges have been widely tested in 2D cell monolayer cultures. However, 3D cell cultures, such as multicellular spheroids (MCS), have various in vivo tissue characteristics including the production of an extracellular matrix [39,40]. This study reports a series of experiments carried out to evaluate the bioeffects of GSH-NS containing two different amounts of disulfide bridges either on 2D cell cultures or on 3D cell cultures of human cancer cells, differing in cancer type and intracellular GSH level, namely, human colorectal, HCT116 and HT-29, and human prostatic, DU145 and PC-3, cancer cell lines.

2. Results

2.1. Characterization of Glutathione Responsive β-cyclodextrin-Based Nanosponges

GSH-NSs have a size of about 200 nm and a negative surface charge, in agreement with our previous papers [35,37,38]. Table 1 reports the average diameters, polydispersity indices, and zeta potentials of blank and 6-coumarin loaded GSH-NS type B and D. The different content of disulfide bridges in the two type of GSH-NS (B and D) did not affect their physico-chemical characteristics. Fluorescent labeling did not significantly alter the values of these parameters. Transmission electron microscopy (TEM) analysis of GSH-NS shows the spherical shape and smooth surfaces of the nanosponges.

Table 1. The physico-chemical characteristics of glutathione-responsive β-cyclodextrin-based nanosponge (GSH-NS) formulations.

	Average Diameter ± SD (nm)	Polydispersity Index (PDI)	Zeta Potential ± SD (mV)
Blank GSH-NS B	183.4 ± 15.3	0.23 ± 0.02	−31.58 ± 3.82
Blank GSH-NS D	180.5 ± 6.7	0.21 ± 0.01	−31.24 ± 3.05
Fluorescent GSH-NS B	188.3 ± 10.2	0.22 ± 0.01	−29.98 ± 2.74
Fluorescent GSH-NS D	185.9 ± 12.5	0.22 ± 0.02	−30.55 ± 2.66

2.2. D Cell Culture Cytotoxicity of Glutathione Responsive β-cyclodextrin-Based Nanosponges

The basal level of reduced glutathione was measured in each of the different cell lines, which were grouped according to cancer type. The results show that HT-29 (Figure 1A) and PC-3 cells (Figure 1B) display significantly lower GSH levels than those in HCT116 (Figure 1A) and DU145 cells (Figure 1B). Dose-response curves were then performed by exposing human colorectal cancer, HCT116 and HT-29, and human prostatic carcinoma, DU145 and PC-3, cell monolayers to various concentrations (0.5, 1.0, 2.0, and 3.0 mg/mL) of the two types of GSH-NSs with increasing disulfide bridge content (GSH-NS B and D), for 24, 48, and 72 h. Table 2 reports the IC_1 and IC_{50} values of GSH-NS, which were determined at 24, 48, and 72 h of exposure.

Figure 2 reports GSH-NS IC_{50} values according to disulfide-bond content and exposure time. These data highlight the remarkable cytotoxicity difference between the two nanosponge formulations in colorectal cancer cell lines, as lower IC_{50} values were observed for the nanosponge formulation with the lower disulfide bridge content (GSH-NS B) at 24, 48, and 72 h (Figure 2A,B). In prostatic cancer cell lines, no significant differences in IC_{50} values were detected for the two nanosponge formulations in DU145 cells (Figure 2C), which were characterized by the highest GSH cell content (Figure 1). Meanwhile, in PC-3 cells, a significant cytotoxicity difference was observed between the two nanosponge formulations, as the IC_{50} values determined by the GSH-NS B formulation were lower compared with what was observed in colorectal cancer cells, even if only at 24 and 48 h (Figure 2D).

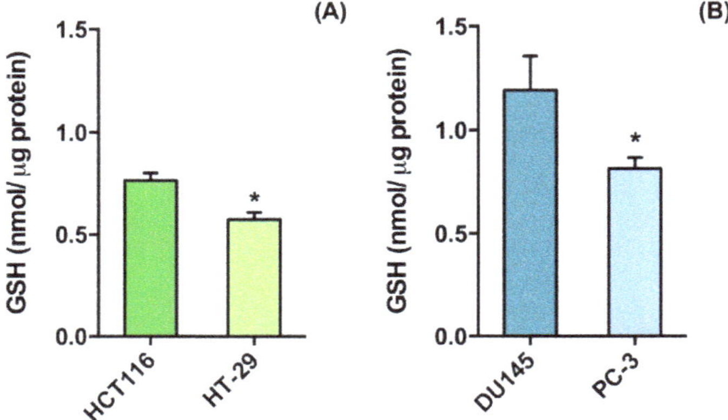

Figure 1. Intracellular glutathione-responsive β-cyclodextrin (GSH) level according to cell type. The reduced glutathione content of human colorectal cancer cell lines, HCT 116 and HT-29 (**A**), and human prostatic carcinoma cell lines, DU145 and PC-3 (**B**), was measured at a basal level, that is, in untreated cells, and was expressed as nmol/µg protein. The results are mean values ± SD of three independent experiments performed in triplicate. Statistically significant difference between cell lines: * $p < 0.05$.

Table 2. Glutathione-responsive β-cyclodextrin-based nanosponge (GSH-NS) inhibition concentration (IC) values on 2D cell cultures.

HCT116 IC Values (mg/mL ± St.Dev) on 2D Cultures								
24 h	GSH-NSB	GSH-NSD	48 h	GSH-NSB	GSH-NSD	72 h	GSH-NS B	GSH-NSD
IC_1	0.27 ± 0.03	0.51 ± 0.04	IC_1	0.47 ± 0.03	0.65 ± 0.03	IC_1	0.50 ± 0.04	0.66 ± 0.05
IC_{50}	1.86 ± 0.31	2.62 ± 0.39 *	IC_{50}	1.60 ± 0.27	2.35 ± 0.40 *	IC_{50}	1.54 ± 0.19	2.38 ± 0.41 *
HT-29 IC Values (mg/mL ± St.Dev) on 2D Cultures								
24 h	GSH-NS B	GSH-NS D	48 h	GSH-NSB	GSH-NS D	72 h	GSH-NSB	GSH-NSD
IC_1	0.05 ± 0.00	0.05 ± 0.00	IC_1	0.15 ± 0.01	0.65 ± 0.03 *	IC_1	0.27 ± 0.02	0.65 ± 0.03 *
IC_{50}	1.71 ± 0.25	2.76 ± 0.35 *	IC_{50}	1.58 ± 0.21	2.80 ± 0.45 *	IC_{50}	1.37 ± 0.30	2.39 ± 0.31 *
DU145 IC Values (mg/mL ± St.Dev) on 2D Cultures								
24 h	GSH-NS B	GSH-NS D	48 h	GSH-NSB	GSH-NS D	72 h	GSH-NSB	GSH-NSD
IC_1	0.01 ± 0.00	0.05 ± 0.00	IC_1	0.05 ± 0.00	0.03 ± 0.00	IC_1	0.17 ± 0.01	0.12 ± 0.01
IC_{50}	1.85 ± 0.40	2.34 ± 0.36	IC_{50}	1.37 ± 0.28	1.57 ± 0.32	IC_{50}	1.38 ± 0.25	1.56 ± 0.30
PC-3 IC Values (mg/mL ± St.Dev) on 2D Cultures								
24 h	GSH-NS B	GSH-NS D	48 h	GSH-NSB	GSH-NSD	72 h	GSH-NSB	GSH-NSD
IC_1	0.04 ± 0.00	0.05 ± 0.00	IC_1	0.06 ± 0.01	0.43 ± 0.03 *	IC_1	0.35 ± 0.02	0.43 ± 0.03
IC_{50}	1.65 ± 0.27	2.56 ± 0.38 *	IC_{50}	1.20 ± 0.22	2.09 ± 0.25 *	IC_{50}	1.37 ± 0.31	1.68 ± 0.27

Statistically significant difference between the two nanosponge formulations: * $p < 0.05$.

Therefore, the nanosponge formulation with the lower disulfide-bond content, GSH-NS B, was found to be more cytotoxic than the nanosponge formulation with the higher disulfide-bond content, GSH-NS D, in all of the cell lines, with the exception of DU145 cell line (Table 2 and Figure 2). Moreover, lower GSH-NS D IC_{50} values were observed on DU145 and PC-3 cells than on HCT116 and HT-29 cells at 48 and 72 h of exposure (Table 2 and Figure 2), suggesting that the nanosponge formulation with the higher disulfide bridge concentration, GSH-NS D, has a higher cytotoxic effect in prostatic cancer cells than in colorectal cancer cells.

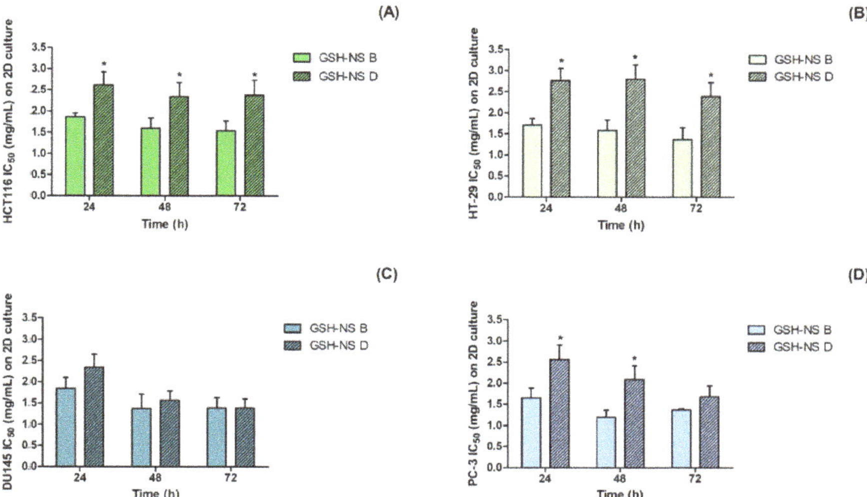

Figure 2. Cytotoxicity of glutathione responsive β-cyclodextrin-based nanosponges in 2D cell cultures. HCT116 (**A**), HT-29 (**B**), DU145 (**C**), and PC-3 (**D**) cells were incubated with GSH-NS B and GSH-NS D at different concentrations (0.5, 1.0, 2.0, 3.0 mg/mL) for 24, 48, and 72 h. Cell proliferation was evaluated by WST-1 assay and the values of the concentration required for a 50% cell growth inhibition (IC_{50}) were determined by the proliferation curves obtained using the CalcuSyn 2.11 software (Biosoft, Cambridge, UK). The results are mean values ± St.Dev of three independent experiments, replicated eight times for each condition. Statistically significant difference between nanosponge formulations: * $p < 0.05$.

2.3. Glutathione Responsive β-cyclodextrin-Based Nanosponge Cellular Uptake on 2D Cell Cultures

The cellular uptake of non-cytotoxic (IC_1) and cytotoxic (IC_{50}) concentrations of fluorescent GSH-NS was analyzed by flow cytometry and fluorescence microscope imaging after 24 h exposure. HCT116, HT-29, DU145, and PC-3 cell monolayers were exposed to 6-coumarin loaded GSH-NS B or 6-coumarin loaded GSH-NS D for 24 h. Significant dose-dependent differences in nanosponge cellular uptake were observed (Figure 3A–D). Interestingly, higher nanosponge cellular uptake was observed in colorectal cancer cells (Figure 3A,B) than in prostatic cancer cells (Figure 3C,D). In particular, the highest nanosponge cellular uptake was observed in HCT116 cells (Figure 3A) and the lowest in PC-3 cells (Figure 3D). As IC_{50} values at 24 h of incubation in colorectal and prostatic cell line were very similar (Table 2), the lower fluorescent GSH-NS intracellular uptake in prostatic cells suggests that they are more highly sensitive to the nanosponge cytotoxic effect than the colorectal cancer cell lines. In particular, Figure 3E–G show images of untreated HCT116 cells with nuclear counterstaining (Figure 3E) and after 24 h exposure to 6-coumarin loaded GSH-NS B (Figure 3F) or 6-coumarin loaded GSH-NS D (Figure 3G).

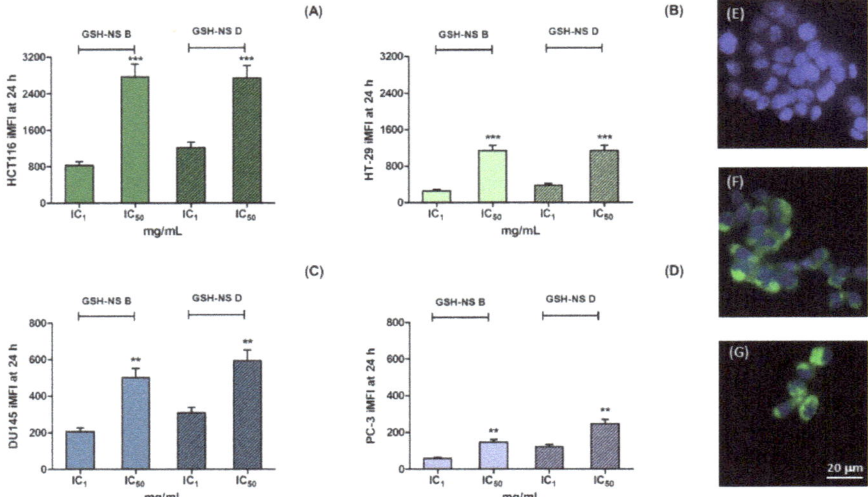

Figure 3. Fluorescent glutathione responsive β-cyclodextrin-based nanosponge cellular uptake. HCT116 (**A**), HT-29 (**B**), DU145 (**C**), and PC-3 (**D**) cells were exposed to the respective IC_1 and IC_{50} of 6-coumarin loaded GSH-NS B and 6-coumarin loaded GSH-NS D for 24 h and analyzed by flow cytometry. Cellular uptake was expressed as integrated mean fluorescence intensity (iMFI). Representative fluorescence images of HCT116 untreated cells (**E**), HCT116 cells exposed to 6-coumarin loaded GSH-NS B IC_{50} (**F**), and 6-coumarin loaded GSH-NS D IC_{50} (**G**) for 24 h using 4′,6-diamidino-2-phenylindole (DAPI) (blue) as nuclear counterstain (63x magnification). Statistically significant difference between IC_1 and IC_{50}: ** $p < 0.01$; *** $p < 0.001$.

2.4. The Effect of Glutathione Responsive β-cyclodextrin-Based Nanosponges on Cell Death and Cell Cycle

As an increase in the number of dead or plasma membrane-damaged cells results in an increase in lactate dehydrogenase (LDH) in the culture supernatant, we measured the LDH leakage of HCT116, HT-29, DU145, and PC-3 cells after 24, 48, and 72 h of incubation with an experimental medium containing different GSH-NS B or GSH-NS D concentrations (0.5, 1.0, 2.0, and 3.0 mg/mL). No significant increase in LDH leakage percentage over untreated control cells was observed under any of the test conditions (data not shown). The lack of apparent plasma membrane-damaged cells in the LDH assay would appear to contrast significantly with the decrease in cell growth observed by WST-1 cell proliferation assay (Table 2 and Figure 2), which was performed at the same incubation times with the same GSH-NS concentrations. This finding prompted us to investigate the cell cycle using flow cytometry to assess any arrest of cell cycle progression.

We then performed cell cycle analyses on the IC_{50} values of each cell line after 24 h of GSH-NS B or GSH-NS D incubation (Table 2). A significant increase in cell percentages in the G_0/G_1 phase and a significant decrease in $S/G_2/M$ cell percentages were observed across the entire cell population after exposure at the respective GSH-NS B or GSH-NS D IC_{50} values (Figure 4). Moreover, the sub-G_0/G_1 peak was absent in all cell lines. This suggests that the observed decrease in cell proliferation (Table 2 and Figure 2) was the result of alterations of cell cycle progression owing to a block in the G_0/G_1 phase. Furthermore, the G_0/G_1 phase cell population percentage was higher in both colorectal cancer (Figure 4A,B) and prostatic cancer (Figure 4C,D) cells after incubation with the lower disulfide-bond content nanosponge (GSH-NS B).

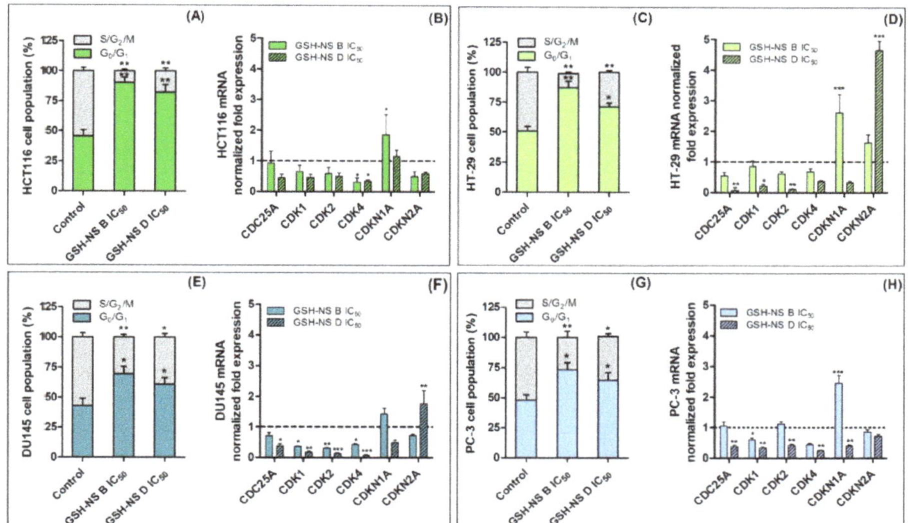

Figure 4. Glutathione responsive β-cyclodextrin-based nanosponge effect on cell cycle and mRNA expression. HCT116 (**A**,**B**), HT-29 (**C**,**D**), DU145 (**E**,**F**), and PC-3 (**G**,**H**) cells were exposed to the respective GSH-NS B or GSH-NS D IC_{50} for 24 h. Cell cycle distribution was analyzed by flow cytometry and the data were expressed as a percentage of cells in the different phases of the cell cycle (**A**,**C**,**E**,**G**). *RRN18S* (ribosomal RNA 18S) was used for the mRNA gene expression analysis as a reference gene to normalize the data and the nanosponge-induced alterations in mRNA levels were compared with those of the controls, that is, untreated cells, fixed at 1 and shown by the dotted line. Statistically significant difference versus control: * $p < 0.05$; ** $p < 0.01$; *** $p < 0.001$.

Thus, to confirm these data, an analysis of mRNA expression of the different cyclin-dependent kinases (CDK) was performed, as they participate in cell cycle regulation, especially during the G_1 to S phase transition. It was observed that *CDK1*, *CDK2*, and *CDK4* mRNA expression was down-regulated in almost all cell lines, as compared with untreated cells (Figure 4B,D,F,H). Interestingly, the mRNA expression of the CDK activator, *CDC25A*, was either unaffected or down-regulated, compared with untreated cells, whereas the mRNA expression of the CDK inhibitors, *CDKN1A* and *CDKN2A*, was up-regulated compared with untreated cells (Figure 4B,D,F,H). No significant differences in the extent of cell cycle arrest were observed between the two nanosponge formulations. However, the formulation with the higher disulfide-bond content, GSH-NS D, seemed to induce a higher down-regulation in CDK mRNA and higher up-regulation in *CDKN2A* mRNA in HT-29 and DU145 cells, whereas GSH-NS B induced a higher *CDKN1A* mRNA expression in HCT116 and PC-3 cells. Furthermore, no significant differences were observed in terms of the extent of cell cycle arrest and gene expressions between cell lines with higher (Figure 4A–F) and lower GSH content (Figure 4C,D,G,H).

2.5. The Effect of Glutathione Responsive β-cyclodextrin-Based Nanosponges on Reactive Oxygen Species Production

After exposure to the respective GSH-NS B and GSH-NS D IC_{50} values at 24 h, the dichlorofluorescein-diacetate (DCFH-DA) assay did not indicate a significant intracellular ROS increase at 1, 12 (data not shown), and 24 h (Figure 5) in all cell lines considered, except for HCT116 cells, where a significant increase in intracellular ROS was observed at 24 h (Figure 5A) after both GSH-NS formulations incubation (Figure 5A).

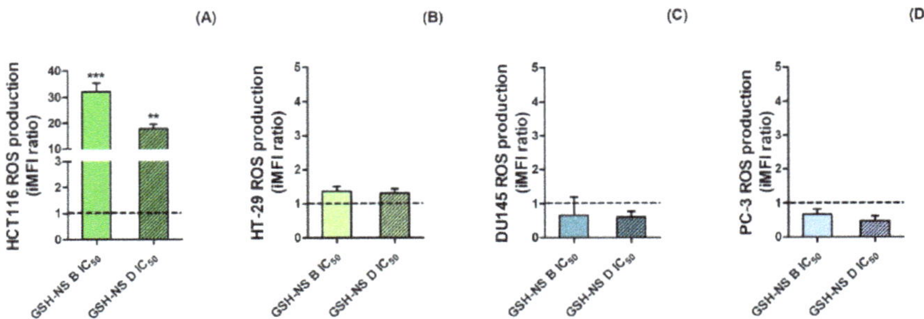

Figure 5. Glutathione responsive β-cyclodextrin-based nanosponge reactive oxygen species (ROS) production. HCT116 (**A**), HT-29 (**B**), DU145 (**C**), and PC-3 (**D**) cells were treated with GSH-NS B or GSH-NS D at the respective IC_{50} for 24 h. ROS levels, detected by dichlorofluorescein-diacetate (DCFH-DA) assay by flow cytometry, were expressed as the integrated median fluorescence intensity (iMFI) ratio and nanosponge-induced ROS levels were compared to those of control, that is, untreated cells, fixed at 1 and shown by the dotted line. Statistically significant difference versus untreated cells: ** $p < 0.01$; *** $p < 0.001$.

2.6. Glutathione Responsive β-cyclodextrin-Based Nanosponge Cytotoxicity in Three-Dimensional Cell Cultures

The next step was the analysis of GSH-NS cytotoxicity in the colorectal and prostatic cancer cell lines with the highest GSH basal level, HCT116 and DU145, which were three-dimensionally cultured as multicellular spheroids (MCSs), cellular aggregates organized in a specific cell-to-cell and cell–matrix interaction, closer to in vivo features [41]. Dose response curves were obtained by exposing HCT116 and DU145 on 3D model to different concentrations (0.5, 2.0, 4.0, and 6.0 mg/mL) of GSH-NS B and GSH-NS D, for 24, 48, and 72 h to obtain the respective IC_1 and IC_{50} values (Table 3 and Figure 6A,G). The MCS uptake of non-cytotoxic (IC_1) and cytotoxic (IC_{50}) concentrations of fluorescent GSH-NS was confirmed by fluorescence microscope imaging after 24 h of exposure (Figure 6B,C,H,I).

Table 3. GSH-NS IC values in the 3D cell cultures.

HCT116 IC Values (mg/mL ± St.Dev) on 3D Cultures								
24 h	GSH-NS B	GSH-NS D	48 h	GSH-NS B	GSH-NS D	72 h	GSH-NS B	GSH-NS D
IC_1	0.01 ± 0.00	0.02 ± 0.00	IC_1	0.01 ± 0.00	0.01 ± 0.00	IC_1	0.01 ± 0.00	0.01 ± 0.00
IC_{50}	3.92 ± 0.95	4.19 ± 0.98	IC_{50}	2.75 ± 0.18	3.05 ± 0.31	IC_{50}	1.98 ± 0.15	2.22 ± 0.21
DU145 IC Values (mg/mL ± St.Dev) on 3D Cultures								
24 h	GSH-NS B	GSH-NS D	48 h	GSH-NS B	GSH-NS D	72 h	GSH-NS B	GSH-NS D
IC_1	0.02 ± 0.00	0.04 ± 0.00	IC_1	0.03 ± 0.00	0.02 ± 0.00	IC_1	0.07 ± 0.01	0.02 ± 0.00
IC_{50}	5.38 ± 1.15	5.80 ± 1.78	IC_{50}	2.68 ± 0.25	3.24 ± 0.34	IC_{50}	2.01 ± 0.24	2.26 ± 0.23

Statistically significant difference between the two nanosponge formulations: ns.

GSH-NS B and D gave similar IC_{50} values in both cell lines (Table 3, Figure 6A,G) and across all incubation times, whereas GSH-NS B gave a higher cytotoxic effect in HCT116 cells than GSH-NS D in the monolayer cultures (Table 2 and Figure 2A). Interestingly, the nanosponge cytotoxic effect was time dependent in both cell lines (Figure 6A,G). This also differs from the results obtained in the monolayer cultures (Figure 2A,C). Moreover, significantly higher GSH-NS B and GSH-NS D IC_{50} values at 24 h were observed in DU145 cells than in HCT116 cells (Figure 6G), suggesting a lower cytotoxic effect of the GSH-NS at 24 h in DU145 cells than in HCT116 cells. These results differ from those observed in monolayer cultures, where the nanosponge formulation with the higher disulfide-bond content, GSH-NS D, was more cytotoxic in DU145 cells than in HCT116 (Figure 2A,C). Figure 6 shows images of 72 h untreated HCT116 MCS (10.51 ± 1.74 μm^3, panel D) and DU145 MCS (11.66 ± 0.92 μm^3, panel J), which are compared to GSH-NS B and GSH-NS D treated HCT116 MCS (4.03 ± 0.09 μm^3, panel E and 4.55 ± 0.77 μm^3, panel F, respectively) ($p < 0.05$) and DU145 MCS (4.63 ± 0.27 μm^3, panel K and 5.12 ± 0.60 μm^3, panel L, respectively) ($p < 0.05$).

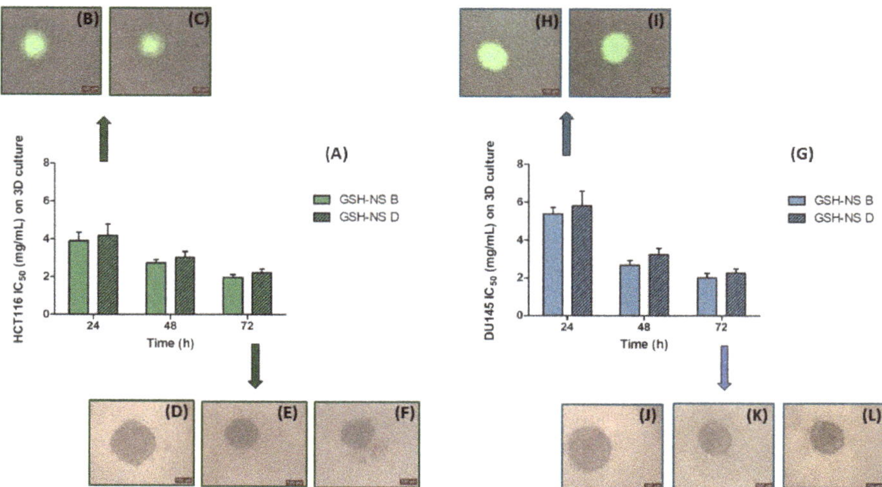

Figure 6. Glutathione responsive β-cyclodextrin-based nanosponge cytotoxicity and uptake on 3D culture. HCT116 (**A**) and DU145 (**G**) multicellular spheroids (MCSs) were incubated with GSH-NS B and GSH-NS D at different concentrations (0.5, 2.0, 4.0, and 6.0 mg/mL) and MCS volume was measured after 24, 48, and 72 h. Representative images at 24 h of HCT116 and DU145 MCS uptake of 6-coumarin loaded GSH-NS B IC$_{50}$ (**B,H**, respectively) and of 6-coumarin loaded GSH-NS D IC$_{50}$ (**C,I**, respectively). Representative phase contrast images at 72 h of HCT116 MCS: untreated (**D**), treated with GSH-NS B IC$_{50}$ (**E**) and GSH-NS D IC$_{50}$ (**F**). Representative phase contrast images at 72 h of DU145 MCS: untreated (**J**), treated with GSH-NS B IC$_{50}$ (**K**), and GSH-NS B IC$_{50}$ (**L**). Images are at 10× magnification.

3. Discussion

Understanding the effect that nanoparticles have on cells is crucial to predict their in vivo toxicity and avoid any undesirable nanoparticle activities. Although there are numerous in vitro cytotoxicity assays that can be applied for the general screening of nanoparticles [42,43], it is of vital importance that the research covers nanoparticle cytotoxicity itself. In this contest the use of strictly controlled in vitro experimental conditions can ensure that the measured effect is the result of nanoparticle toxicity and not unstable culturing conditions [44]. Moreover, up to nowadays, there has been no single analysis able to provide sufficient information to correlate the biomaterial chemistry and surface with biological response [45]. Herein, we investigated the in vitro biological effects of a stimuli-responsive nanosystem, that is, glutathione responsive β-cyclodextrin-based nanosponges (GSH-NS), in various cancer cell lines, characterized by their GSH basal content, as this nanosystem is designed to be a GSH responsive anticancer drug carrier.

GSH plays a key role in cellular defense against oxidative stress [46] and its increased redox capacity in cancer cells is well-known [34,47]. Consequently, GSH has been recognized to be an ideal intracellular trigger for selective drug delivery by responsive nanocarriers, as many compounds exert their therapeutic effects only inside cells. As disulfide chemistry is particularly versatile, a wide range of GSH-responsive nano-vehicles, such as micelles, nanoparticles, and nanogels, have been recently developed [28]. Among them, glutathione responsive β-cyclodextrin-based nanosponges incorporate high payload and provide controlled drug release over time, with the further advantage of triggered intracellular drug delivery in response to cell GSH content. In addition, GSH-NSs are able to protect degradable drugs from the external environment. It is foreseen that β cyclodextrin-based nanosponges will have a significant positive impact on anticancer therapeutic scenarios [13,24,37,38]. Taking into account the promising results concerning the efficacy of GSH-NSs as an anticancer drug delivery system [37,38], the biological safety of the nanosponge itself is a critical parameter for their future clinical application.

β-cyclodextrin toxicology has been evaluated in in vitro and in vivo studies that have reported it as non-toxic and well tolerated even at very high doses [48]. Previous in vitro studies showed no signs of cytotoxicity after cell exposure to unloaded nanosponges in the 10–100 µg/mL concentration range used for the delivery of therapeutic drugs [23,49,50]. In addition, in vivo experiments have shown that β-cyclodextrin-based nanosponges prepared with pyromellitic dianhydride as a cross-linking agent have been orally administered to rats without showing any toxic side effects at selected doses in an acute and repeated dose toxicity study [51]. Previously, GSH-NSs have been investigated as doxorubicin carrier. No acute cardiotoxic effects were observed in mice after the in vivo administration of doxorubicin-loaded GSH-NS [37]. Recently, the hepatotoxicity of this nanoformulation was investigated either in vitro on human HepG2 cell line or ex vivo on rat precision-cut liver slices (PCLSs), where a good nanosponge safety profile was demonstrated, showing a comparable hepatotoxicity to that of free doxorubicin [52].

As no reports have been published on the effects at a cellular level of GSH-NS as such, it was decided to study the effect of GSH-NS per se on HCT116, HT-29, DU145, and PC-3 cancer cell lines with various GSH content in a concentration range that is about fifty times higher than that used in the above mentioned studies to ensure the use of cytotoxic concentrations. HCT116 and DU145 cells showed the highest GSH values in colorectal and prostatic cancer cell lines, respectively; previous research studies have shown that DU145 cells have the highest GSH content [53]. Non-toxic (IC_1) and cytotoxic (IC_{50}) GSH-NS concentrations were determined by a 2D cell assay, which measured mitochondrial activity. A decrease in cell growth with significantly different IC_1 or IC_{50} values was observed when the two nanosponge formulations were compared in all cell lines, except in DU145 cell line, where no statically significant difference was observed.

DU145 cell line was the most sensitive to the GSH-NS D cytotoxic effect among all cell lines tested. Notably, DU145 cells are more resistant to electrophilic toxicity than other cells owing to their high levels of redox-sensitive transcription factor, nuclear factor erythroid 2-related factor-2 (Nrf2), which activates cytoprotective pathways against oxidative injury, such as GSH synthesis [54,55]. As this nanosystem has the ability to disrupt itself in the presence of GSH, we can hypothesize that it is the high GSH content in DU145 cells that allows GSH-NS to exert their cytotoxic effect, whatever the disulfide-bond concentration. Further studies are needed to investigate whether agents able to modulate intracellular GSH, such as *N*-acetyl cysteine or buthionine sulfoximine [56] could affect nanosponge intracellular drug release and cytotoxicity.

Our study shows that colorectal cancer cells, in particular HCT116 cells, have the most pronounced GSH-NS B and D cellular uptake. This difference in nanosponge cellular uptake in this cell line may be owing to differing uptake mechanisms, as cell surface thiols have been reported to affect disulfide-conjugated peptide cell entry [57]. Indeed, disulfide bridge cleavage may start at the cell surface via thiol/disulfide exchange reactions catalyzed by redox proteins such as thioredoxines [58]. Therefore, the 2D data on IC_{50} would appear to indicate that prostatic cancer cell lines are more sensitive to GSH-NS cytotoxic effects.

Worthy of note is that cell cycle analyses revealed a significant cell cycle arrest in the G_0/G_1 phase in all cell lines at 24 h IC_{50} values. Thus, to further investigate this cell cycle arrest, we analyzed a panel of genes that are involved in cell cycle regulation. Notably, the results show significant mRNA over-expression in the cell cycle progression regulators at G_1, *CDKN1A*, and *CDKN2A*, which code for p21 and p16 that inhibit the cyclin-CDK2 and -CDK4 complexes, respectively. Apart from this, the mRNA expression of *CDC25A*, *CDK1*, *CDK2*, and *CDK4* was either unaffected or down-regulated in all cell lines. These data demonstrate that GSH-NS inhibition of cell proliferation is essentially owing to G_1 cell cycle arrest, in agreement with previous reports by Choi et al. [59]. Interestingly, only HTC116 cells showed significant ROS production after GSH-NS exposure, which is most likely owing to their high GSH-NS cellular uptake.

Lastly, the investigation of nanosponges effects on MCS growth was carried out. The results were interesting as differences in the 2D study were observed. There were no significant differences between the two GSH-NS formulations in HCT116 and DU145 MCS, whereas there was a significant difference in the 2D HCT-116 culture. Indeed, IC_{50} values were significantly lower in the 2D cultures than in the 3D cultures, especially after 24 h incubation, where similar values were reached only after 72 h of incubation. For example, IC_{50} was twofold higher after 24 h in 3D cultures for HCT116 and three-fold higher in DU145 than in their respective monolayers. On the other hand, IC_1 concentrations were significantly lower in HCT116 MCS than in HCT116 cell monolayers, whereas IC_1 was quite similar both in DU145 spheroids and monolayers.

GSH-NS cytotoxicity might appear to be linked to disulfide-bond content in 2D cell monolayers as the formulation with the higher disulfide-bond content, GSH-NS D, had the lowest cytotoxic effect in all cell lines, except for the DU145 cell line. On the other hand, GSH-NS cytotoxicity was not influenced by the disulfide-bond content in MCS and the most pronounced cell growth decrease was observed in the colorectal cancer cell line, HCT116, after 24 h of exposure to GSH-NS. Tissue-like morphology and phenotypic change may be identified as the major factors in diminishing toxicity on MCS. This means that in vitro 3D cell culture models could act as an intermediate stage and bridge the gap between in vitro 2D and in vivo studies, which would extend current cellular level cytotoxicity to the tissue level and improve the predictive power of in vitro nanoparticle toxicology [60]. Finally, GSH-NSs showed a limited toxicity, leading to G_1 cell cycle arrest, without membrane damage or oxidative stress generation at significantly higher concentrations about fifty times those used for the delivery of anticancer drugs.

4. Methods

4.1. Synthesis of Glutathione Responsive β-cyclodextrin-Based Nanosponges

Glutathione-responsive β-cyclodextrin-based nanosponges (GSH-NSs) were synthetized according to the method developed by Trotta et al. [35].

Briefly, GSH-NSs were obtained using a one-step synthetic route by reacting β-cyclodextrin and pyromellitic dianhydride, in the presence of 2-hydroxyethyl disulfide to insert disulfide bridges in the NS nanostructure [35]. Varying amounts of 2-hydroxyethyl disulfide were used to obtain a series of GSH-NS with different disulfide bridge percentages in the polymer matrix. In particular, 2-hydroxyethyl disulfide/β-cyclodextrin molar ratios of 1:20 and 1:5 were used for the synthesis of GSH-NS with low (GSH-NS B) and high disulfide-bridge content (GSH-NS D), respectively. The reaction was estimated at room temperature under stirring for 24 h. The nanosponges were then purified by Soxhlet extraction with acetone for a few hours. The percentage of sulfur in the two types of GSH-NS was measured by elemental analysis and was 0.62 and 1.90 for GSH-NS B and GSH-NS D, respectively.

4.2. Preparation of Glutathione Responsive β-cyclodextrin-Based Nanosponge Nanosuspension

GSH-NS nanosuspensions were prepared following the preparation protocol previously reported [35,38]. A weighed amount of GSH-NSs was suspended in a saline solution (NaCl 0.9%) at a concentration of 10 mg/mL. The suspension was homogenized by a high shear homogenizer (Ultraturrax®, IKA, Konigswinter, Germany) for 5 min at 24,000 rpm. The sample was then homogenized on a high-pressure homogenizer (EmulsiFlex C5, Avastin, Mannheim, Germany) for 90 min at a back pressure of 500 bar to further reduce the size of the nanosponges. The aqueous nanosponge nanosuspension was subsequently purified by dialysis (membrane cutoff 12,000 Da) to eliminate potential synthesis residues. The nanosuspension was stored at +4 °C and used for all experiments.

4.3. Preparation of Fluorescent Glutathione Responsive β-cyclodextrin-Based Nanosponges

Fluorescent GSH-responsive nanosponges were obtained by adding 6-coumarin (0.1 mg/mL) to the aforementioned aqueous GSH-NS nanosuspensions (previously described) (10 mg/mL) under stirring for 24 h at room temperature in the dark.

4.4. Characterization of Glutathione Responsive β-cyclodextrin-based Nanosponges

The two types of GSH-NS (GSH-NS B and D), either blank or 6-coumarin loaded, were characterized in vitro to measure their physico-chemical parameters. The average diameters, polydispersity indices, and zeta potential values were determined by photon correlation spectroscopy (PCS) and electrophoretic mobility using a 90 Plus Instrument (Brookhaven, NY, USA) at a fixed angle of 90° and a temperature of +25 °C. The analyses were performed on diluted GSH-NS samples (1:30 *v/v*). For zeta potential determination, the samples were placed in an electrophoretic cell where an electric field of approximately 15 V/cm was applied. Three batches were analyzed for each NS type and each measured value was the average of ten repetitions. Nanosponge morphology was evaluated by transmission electron microscopy (TEM) (Philips CM10 instrument, Eindhoven, Netherlands) after the diluted aqueous nanosponge nanosuspensions were sprayed onto a Form war-coated copper grid and air-dried.

4.5. Cell Culture and Treatment with Glutathione Responsive β-cyclodextrin-Based Nanosponges

Human colorectal cancer cell lines, HCT116 and HT-29 (ICLC, Interlab Cell Line Collection, Genova, Italy), and human prostatic carcinoma cell lines, DU145 and PC-3 (ICLC), were cultured in McCoy's 5A Medium and RPMI-1640 Medium, respectively. These media were supplemented with 2 mM L-glutamine, 100 UI/mL penicillin, 100 μg/mL streptomycin, and 10% (*v/v*) heat-inactivated fetal calf serum (Sigma, ST Louis, MO, USA) in a humidified atmosphere of 5% CO_2 air at 37 °C. At 85% confluence, cells were harvested with 0.25% trypsin and sub-cultured into 75 cm^2 flasks, 6-well plates or 96-well plates according to need. Cells were allowed to attach to the surface for 24 h prior to treatment. GSH-NS B and D were then suspended in a cell culture medium and diluted to the appropriate concentrations. After treatment, the cells were harvested to determine cytotoxicity, cell cycle distribution and ROS production. The cells that were not exposed to GSH-NS were used as control conditions for each experiment.

4.6. Measurement of Basal Intracellular Reduced Glutathione Levels

The total glutathione level (GSSG + GSH) in HT-29, HCT116, DU154, and PC-3 cells were assayed by the Glutathione Assay Kit (Sigma, Milano, Italy), according to manufacturer's instructions. The protein concentration (μg/mL) was quantified by the Qubit fluorometer (Invitrogen, Milan, Italy) and the Quant-IT Protein Assay Kit (Invitrogen, Milano, Italy). Calibration was performed by the application of a two-point standard curve, according to the manufacturer's instructions.

Briefly, reduced glutathione (GSH) reacts with 5,5′-dithiobis(2-nitrobenzoic acid) (DTNB) in a recycling assay and produces glutathione disulfide (GSSG) and the 1,3,5-trinitrobenzene (TNB) anion, which can be detected by absorbance. In turn, the enzyme glutathione reductase then reduces GSSG, which release GSH that can react with another DTNB molecule. Therefore, the rate of TNB production is measured rather than a single determination of how much DTNB react with GSH, as it is proportional to the initial amount of GSH [61]. The plate was read at 412 nm on a microplate reader Asys UV 340 (Biochrom, Cambridge, UK) and the amount of GSH was expressed in nmol/μg protein.

4.7. Cell Proliferation Assay

The effect that GSH-NS B and D had on HCT116, HT-29, DU145, and PC-3 cell growth was evaluated by WST-1 cell proliferation assay (Roche Applied Science, Penzberg, Germany). Briefly, 2.0×10^3 HT-29, 1.5×10^3 HCT116, 5.0×10^2 DU145, and 1.2×10^3 PC-3 cells were seeded in 100 μL of growth medium in replicates (*n* = 8) in 96-well culture plates; the seeding density of each cell line was chosen according to the best proliferation rate. The medium was removed after 24 h and the cells were incubated with in an experimental medium containing differing GSH-NS B or GSH-NS D concentrations (0.5, 1.0, 2.0, and 3.0 mg/mL). At 24, 48, and 72 h, WST-1 reagent (10 μL) was added and the plates were incubated at 37 °C in 5% CO_2 for 1.5 h. Well absorbance was measured at 450 and 620 nm (reference wavelength) on a microplate reader Asys UV 340.

Cell proliferation data were expressed as a percentage of control, that is, untreated cells. At 24, 48, and 72 h, the inhibition concentration 50% (IC_{50}), defined as the dose of compound that inhibited 50% of cell growth, was interpolated from the growth curves, as was the inhibition concentration 1% (IC_1), defined as the dose of compound that inhibited 1% of cell growth. Thus, to compare the effects of GSH-NS on the different cell lines, the IC_1 and IC_{50} values obtained were used to carry out the following experiments.

4.8. Nanosponge Cellular Uptake Assays

Coumarin 6-loaded GSH-NS cellular uptake was assessed by cytofluorimetric analysis using a C6 flow cytometer (Accuri Cytometers, Ann Arbor, MI, USA) and imaging analysis using a DMI4000B fluorescence microscopy (Leica, Wetzlar, Germany). For flow cytometry analysis, 5.0×10^4 cells were seeded in a six-well culture plate. Forty-eight hours after seeding, HT-29, HCT116, DU145, and PC-3 cells were treated with the respective not-cytotoxic (IC_1) and cytotoxic (IC_{50}) concentrations of either fluorescent GSH-NS B or fluorescent GSH-NS D at 24 h. After a 24 h incubation, the cells were washed three times with phosphate-buffered saline PBS, suspended in 250 μL PBS, and run on the flow cytometer with 488 nm excitation. Intracellular fluorescence was expressed as integrated mean fluorescence intensity (iMFI), which was the product of the frequency of 6-coumarin-loaded GSH-NS positive cells and the mean fluorescence intensity.

Microscopy observation was carried out after glass coverslips were placed in 24-well plates and the cells seeded at a density of 5.0×10^4 cells/coverslip for 48 h of incubation. The coumarin 6-loaded nanosponges were then added at the respective IC_{50} values for GSH-NS B and D, and incubated for 24 h. The cells were incubated with 1 μg/mL of 4',6-diamidino-2-phenylindole (DAPI) for nuclear counterstaining 30 min before the programmed stop time. After the cells were washed with PBS, the cells on the coverslip were mounted on a glass slide, observed under a fluorescence microscope, and photographed.

4.9. Lactate Dehydrogenase Leakage Assay

Lactate dehydrogenase (LDH) is an enzyme that is widely present in cytosol and catalyzes the conversion of lactate to pyruvic acid. If plasma membrane integrity is disrupted, the LDH leaks into culture media, increasing its extracellular level, and the amount of LDH release is proportional to the number of damaged cells [62]. The LDH leakage was evaluated by the LDH-Cytotoxicity Detection Kit (Roche Diagnostic, Indianapolis, USA), according to manufacturer's instructions. Briefly, 96-well plates were seeded with HT-29, HCT116, DU145, and PC-3 cell lines at a density of 2.0×10^3, 1.5×10^3, 5.0×10^2, and 1.2×10^3 cells/100 μL culture medium, respectively. Twenty-four hours after the seeding, 100 μL of different concentrations (0.5, 1.0, 2.0, and 3.0 mg/mL) of GSH-NS B or GSH-NS D was added to the wells. The plates were then incubated for 24, 48, and 72 h, at 37 °C, in a humidified atmosphere of 5% CO_2 air. Cell-free culture media were then collected and incubated with the same volume of reaction mixture for 30 min. LDH activity was measured at 490 nm on a microplate reader Asys UV 340. The background control was obtained by measuring the LDH activity of the assay medium, the untreated control by measuring the LDH activity of untreated cells, and the positive control by measuring the maximum releasable LDH activity after the treatment with the lysis buffer. The LDH leakage percentage was calculated as follows: LDH leakage (%) = (experimental value−untreated control)/(positive control−untreated control) × 100, and is the mean of three independent wells.

4.10. Cell Cycle Analysis

Cell cycle distribution was evaluated 24 h after cell treatment with the respective IC_{50} of GSH-NS B or GSH-NS D. The occurrence of the so-called sub-G_0/G_1 peak, which is a distinct cell population characterized by subdiploid DNA fluorescence and might correlate with the internucleosomal DNA fragmentation typical of apoptosis (Pozarowski and Darzynkiewicz, 2004), was also evaluated. Briefly, 1×10^6 HCT116, 1×10^6 HT-29, 1×10^6 DU145, and 1×10^6 PC-3 cells were incubated with 2 µM of the live cell staining Vybrant Dye Cycle Green (Invitrogen) for 30 min at 37 °C. The samples were run on a flow cytometer with 488 nm excitation to measure Vybrant Dye Cycle Green staining and data analysis was performed by FCS Express software version 4 (BD Bioscience, Milano, Italy).

4.11. Real Time Reverse Transcriptase-Polymerase Chain Reaction (RT-PCR)

Total RNA was isolated from the HCT116, HT-29, DU145, and PC-3 cells, 24 h after incubation with the respective GSH-NS B or GSH-NS D IC_{50}. Briefly, the cells were collected in RNA Cell Protection Reagent (Qiagen, Milano, Italy) and stored at −80 °C. Total RNA was obtained by the RNeasy Plus Mini Kit (Qiagen Milano, Italy). Total RNA concentration (µg/mL) was determined using the fluorometer Qubit (Invitrogen) and the Quant-IT RNA Assay Kit (Invitrogen). Calibration was carried out by applying a two points standard curve, according to the manufacturer's instructions. RNA sample integrity was determined by the Total RNA 6000 Nano Kit (Agilent Technologies, Milano, Italy) using the Agilent 2100 Bioanalyzer (Agilent Technologies, Milano, Italy).

Real-time RT-PCR analysis was carried out using 1 µg of total RNA, which was reverse transcribed in a 20 µL cDNA reaction volume using the QuantiTect Reverse Transcription Kit (Qiagen, Milano, Italy). Each 10 µL real-time RT-PCR reaction was obtained using 12.5 ng of cDNA, according to the manufacturer's instructions. Quantitative RT-PCR was performed by the SsoFast EvaGreen (Bio-Rad, Milan, Italy) and the QuantiTect Primer Assay (Qiagen, Milano, Italy) was used as the gene-specific primer pair for the studied gene panel (Table 4).

Table 4. Gene description.

Gene	Primer Codes	Description
CDC25A	QT00001078	cell division cycle 25 homolog A
CDK1	QT00042672	cyclin-dependent kinase 1
CDK2	QT00005586	cyclin-dependent kinase 2
CDK4	QT00016107	cyclin-dependent kinase 4
CDKN1A	QT00031192	cyclin-dependent kinase inhibitor 1A, p21
CDKN2A	QT00089964	cyclin-dependent kinase inhibitor 2A, p16
RRN18S	QT00199367	18S ribosomal RNA

The transcript of the reference gene 18S ribosomal RNA (*RRN18S*) was used to normalize mRNA data and real-time RT-PCR was performed by the MiniOpticon Real Time PCR system (Bio-Rad, Milan, Italy). The PCR protocol conditions were as follows: a HotStarTaq DNA polymerase activation step at +95 °C for 30 s, followed by 40 cycles at +95 °C for 5 s and +55 °C for 10 s. All runs were performed on at least three independent cDNA preparations per sample and all samples were run in duplicate. At least two non-template controls were included in each PCR run. Quantification data analyses were performed by the Bio-Rad CFX Manager software version 1.6 (Bio-Rad, Milan, Italy), according to the manufacturer's instructions. These analyses were performed in compliance with MIQE guidelines (Minimum Information for Publication of Quantitative Real-time PCR Experiments) [63].

4.12. Reactive Oxygen Species Production Assay

The production of intracellular reactive oxygen species (ROS) was measured by flow cytometry using dichlorofluorescein-diacetate (DCFH-DA) (Sigma, Milano, Italy) as the oxidation-sensitive probe. Briefly, after 1, 12, and 24 h cell exposure to the respective GSH-NS B or GSH-NS D IC_1 and IC_{50}

at 24 h, HT-29, HCT116, DU145, and PC-3 cells were washed twice with PBS in six-well plates and incubated with 10 µM DCFH at 37 °C in the dark for 30 min. The cells were then washed with PBS, trypsinized, collected in 500 µL of PBS, and analyzed. ROS production was expressed as iMFI ratio, that is, the difference between the iMFI of treated and untreated cells over the iMFI of untreated cells (iMFI is the product of the frequency of ROS-producing cells and the median fluorescence intensity).

4.13. Cell Growth and Nanosponge Cellular Uptake Assays on Three-Dimensional Cell Culture

Cell suspensions (250-cell spheroids) 40 µL were dispensed into the access hole at each cell culture site to form a hanging drop on a Perfecta3D® 96-well Hanging Drop Plate (3D Biomatrix, Ann Arbor, MI, USA). On day 8 of the HCT116 and DU145 spheroid culture, 15 µL of different GSH-NS B or GSH-NS D concentrations (0.5, 2.0, 4.0, and 6.0 mg/mL) was added to each cell hanging drop and MCS growth was analyzed at 24, 48, and 72 h after nanosponge incubation. Noteworthy is the fact that we had to use a different concentration range for 3D cell growth assay (0.5, 2.0, 4.0, and 6.0 mg/mL) to obtain the dose-response data necessary to calculate the IC_{50} values from the one used in the 2D cell growth assay (0.5, 2.0, 4.0, and 3.0 mg/mL). Phase contrast photographs were taken by the DMI4000B microscope (Leica, Milano, Italy) and the diameter of each MCS was measured by Leica Application Suite Software (Leica) and the volume (V) was calculated using the equation $V = 4/3\pi r^3$. Coumarin 6-loaded nanosponge uptake by MCS at the respective IC_{50} at 24 h for GSH-NS B or GSH-NS D was analyzed by fluorescence microscopy using a DMI4000B microscope (Leica).

4.14. Statistical Analysis

The results are expressed as the average value ± standard deviation (St.Dev) of three independent experiments. Median-effect analysis was performed by CalcuSyn software version 2.11 (Biosoft, Cambridge, UK) to calculate the values of the concentration required to cause a 1% inhibition of cell growth (IC_1) and for a 50% inhibition of cell growth (IC_{50}) for each nanosponge formulation. Statistical analyses were performed on Prism software version 6 (Graph-Pad, La Jolla, CA, USA) using a Student's *t*-test and one-way analysis of variance (ANOVA) to calculate the threshold of significance as appropriate. Statistical significance was set at $p < 0.05$.

Author Contributions: Conceptualization, R.C. (Roberto Canaparo), F.T., L.S., and R.C. (Roberta Cavalli); Data curation, M.A. and F.F.; Investigation, M.A., F.F., R.S., and F.C.; Supervision, R.C. (Roberto Canaparo), C.D.P., F.T., L.S., and R.C. (Roberta Cavalli); Writing—original draft, M.A. and F.F.; Writing—review & editing, R.C.(Roberta Cavalli), L.S., and R.C. (Roberto Canaparo). All authors have read and agreed to the published version of the manuscript.

Funding: The authors gratefully acknowledge funding from the University of Torino (Grant "Progetti di Ricerca di Ateneo 2011" and Grant "Ricerca Locale 2019") and from MIUR (PRIN 2010–2011 NANOMED, code 2010FPTBSH).

Conflicts of Interest: The authors declare no conflict of interest.

References

1. Bawarski, W.E.; Chidlowsky, E.; Bharali, D.J.; Mousa, S.A. Emerging nanopharmaceuticals. *Nanomedicine* **2008**, *4*, 273–282. [CrossRef] [PubMed]
2. Rivera, P.; Huhn, D.; del Mercato, L.L.; Sasse, D.; Parak, W.J. Nanopharmacy: Inorganic nanoscale devices as vectors and active compounds. *Pharm. Res.* **2010**, *62*, 115–125. [CrossRef] [PubMed]
3. Ferrari, M. Nanotechnology-enabled medicine. *Discov. Med.* **2005**, *5*, 363–366. [PubMed]
4. Malam, Y.; Loizidou, M.; Seifalian, A.M. Liposomes and nanoparticles: Nanosized vehicles for drug delivery in cancer. *Trends Pharm. Sci.* **2009**, *30*, 592–599. [CrossRef] [PubMed]
5. Wang, M.D.; Shin, D.M.; Simons, J.W.; Nie, S. Nanotechnology for targeted cancer therapy. *Expert Rev. Anticancer Ther.* **2007**, *7*, 833–837. [CrossRef]
6. Weissig, V.; Pettinger, T.K.; Murdock, N. Nanopharmaceuticals (part 1): Products on the market. *Int. J. Nanomed.* **2014**, *9*, 4357–4373. [CrossRef]

7. Dilnawaz, F.; Acharya, S.; Sahoo, S.K. Recent trends of nanomedicinal approaches in clinics. *Int J. Pharm.* **2018**, *538*, 263–278. [CrossRef]

8. Alvarez-Lorenzo, C.; Blanco-Fernandez, B.; Puga, A.M.; Concheiro, A. Crosslinked ionic polysaccharides for stimuli-sensitive drug delivery. *Adv. Drug Deliv. Rev.* **2013**, *65*, 1148–1171. [CrossRef]

9. Curcio, M.; Diaz-Gomez, L.; Cirillo, G.; Concheiro, A.; Iemma, F.; Alvarez-Lorenzo, C. pH/redox dual-sensitive dextran nanogels for enhanced intracellular drug delivery. *Eur. J. Pharm Biopharm.* **2017**, *117*, 324–332. [CrossRef]

10. Moya-Ortega, M.D.; Alvarez-Lorenzo, C.; Concheiro, A.; Loftsson, T. Cyclodextrin-based nanogels for pharmaceutical and biomedical applications. *Int. J. Pharm.* **2012**, *428*, 152–163. [CrossRef]

11. Ahmed, R.Z.; Patil, G.; Zaheer, Z. Nanosponges–A completely new nano-horizon: Pharmaceutical applications and recent advances. *Drug Dev. Ind. Pharm.* **2013**, *39*, 1263–1272. [CrossRef] [PubMed]

12. Selvamuthukumar, S.; Anandam, S.; Krishnamoorthy, K.; Rajappan, M. Nanosponges: A novel class of drug delivery system–review. *J. Pharm. Pharm. Sci.* **2012**, *15*, 103–111.

13. Chilajwar, S.V.; Pednekar, P.P.; Jadhav, K.R.; Gupta, G.J.; Kadam, V.J. Cyclodextrin-based nanosponges: A propitious platform for enhancing drug delivery. *Expert Opin. Drug Deliv.* **2014**, *11*, 111–120. [CrossRef] [PubMed]

14. Tejashri, G.; Amrita, B.; Darshana, J. Cyclodextrin based nanosponges for pharmaceutical use: A review. *Acta. Pharm.* **2013**, *63*, 335–358. [CrossRef]

15. Trotta, F.; Zanetti, M.; Cavalli, R. Cyclodextrin-based nanosponges as drug carriers. *Beilstein J. Org. Chem.* **2012**, *8*, 2091–2099. [CrossRef]

16. Sherje, A.P.; Dravyakar, B.R.; Kadam, D.; Jadhav, M. Cyclodextrin-based nanosponges: A critical review. *Carbohydr. Polym.* **2017**, *173*, 37–49. [CrossRef]

17. Allahyari, S.; Trotta, F.; Valizadeh, H.; Jelvehgari, M.; Zakeri-Milani, P. Cyclodextrin-based nanosponges as promising carriers for active agents. *Expert Opin. Drug Deliv.* **2019**, *16*, 467–479. [CrossRef]

18. Pandey, P.; Purohit, D.; Dureja, H. Nanosponges -A Promising Novel Drug Delivery System. *Recent Pat. Nanotechnol.* **2018**, *12*, 180–191. [CrossRef]

19. Argenziano, M.; Haimhoffer, A.; Bastiancich, C.; Jicsinszky, L.; Caldera, F.; Trotta, F.; Scutera, S.; Alotto, D.; Fumagalli, M.; Musso, T.; et al. In Vitro Enhanced Skin Permeation and Retention of Imiquimod Loaded in β-Cyclodextrin Nanosponge Hydrogel. *Pharmaceutics* **2019**, *11*, 138. [CrossRef]

20. Clemente, N.; Boggio, E.; Gigliotti, L.C.; Raineri, D.; Ferrara, B.; Miglio, G.; Argenziano, M.; Chiocchetti, A.; Cappellano, G.; Trotta, F.; et al. Immunotherapy of experimental melanoma with ICOS-Fc loaded in biocompatible and biodegradable nanoparticles. *J. Control. Release* **2020**, *320*, 112–124. [CrossRef]

21. Clemente, N.; Argenziano, M.; Gigliotti, C.L.; Ferrara, B.; Boggio, E.; Chiocchetti, A.; Caldera, F.; Trotta, F.; Benetti, E.; Annaratone, A.; et al. Paclitaxel-Loaded Nanosponges Inhibit Growth and Angiogenesis in Melanoma Cell Models. *Front. Pharm.* **2019**, *10*, 776. [CrossRef] [PubMed]

22. Argenziano, M.; Gigliotti, C.L.; Clemente, N.; Boggio, E.; Ferrara, B.; Trotta, F.; Pizzimenti, S.; Barrera, G.; Boldorini, R.; Bessone, F.; et al. Improvement in the Anti-Tumor Efficacy of Doxorubicin Nanosponges in In Vitro and in Mice Bearing Breast *Tumor Models. Cancers (Basel)* **2020**, *12*, 162. [CrossRef] [PubMed]

23. Swaminathan, S.; Pastero, L.; Serpe, L.; Trotta, F.; Vavia, P.; Aquilano, D.; Trotta, M.; Zara, G.P.; Cavalli, R. Cyclodextrin-based nanosponges encapsulating camptothecin: Physicochemical characterization; stability and cytotoxicity. *Eur. J. Pharm. Biopharm.* **2010**, *74*, 193–201. [CrossRef] [PubMed]

24. Trotta, F.; Dianzani, C.; Caldera, F.; Mognetti, B.; Cavalli, R. The application of nanosponges to cancer drug delivery. *Expert Opin. Drug Deliv.* **2014**, *11*, 931–941. [CrossRef] [PubMed]

25. Hoffman, A.S. Stimuli-responsive polymers: Biomedical applications and challenges for clinical translation. *Adv. Drug Deliv. Rev.* **2013**, *65*, 10–16. [CrossRef]

26. Caldera, F.; Tannous, M.; Cavalli, R.; Zanetti, M.; Trotta, F. Evolution of Cyclodextrin Nanosponges. *Int. J. Pharm.* **2017**, *531*, 470–479. [CrossRef]

27. Saravanakumar, K.; Hu, X.; Ali, D.M.; Wang, M.H. Emerging Strategies in Stimuli-Responsive Nanocarriers as the Drug Delivery System for Enhanced Cancer Therapy. *Curr. Pharm. Des.* **2019**, *25*, 2609–2625. [CrossRef]

28. Cheng, R.; Feng, F.; Meng, F.; Deng, C.; Feijen, J.; Zhong, Z. Glutathione-responsive nano-vehicles as a promising platform for targeted intracellular drug and gene delivery. *J. Control. Release.* **2011**, *152*, 2–12. [CrossRef]

29. Guo, X.; Cheng, Y.; Zhao, X.; Luo, Y.; Chen, J.; Yuan, W.E. Advances in redox-responsive drug delivery systems of tumor microenvironment. *J. Nanobiotechnol.* **2018**, *16*, 74. [CrossRef]

30. Raza, A.; Hayat, U.; Rasheed, T.; Bilal, M. Iqbal HMN. Redox-responsive nano-carriers as tumor-targeted drug delivery systems. *Eur. J. Med. Chem.* **2018**, *157*, 705–715. [CrossRef]

31. Klaunig, J.E. Oxidative Stress and Cancer. *Curr. Pharm. Des.* **2018**, *24*, 4771–4778. [CrossRef] [PubMed]

32. Kim, S.J.; Kim, H.S.; Seo, Y.R. Understanding of ROS-Inducing Strategy in Anticancer Therapy. *Oxid. Med. Cell Longev.* **2019**, *2019*, 5381692. [CrossRef]

33. Traverso, N.; Ricciarelli, R.; Nitti, M.; Marengo, B.; Furfaro, A.L.; Pronzato, M.A.; Marinari, U.M.; Domenicotti, C. Role of glutathione in cancer progression and chemoresistance. *Oxid. Med. Cell Longev.* **2013**, *2013*, 972913. [CrossRef] [PubMed]

34. Wang, J.; Yi, J. Cancer cell killing via ROS: To increase or decrease; that is the question. *Cancer Biol.* **2008**, *7*, 1875–1884. [CrossRef]

35. Trotta, C.F.; Dianzani, C.; Argenziano, M.; Barrera, G.; Cavalli, R. Glutathione bioresponsive cyclodextrin nanosponges. *Chem. Plus Chem.* **2016**, *81*, 5.

36. Yang, D.; Chen, W.; Hu, J. Design of controlled drug delivery system based on disulfide cleavage trigger. *J. Phys. Chem. B* **2014**, *118*, 12311–12317. [CrossRef]

37. Daga, M.; Ullio, C.; Argenziano, M.; Dianzani, C.; Cavalli, R.; Trotta, F.; Ferretti, C.; Zara, G.P.; Gigliotti, C.L.; Ciamporcero, E.S.; et al. GSH-targeted nanosponges increase doxorubicin-induced toxicity "in vitro" and "in vivo" in cancer cells with high antioxidant defenses. *Free Radic. Biol. Med.* **2016**, *97*, 24–37. [CrossRef]

38. Argenziano, M.; Lombardi, C.; Ferrara, B.; Trotta, F.; Caldera, F.; Blangetti, M.; Koltai, H.; Kapulnik, Y.; Yarden, R.; Gigliotti, L.; et al. Glutathione/pH-responsive nanosponges enhance strigolactone delivery to prostate cancer cells. *Oncotarget* **2018**, *9*, 35813–35829. [CrossRef]

39. Kimlin, L.C.; Casagrande, G.; Virador, V.M. In vitro three-dimensional (3D) models in cancer research, an update. *Mol. Carcinog.* **2013**, *52*, 167–182. [CrossRef]

40. Wang, X.; Zhen, X.; Wang, J.; Zhang, J.; Wu, W.; Jiang, X. Doxorubicin delivery to 3D multicellular spheroids and tumors based on boronic acid-rich chitosan nanoparticles. *Biomaterials* **2013**, *34*, 4667–4679. [CrossRef]

41. Kunz-Schughart, L.A.; Kreutz, M.; Knuechel, R. Multicellular spheroids: A three-dimensional in vitro culture system to study tumour biology. *Int. J. Exp. Pathol.* **1998**, *79*, 1–23. [CrossRef]

42. Clemedson, C.; Ekwall, B. Overview of the Final MEIC Results: I. The In Vitro-In Vitro Evaluation. *Toxicol. In Vitro* **1999**, *13*, 657–663. [CrossRef]

43. Scheers, E.M.; Ekwall, B.; Dierickx, P.J. In vitro long-term cytotoxicity testing of 27 MEIC chemicals on Hep G2 cells and comparison with acute human toxicity data. *Toxicol. In Vitro* **2001**, *15*, 153–161. [CrossRef]

44. Lewinski, N.; Colvin, V.; Drezek, R. Cytotoxicity of nanoparticles. *Small* **2008**, *4*, 26–49. [CrossRef] [PubMed]

45. Jones, C.F.; Grainger, D.W. In vitro assessments of nanomaterial toxicity. *Adv. Drug Deliv. Rev.* **2009**, *61*, 438–456. [CrossRef]

46. Ballatori, N.; Krance, S.M.; Notenboom, S.; Shi, S.; Tieu, K.; Hammond, C.L. Glutathione dysregulation and the etiology and progression of human diseases. *Biol. Chem.* **2009**, *390*, 191–214. [CrossRef]

47. Kramer, R.A.; Zakher, J.; Kim, G. Role of the glutathione redox cycle in acquired and de novo multidrug resistance. *Science* **1988**, *241*, 694–697. [CrossRef]

48. Park, J.H.; Choi, K.H.; Kwak, H.S. Single- and 14-day repeat-dose toxicity of cross-linked beta-cyclodextrin in rats. *Int. J. Toxicol.* **2011**, *30*, 700–706. [CrossRef]

49. Minelli, R.; Cavalli, R.; Ellis, L.; Pettazzoni, P.; Trotta, F.; Ciamporcero, E.; Barrera, G.; Fantozzi, R.; Dianzani, C.; Pili, R.; et al. Nanosponge-encapsulated camptothecin exerts anti-tumor activity in human prostate cancer cells. *Eur. J. Pharm. Sci.* **2012**, *47*, 686–694. [CrossRef]

50. Torne, S.; Darandale, S.; Vavia, P.; Trotta, F.; Cavalli, R. Cyclodextrin-based nanosponges: Effective nanocarrier for tamoxifen delivery. *Pharm. Dev. Technol.* **2013**, *18*, 619–625. [CrossRef]

51. Shende, P.; Kulkarni, Y.A.; Gaud, R.S.; Deshmukh, K.; Cavalli, R.; Trotta, F.; Caldera, F. Acute and repeated dose toxicity studies of different beta-cyclodextrin-based nanosponge formulations. *J. Pharm. Sci.* **2015**, *104*, 1856–1863. [CrossRef] [PubMed]

52. Daga, M.; de Graaf, I.A.M.; Argenziano, M.; Barranco, A.S.M.; Loeck, M.; Al-Adwi, Y.; Cucci, M.A.; Caldera, F.; Trotta, F.; Barrera, G.; et al. Glutathione-responsive cyclodextrin-nanosponges as drug delivery systems for doxorubicin: Evaluation of toxicity and transport mechanisms in the liver. *Toxicol. In Vitro* **2020**, *65*, 104800. [CrossRef] [PubMed]

53. Jayakumar, S.; Kunwar, A.; Sandur, S.K.; Pandey, B.N.; Chaubey, R.C. Differential response of DU145 and PC3 prostate cancer cells to ionizing radiation: Role of reactive oxygen species; GSH and Nrf2 in radiosensitivity. *Biochim. Biophys. Acta* **2014**, *1840*, 485–494. [CrossRef] [PubMed]

54. Nguyen, T.; Nioi, P.; Pickett, C.B. The Nrf2-antioxidant response element signaling pathway and its activation by oxidative stress. *J. Biol. Chem.* **2009**, *284*, 13291–13295. [CrossRef] [PubMed]

55. Wakabayashi, N.; Slocum, S.L.; Skoko, J.J.; Shin, S.; Kensler, T.W. When NRF2 talks, who's listening? *Antioxid. Redox Signal.* **2010**, *13*, 1649–1663. [CrossRef]

56. Franco, R.; Cidlowski, J.A. Apoptosis and glutathione: Beyond an antioxidant. *Cell Death Differ.* **2009**, *16*, 1303–1314. [CrossRef]

57. Aubry, S.; Burlina, F.; Dupont, E.; Delaroche, D.; Joliot, A.; Lavielle, S.; Chassaing, G.; Sagan, S. Cell-surface thiols affect cell entry of disulfide-conjugated peptides. *Faseb. J.* **2009**, *23*, 2956–2967. [CrossRef]

58. Feener, E.P.; Shen, W.C.; Ryser, H.J. Cleavage of disulfide bonds in endocytosed macromolecules. A processing not associated with lysosomes or endosomes. *J. Biol. Chem.* **1990**, *265*, 18780–18785.

59. Choi, Y.A.; Chin, B.R.; Rhee, D.H.; Choi, H.G.; Chang, H.W.; Kim, J.H.; Baek, S.H. Methyl-beta-cyclodextrin inhibits cell growth and cell cycle arrest via a prostaglandin E(2) independent pathway. *Exp. Mol. Med.* **2004**, *36*, 78–84. [CrossRef]

60. Lee, J.; Lilly, G.D.; Doty, R.C.; Podsiadlo, P.; Kotov, N.A. In vitro toxicity testing of nanoparticles in 3D cell culture. *Small* **2009**, *5*, 1213–1221. [CrossRef]

61. Vandeputte, C.; Guizon, I.; Genestie-Denis, I.; Vannier, B.; Lorenzon, G. A microtiter plate assay for total glutathione and glutathione disulfide contents in cultured/isolated cells: Performance study of a new miniaturized protocol. *Cell Biol. Toxicol.* **1994**, *10*, 415–421. [CrossRef] [PubMed]

62. Haslam, G.; Wyatt, D.; Kitos, P.A. Estimating the number of viable animal cells in multi-well cultures based on their lactate dehydrogenase activities. *Cytotechnology* **2000**, *32*, 63–75. [CrossRef] [PubMed]

63. Bustin, S.A.; Benes, V.; Garson, J.A.; Hellemans, J.; Huggett, J.; Kubista, M.; Mueller, R.; Nolan, T.; Pfaffl, M.W.; Shipley, G.; et al. The MIQE guidelines: Minimum information for publication of quantitative real-time PCR experiments. *Clin. Chem.* **2009**, *55*, 611–622. [CrossRef] [PubMed]

Sample Availability: Samples of the GSH-NS are to be requested to Professor Francesco Trotta.

 *molecules*

Article

Thermodynamic Characterization of the Interaction between the Antimicrobial Drug Sulfamethazine and Two Selected Cyclodextrins

Hiba Mohamed Ameen [1,2], Sándor Kunsági-Máté [2,3], Balázs Bognár [2], Lajos Szente [4], Miklós Poór [3,5] and Beáta Lemli [2,3,*]

1. Department of General and Physical Chemistry, Faculty of Sciences, University of Pécs, Ifjúság 6, H-7624 Pécs, Hungary; hiba83@gamma.ttk.pte.hu
2. Institute of Organic and Medicinal Chemistry, Medical School, University of Pécs, Szigeti 12, H-7624 Pécs, Hungary; kunsagi-mate.sandor@gytk.pte.hu (S.K.-M.); balazs.bognar@aok.pte.hu (B.B.)
3. János Szentágothai Research Center, University of Pécs, Ifjúság 20, H-7624 Pécs, Hungary; poor.miklos@pte.hu
4. CycloLab Cyclodextrin Research & Development Laboratory, Ltd., Illatos 7, H-1097 Budapest, Hungary; szente@cyclolab.hu
5. Department of Pharmacology, Faculty of Pharmacy, University of Pécs, Szigeti 12, H-7624 Pécs, Hungary
* Correspondence: beata.lemli@aok.pte.hu; Tel.: +36-72-503-600 (ext. 35462)

Academic Editors: Marina Isidori, Margherita Lavorgna and Rosa Iacovino
Received: 17 October 2019; Accepted: 11 December 2019; Published: 13 December 2019

Abstract: Sulfamethazine is a representative member of the sulfonamide antibiotic drugs; it is still used in human and veterinary therapy. The protonation state of this drug affects its aqueous solubility, which can be controlled by its inclusion complexes with native or chemically-modified cyclodextrins. In this work, the temperature-dependent (298–313 K) interaction of sulfamethazine with native and randomly methylated β-cyclodextrins have been investigated at acidic and neutral pH. Surprisingly, the interaction between the neutral and anionic forms of the guest molecule and cyclodextrins with electron rich cavity are thermodynamically more favorable compared to the cationic guest. This property probably due to the enhanced formation of zwitterionic form of sulfamethazine in the hydrophobic cavities of cyclodextrins. Spectroscopic measurements and molecular modeling studies indicated the possible driving forces (hydrophobic interaction, hydrogen bonding, and electrostatic interaction) of the complex formation, and highlighted the importance of the reorganization of the solvent molecules during the entering of the guest molecule into the host's cavity.

Keywords: cyclodextrin; sulfamethazine; zwitterion; host-guest complex; thermodynamics

1. Introduction

Formation of host-guest type inclusion complexes typically occurs when the host molecule uses its cavity to encapsulate a guest through noncovalent interactions. According to the significant practical utility of macrocyclic molecules, such as calixarenes [1,2], cavitands [3,4], and cyclodextrins (CDs) [5,6] in host-guest complex formations, chemists, biologists, and material scientists got interested the physical properties, chemical nature, and related biological activity of these molecules. However, utilization of these noncovalent interactions (hydrogen-bonding, π-stacking, electrostatic interaction, van der Waals force, and hydrophobic/hydrophilic attraction) are still a great challenge [7–10]. CDs, a fascinating class of macrocycles, are composed of six, seven, or eight glucose units, called α-, β-, and γ-CDs, respectively. CDs are used as host components for the construction of various interesting supramolecular structures [11,12]. Sulfonamide antibiotics are widely used in both human medicine and livestock production to treat some bacterial infections of the urinary tract, ears, lungs, skin, and soft tissues [13,14]. Furthermore, sulfonamides can appear as contaminants in various foods, which may cause adverse health effects [15–17]. The host-guest type complex formation of these antibiotics with CDs is an extensively studied field [18–24]. Zoppi et al. focus on the increased water solubility of sulfonamide drugs in the presence of native and methylated β-CD [23,24]. In the case of sulfamethazine (SMT), their nuclear magnetic resonance (NMR) and molecular modeling results demonstrate that SMT included the substituted pyrimidine ring into the β-CD cavity. Contradictory, NMR and quantum chemical results of Bani-Yaseen and Mo'ala revealed that complex formation is favorable with inclusion of the aniline moiety through the β-CD cavity [18].

Several studies have been performed to get an insight into the factors which affect the thermodynamic and kinetic stability or selectivity of host-guest complexes [25,26], because the deeper understanding of these interactions has high importance. The pH-responsive host-guest encapsulation is also a highly studied field in material sciences [27] and in pharmacology [28,29]. Therefore, besides the complex stability and stoichiometry of SMT – β-CD complex and along the contradictory description of the related structures [18,23,24], the investigation of the pH dependence interaction of SMT with CDs is also reasonable.

In our recent study [30], we demonstrated the importance of pH-dependent dipole moment of SMT molecule, which phenomenon can affect the complex geometry formed with β-CD (BCD) and randomly methylated β-CD (RAMEB) (Figure 1). Now we focus on the thermodynamic properties of the formation of inclusion complexes at different pH values. Our aim is to analyze the weak interactions between the pH dependent ionic and neutral forms of SMT and native or methylated CDs at molecular level to clarify the previous contradictory results. In this way, the involvement of weak molecular interactions (electrostatic forces and hydrogen bonds) have been tested by the temperature-dependent measurements and molecular modeling studies.

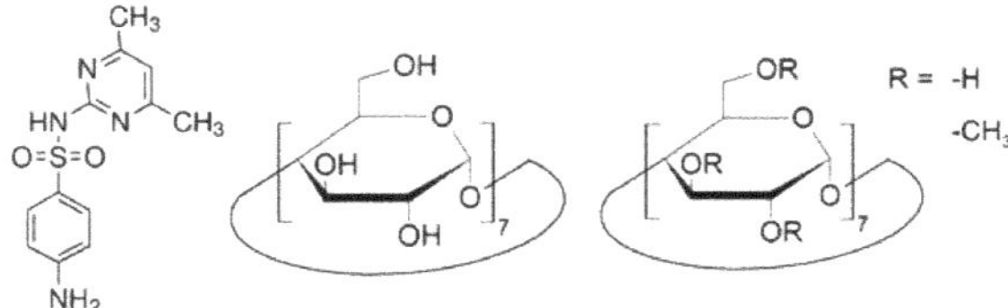

Figure 1. Chemical structures of sulfamethazine (SMT), native β-cyclodextrin (BCD), and randomly methylated β-cyclodextrin (RAMEB).

2. Results and Discussion

2.1. Temperature Dependence of the Association Constants of Sulfamethazine-CD Complexes at Different pH

Figure 2 shows the van't Hoff plot of SMT-CD complexes, based on association constants determined at different temperatures. In accordance with our earlier findings [30], significant difference between the association constants at elevation pH (pH = 5 and pH = 7) and at strong acidic environment (pH = 2) has been found. The slight dependence of complex stabilities on the temperature reflects low enthalpy changes, i.e., weak interactions between the molecules. At pH 7, where the nonionic and anionic guest molecules are dominant, higher stability is associated to the complexes at decreased temperatures. In contrast, in the presence of considerable amount of cationic guest at pH 2, the complex stability increases with the elevation of the temperature. Although, only one form of the guest molecule (nonionic SMT) is available at pH 5, the substitution of the β-CD affects the change of the association constants with the temperature. The association constants of CD complexes generally decrease with the elevation of the temperature [19,31]. However, one of our earlier work showed an opposite example [32]. The thermodynamic parameters have been also determined to analyze further the related processes.

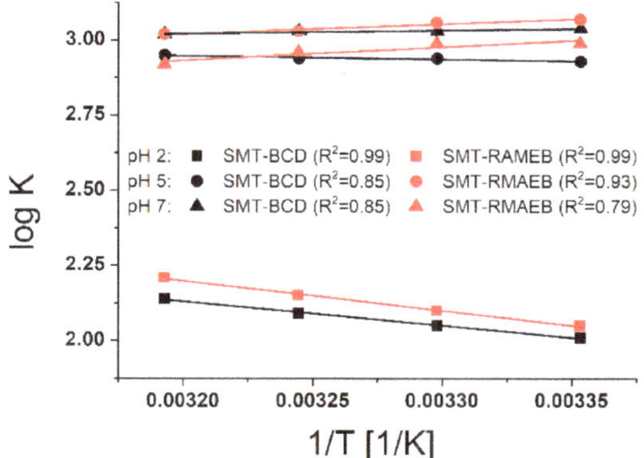

Figure 2. The van't Hoff plots of SMT-BCD and SMT-RAMEB complex formations at different pH values.

Thermodynamic parameters (Table 1) were calculated from the slopes and the intercepts of the lines fitted to the experimental data based on the van't Hoff plot (Equation (1), see Figure 2). The negative ΔG values yield spontaneous complex formation between SMT and CDs. Results showed exothermic association at high pH (pH = 7), while an endothermic molecular association was obtained at low pH (pH = 2). At pH 5, the endothermic character of the complex formation was just changed to exothermic as a result of the methyl substitution of BCD. In each interaction, an entropy gain was observed; however, the entropy increase during the complex formation correlates with the enthalpy change. The entropy increase during the association reaction was probably due to the process when SMT enters the CD cavity (it releases its solvation shell). Furthermore, higher entropy gain associated with positive or less negative enthalpy, which property reflects to the removal of more or less water molecules from the solvation shell regarding the molecules interacted during formation of complexes. Decreased ΔS at higher pH values suggest the release of less water molecules from the solvation shell of SMT molecules, because the stabilization is also supported by the attractive coulomb forces between the negatively charged SMT and the dipole moments of the solvent molecules. The correlation between the enthalpy and the entropy changes can be described by the changes of the solvation shell of guests, since the removal of less water molecules from the solvation shell costs less energy. This description agrees with the enthalpy-entropy compensation and highlights that the exothermicity of molecular association usually restricts the movement of the constituents, thereby causing growing entropy loss.

Table 1. Thermodynamic parameters associated to the formation of SMT-CD complexes. Data are determinate based on temperature-dependent fluorescence spectroscopic measurements. (ΔH [kJ mol^{-1}], ΔS [J K^{-1} mol^{-1}] $\Delta G298K$ [kJ mol^{-1}]).

Host Species	pH								
	2			5			7		
	ΔH	ΔS	ΔG_{298K}	ΔH	ΔS	ΔG_{298K}	ΔH	ΔS	ΔG_{298K}
BCD	15.4 ± 0.8	90.0 ± 2.5	−11.4 ± 1.5	2.2 ± 0.5	63.3 ± 1.7	−16.7 ± 1.0	−2.2 ± 0.5	51.0 ± 1.7	−17.3 ± 1.0
RAMEB	18.9 ± 0.8	102.7 ± 2.6	−11.7 ± 1.6	−6.4 ± 1.0	37.2 ± 3.3	−17.5 ± 2.0	−8.5 ± 1.2	28.8 ± 3.8	−17.1 ± 2.3

2.2. Modeling Studies

To get a deeper insight into the complex formation processes, molecular modeling studies were performed at semi-empirical level. During these calculations, the energetically favorable deprotonation route of SMT molecule was determined first in aqueous solutions considering the presence of other ions as described in the Materials and Methods section. Sulfamethazine exists as cationic (SMT$^+$), anionic (SMT$^-$), nonionic (SMT^0) and zwitterionic (SMT$^{+/-}$) forms in aqueous solutions. Figure 3 shows that the aromatic amine moiety, which is protonated at low pH loses first the proton while the second deprotonation occurs at the sulfonamide nitrogen. The associated experimental pKa$_1$ and pKa$_2$ values at room temperature were 2.07 and 7.49, respectively. We should mention here that the Gibbs free energy difference between the nonionic and zwitterionic forms of SMT was found to be 12.3 kJ mol^{-1} in this environment. This result suggests presence of SMT in nonionic rather than zwitterionic form in the solutions, however, it is known that zwitterionic form can stabilized e.g., in adsorbed state [33]. Then the interactions of these three forms of SMT (cationic, nonionic and anionic) were examined with BCD and RAMEB host molecules in the aqueous buffer. Due to the huge computation time of the large systems, in the case of RAMEB the electron releasing property of the methyl groups was considered as negatively charged specie of the native BCD molecules. Thus, the repulsive Coulomb interaction between the negatively charged RAMEB cavity (simulated by −1 BCD) and the deprotonated SMT species will reduce the secondary interactions between the host and guest molecules. In contrast, the charged SMT species showed stronger interactions with the negatively charged cavity of RAMEB. Furthermore, the host molecules formed even more stable complexes with the anionic form of the guest. From the point of view the enthalpy (Table 2), the following process is responsible for these unexpected results: at low pH the cationic SMT molecule enters into the host cavity with its aromatic amine moiety. However, at higher pH, SMT molecule enters with its methyl substituents. In the former cases, hydrogen bridges between the (guest amine) N-H ⋯ O (host hydroxyl), while in the latter cases, the hydrogen bridges between the (guest methyl) C-H ⋯ O (host hydroxyl) are moderate the weak interactions between the host and guest (Figure 4). Noted here, that this pH dependent orientation of guest molecule in the complexes support the earlier described structures based on the inclusion of the aniline moiety [18] as well as the pyrimidine ring [24] through the CD cavity.

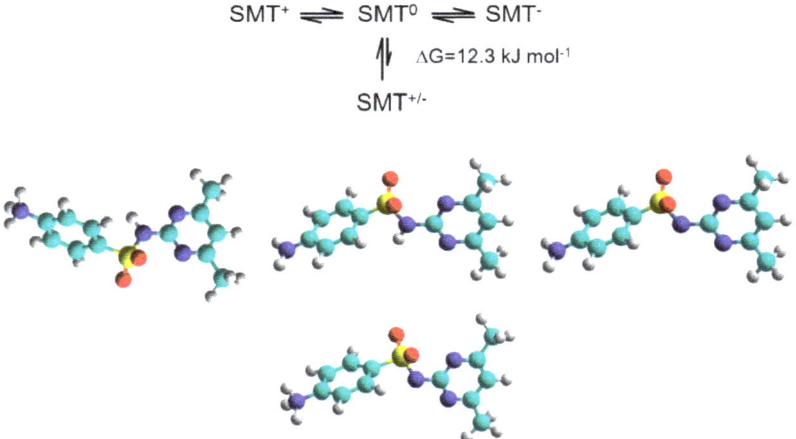

Figure 3. The energetically most favorable deprotonation routes of SMT (cationic: left, nonionic: middle, anionic: right, zwitterionic: bottom) determined by MINDO/3 approximation using the TIP3P solvation model for the buffer [34]. Gibbs free energy between the nonionic and zwitterionic forms suggest presence preferably of nonionic form in the solution.

Table 2. Thermodynamic parameters associated to the formation of SMT-CD complexes. Semiempirical MINDO/3 method with TIP3P solvation model is applied. (ΔH [kJ mol^{-1}], ΔS [J K^{-1} mol^{-1}]).

Host Specie	Host Simulated as	Guest's Charges							
		+1 (Cationic)		0 (Nonionic)		0 (Zwitterionic)		−1 (Anionic)	
		ΔH	ΔS	ΔH	ΔS	ΔH	ΔS	ΔH	ΔS
BCD	0 BCD	16.3	93.0	9.3	78.2	5.4	68.4	−3.7	47.5
RAMEB	−1 BCD	19.1	105.4	14.3	99.7	−8.7	35.2	−9.4	26.4

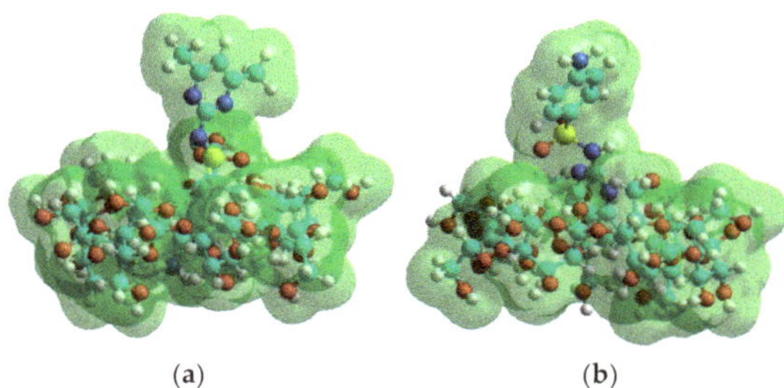

(a)   (b)

Figure 4. Equilibrium conformation of SMT-BCD complexes. (**a**) SMT molecules with their aromatic amine moiety and (**b**) with their methyl groups enter into the cavities of hosts.

Furthermore, the inclusion of SMT by its aromatic amine moiety in case of RAMEB host enhances formation of zwitterionic form of SMT in the cavity. This is due to the tautomerization of the proton from the sulfonamide to the aromatic amine moiety enhanced by the Coulomb interaction of the proton with the negatively charged cavity of RAMEB. With the aim to justify this conception of the SMT's zwitterion formation in the RMAEB's cavity, simultaneous analysis of complexation behavior has been done using infrared (IR) spectroscopy. In general, our results are in agreement with the IR analyses of SMT-BCD complexes prepared by a freeze-drying method [23], the characteristic bands of SMT shifted and are more or less intense in the presence of CD. Moreover, IR spectra of the SMT-RAMEB complexes and the species interacted support our idea described above (Figure 5): significant changes of two characteristic vibrations of SMT molecules were observed upon complexation by the RAMEB host as follows. Quantum chemical analysis revealed that belting vibration of SNH bond angle at sulfonamide moiety (1103 cm^{-1}) disappeared while the bond stretching associated to the aromatic NH$_3$ is appeared at 2812 cm^{-1} in the experimental IR spectra of the complexes. These changes in the experimental IR spectra indicate the stabilization of the zwitterionic form of SMT in the RAMEB cavity. This phenomenon has not been observed in the case of the BCD host.

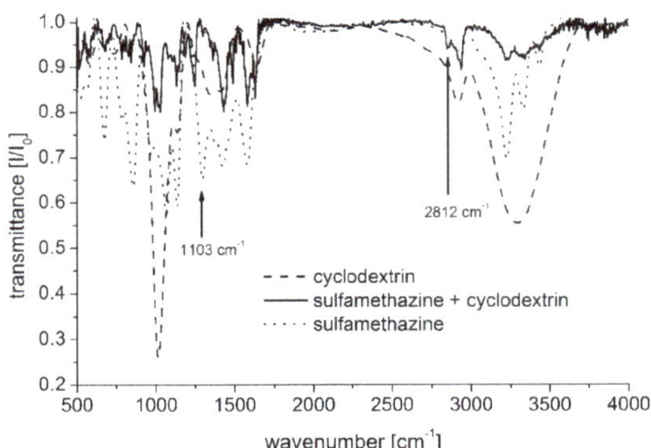

Figure 5. Infrared spectra of SMT – RAMEB complexes.

In all eight situations, the interactions show an increased entropy term (Table 2). This property is associated with two facts: the solvent water molecules leave the host's cavity prior to the complex formation and the guest molecules (at least partly) lose their hydration shell. Both processes increase the entropy. In particular, the entropy gain decreases by the second deprotonation step. This is probably due to the increase in the stability of the hydration shell regarding the anionic SMT molecules.

Considering that the formation of hydrogen bridges between the host and guest always assumes dehydration of the appropriate part of the host and guest molecules, the energy cost of dehydration compensated by the entropy gain associated to the increased freedom of the water molecules after the dehydration. This assumption is supported by the good agreement between the measured and calculated thermodynamic parameters.

2.3. Driving Forces of the SMT-CD Complex Formations

Taking into account the binding conformations suggested by theoretical modeling, we can discuss in detail the thermodynamic parameters of complex formation between CDs and SMT. However, it should be noted here, that thermodynamic parameters derived from temperature- dependent spectroscopic measurements assume that these parameters are constants within the temperature range of investigation. Furthermore, these data reflect not only for the temperature-dependent change of the association constants, but also for the way how the association constants have been determined. Spectroscopic identification of association constant based on changes of the environment around the guest when the molecule enters from the polar aqueous media the hydrophobic cavity of the CD. Therefore, the related enthalpy changes and entropy changes describe the complex formation without solvent interaction in the bulk phase. Isothermal titration calorimetry is the accurate technique to solve this problem and to measure directly thermodynamic properties of host-guest complex formation. However, in this work, the thermodynamic parameters were determined based on fluorescence spectroscopic measurements, using the van't Hoff equation. Relevant experiments showed [19] that the results of calorimetric studies are similar to the spectroscopic findings regarding host-guest type CD complexes, and the data of thermodynamic parameters only slightly differs between the two methods. This property supports our conclusions made on the spectroscopic data.

The possible driving forces which stabilize the host-guest complexes of CDs are electrostatic interaction, van der Waals interaction, hydrophobic interaction, hydrogen bonding, relief of conformational strain, charge transfer interaction, and release of water molecules from the hydrophobic cavity of the host to the bulk phase [35]. The values of thermodynamic parameters consist the contribution of the species' desolvation and the different kind of noncovalent interactions listed above. In general, the combination of both negative or positive enthalpy and entropy changes indicate that van der Waals forces and hydrogen bonding or hydrophobic interaction take places in complex formation, respectively. While higher negative values of ΔH combined with positive ΔS have found for the electrostatic driving forces combined with hydrogen bonds of ionized groups [36]. However, the given values can be strongly affected by intensive dehydration and solvent reorganization. In the discussion of the present experimental data (Table 1) we focus on two tendencies observed in the thermodynamic parameters: both the enthalpy and entropy changes associated to the complex formation decrease while the charge of the guest SMT molecules varies from +1, 0 to −1. On this base, considering the attractive forces between the anionic cavity of the host and the cationic guest at pH = 2, highly negative enthalpy changes should be observed in vacuo. However, the desolvation of the guest costs more energy than it is causes during the association of SMT with BCD, therefore a positive enthalpy change can be observed. The ordered structure of solvent molecules in the solvation shell is destroyed after the complex is formed and the free solvent molecules gain the entropy. Results related to the complexation of the neutral form at pH = 5 suggest preference of the latest effect: weaker stability of the solvation shell assumes much lower energy costs for its destroying, therefore the enthalpy change lowered instead the weaker contribution of the attractive coulomb forces. As parallel effect on the entropy, weaker stability of solvent molecules in the solvation shell of the guest assumes higher entropy content of the solvent molecules prior complex formation which property causes lower entropy gain during the interaction with the CD hosts. The complex formation, however, is also affected by the formation of zwitterion of the guest at pH = 5 and this property enhances the decrease of the enthalpy when the positively charged NH_3 group of SMT interact the more negatively charged cavity of the RAMEB while the negative sulfonamide nitrogen of SMT interact with the positively charged methyl groups of the host. These three processes (coulomb interaction, desolvation of the guest prior formation of the complex and the formation of zwitterionic derivative of the SMT) will then compete. At pH = 7 comparable amount of neutral and anionic form of SMT are presented in aqueous solution. The further decrease of both the enthalpy and entropy changes associated to the complex formation highlighted the complex stabilization effect of the deprotonated sulfonamide nitrogen. Presence of competition of the processes above was then confirmed by the analysis of enthalpy – entropy compensation.

The enthalpy-entropy compensation is still a widely observed and unresolved phenomenon in chemical thermodynamics [37–39]. The linear correlation when the experimentally found ΔH and ΔS values are plotted against each other is believed to play an important role in the formation of weak interactions. However, for similar systems, the Gibbs free energy remains the same. Figure 6 shows the ΔH vs. ΔS plots for SMT-CD complexes analyzed in the present work. Both experimentally and theoretically determined data are presented. Although the processes have been investigated in a small temperature range (298–313 K), the compensation temperature determined from the slope of the good straight line (387 K and 374 K experimental and theoretical data, respectively) are far to the average temperature. This small difference could arise from the indirect determination of the thermodynamic parameters based on spectroscopic measurements. The difference between the ΔG values (~6.1 kJ mol^{-1} in the present systems) brings the experimental and compensation temperature farther [38]. In biological supramolecular systems, also in CD chemistry, the studies of enthalpy-entropy compensation have been started early and it has been widely investigated. Twenty years ago, a review comprises more than 1000 thermodynamic data of the inclusion complexes of native and chemically-modified CDs [40]. Based on the analyzes of the enthalpy-entropy compensation plot of native and modified CDs or the α-, β- and γ-CDs, it was found that the linearity and the slope of the straight line could be affected by the difference between the conformational change of the native and modified CDs, by the desolvation of both host and guest molecules, and by the ring size and flexibility. However, in recent studies [35,37], the compensation effect is mainly interpreted based on the changes in the level of hydration and reorganization of the solvent molecules. The considerable effect of the solvent is not surprising, since solvation known to affect the electronic structure of molecules, so affects the interactions between electrons of different atomic or molecular orbitals. Therefore, it affects also the molecular interactions, especially when they are weak [41,42]. In the present case, if the anionic guest molecule keeps the part of its solvation shell, then the higher ordered structure of the complexes (included by its solvation shell) explain the deprotonation enhanced entropy gain decreases. Because there is no significant difference between the cavity size and flexibility of BCD and RAMEB and the enhanced electron rich character of the methylated CD should result in opposite effect than we have found, the small entropy term differences can be explained by poor solubility of native BCD (owing to the highly ordered water molecules in its solvation shell) [43]. When the guest molecule enters into the CD cavity, the interaction (at least partly) destroys the solvation shell of the host and weakens the CD-solvent interaction. Similarly to our earlier findings [32], when the solvent molecules leave the host's cavity, reorganization of the more ordered BCD-water structure results in a higher entropy change vs. the less ordered RAMEB-water system.

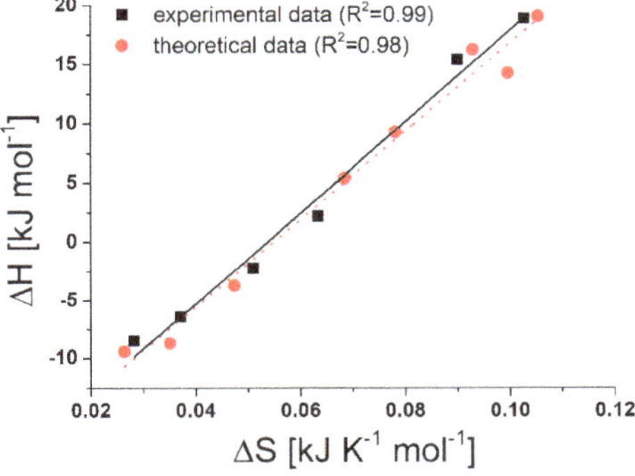

Figure 6. Enthalpy-entropy compensation plot of SMT-BCD and SMT-RAMEB complexes.

3. Materials and Methods

3.1. Reagents

Sulfamethazine (SMT) was purchased from Alfa Aesar (Kandel, Germany). Stock solutions of SMT (5000 µM) were prepared in methanol (spectroscopic grade, Reanal, Budapest, Hungary). Diluted solutions of SMT were prepared by evaporating the methanol under relatively low pressure, then SMT was dissolved in appropriate volumes of the phosphate buffer of interest. CDs, including β-cyclodextrin (BCD) and randomly methylated β-cyclodextrin (RAMEB) were obtained from CycloLab Cyclodextrin Research and Development Laboratory, Ltd. (Budapest, Hungary). All the other analytical grade chemicals were purchased from VWR International Ltd. (Debrecen, Hungary). Phosphate buffer solutions have been prepared by mixing (0.1 M) H_3PO_4 and (0.1M) Na_2HPO_4, (0.01 M) H_3PO_4 and (0.01 M) Na_2HPO_4 or (0.01 M) KH_2PO_4 and (0.01 M) Na_2HPO_4 stock solutions until the requested pH 2, 5 or 7 were reached, respectively. Ultrapure water (conductivity < 0.1 µS/cm,) were prepared by an Adrona (Riga, Latvia) water purification system.

3.2. Fluorescence Spectroscopic Studies

Highly sensitive Fluorolog tau3 spectrofluorometer (Jobin-Yvon/SPEX, Longjumeau, France) was used to investigate the fluorescence spectra of the different solutions. For data collection, the photon counting method with 0.1 s integration time was used. Excitation and emission bandwidths were set to 4 nm. A 10 mm thickness of the fluorescent probes with right-angle detection was applied. Temperature-dependent steady-state fluorescence spectroscopic measurements were carried out at different temperatures: 298.2 K, 303.2 K, 308.2 K, and 313.2 K. The fluorescence emission spectra of SMT (30 µM) was recorded in the absence and presence of increasing concentration of BCD or RAMEB (0–3 mM) in different phosphate buffers, using 280 nm excitation wavelength. Similarly, to our previous studies [30,44,45], overall and stepwise association constants of the complex formation were calculated by non-linear fitting, based on the fluorescence emission data obtained, employing the HyperQuad2006 program package [46].

To determine the thermodynamic parameters, temperature dependence of the complex stabilities was examined. According to the van't Hoff Equation (1) the temperature-dependence of the association constants offers possibility to determine the thermodynamic parameters related to the formation of the SMT-BCD and SMT-RAMEB complexes:

$$lnK = -\frac{\Delta G}{RT} = -\frac{\Delta H}{RT} + \frac{\Delta S}{R}, \tag{1}$$

where the ΔH and ΔS stand for the enthalpy and entropy changes of the complex formation, while ΔG is the Gibbs free energy change. R stands for the gas constant, while T is the temperature in Kelvin.

3.3. Infrared Spectroscopy

Fourier transform infrared spectra of SMT, BCD, RAMEB and SMT-BCD and SMT-RAMEB complexes were recorded on Platinum Alpha T FT-IR Spectrometer (Bruker, Ettlingen, Germany). Droplets of samples is used for these measurements. Average of ten scans with 5 cm^{-1} resolution is applied.

3.4. Modeling Studies

The thermodynamic parameters of the SMT-BCD or SMT-RAMEB complexes were determined at 298 K as follows: The enthalpy change considered as the energy change calculated by subtracting the total energies of the reactants from the total energies of the products. Similarly, the entropy changes were calculated by subtracting the entropy terms of the reactants from the entropy terms of the products. The entropy terms of the species interacted were calculated applying Boltzmann statistics.

The higher contribution to the entropy comes from the vibrational motions. Therefore, after calculating the vibrational frequencies using the harmonic approximation, the entropy was then determined as the following equation implemented in the HyperChem code:

$$S_{vib} = R \sum_i \left\{ \frac{h\nu_i/kT}{e^{(h\nu_i/kT)} - 1} - \ln\left[1 - e^{(-h\nu_i/kT)}\right] \right\} \tag{2}$$

where the ν_i is the frequency of vibration and T is the temperature (298.16 K).

The total energies of the species interacted have been calculated at semi-empirical MINDO/3 level using HyperChem 8 code. After the geometry optimization at MINDO/3 level the vibrational-rotational analysis was performed in harmonic approximation using AM1 approximation. Neutral aqueous environment was considered by the TIP3P solvation model implemented in HyperChem code [47]. Considering that in the present studies ionic species are interacted, the ionic strength of the buffer were considered by the additional PO_4^{3-}, K^+, Na^+ and H_3O^+ ions as described in an earlier study [34]. Accordingly, the final cube for representing solvents has 30 Å × 30 Å × 30 Å sizes and contained water, PO_4^{3-}, HPO_4^{2-}, K^+, Na^+ and 9 H_3O^+ according to the composition of the buffer solution while the pH varied from 7, 5 and 2. After 10 ps MD simulation to equilibrate the system at room-temperature at MM+ level, the calculations for the complexes and the separated species interacted were performed at MINDO/3 level. To reduce the huge computational time, the random-methylated CD derivative (RAMEB, which have electron-rich cavity) was considered as negatively charged species of the native BCD [48,49].

4. Conclusions

The complex formation between different sulfonamides and cyclodextrins still has attract much attention. Previous studies described the ability of cyclodextrin to increase the solubility of these drugs in water [23,24]. Subsequently, efforts were made to study the structure of the complexes by experimental and molecular modelling techniques [18,23,24]. Earlier studies focus on the buffer free solution, suspension or freeze-dried solid state complexes. According to our present knowledge our work is the first study to describe the sulfamethazine–β-cyclodextrin and sulfamethazine–randomly methylated β-cyclodextrin complexes in aqueous solution at different pH and temperature values using combined experimental and theoretical techniques. Both spectroscopic measurements and molecular modeling studies highlight the importance of the reorganization of the solvent molecules during the guest molecule enters the host's cavity. Results highlight formation of zwitterionic sulfamethazine molecule in the cyclodextrin cavity which affect significantly the stability of SMT-CD complexes. The pH-affected structures of the complexes investigated explain the previous contradictory findings. The presented results might relevant for the preparation of orally administreted products of sulfamethazine-cyclodextrin complexes.

Author Contributions: Conceptualization, B.L. and S.K.-M.; methodology, B.L. and S.K.-M.; formal analysis, H.M.A.; investigation, H.M.A.; L.S.; M.P.; B.L. B.B. and S.K.-M.; resources, S.K.-M. L.S.; data curation, H.M.A.; B.L. and S.K.-M.; writing—original draft preparation, H.M.A.; L.S.; M.P.; B.L. and S.K.-M.; writing—review and editing B.L. and S.K.-M.

Funding: This research received no external funding.

Acknowledgments: This work was supported by the GINOP-2.3.2-15-2016-00049 grant.

Conflicts of Interest: The authors declare no conflict of interest.

References

1. Liu, C.; Zhang, D.; Zhao, L.; Zhang, P.; Lu, X.; He, S. Extraction property of *p-tert*-butylsulfonylcalix[4]- arene possessing irradiation stability towards cesium(I) and strontium(II). *Appl. Sci.* **2016**, *6*, 212. [CrossRef]
2. Galindo-Murillo, R.; Olmedo-Romero, A.; Cruz-Flores, E.; Petrar, P.M.; Kunsagi-Mate, S.; Barroso-Flores, J. Calix[n]arene-based drug carriers: A DFT study of their electronic interactions with a chemotherapeutic agent used against leukemia. *Comput. Theor. Chem.* **2004**, *1035*, 84. [CrossRef]
3. Brancatelli, G.; Dalcanale, E.; Pinalli, R.; Geremia, S. Probing the structural determinants of amino acid recognition: X-Ray studies of crystalline ditopic host-guest complexes of the positively charged amino acids, Arg, Lys, and His with a cavitand molecule. *Molecules* **2018**, *23*, 3368. [CrossRef] [PubMed]
4. Czibulya, Z.; Horvath, E.; Nagymihaly, Z.; Kollar, L.; Kunsagi-Mate, S. Competitive processes associated to the interaction of a cavitand derivative with caffeic acid. *Supramol. Chem.* **2016**, *28*, 582. [CrossRef]
5. Hahm, E.; Kang, E.J.; Pham, X.-H.; Jeong, D.; Jeong, D.H.; Jung, S.; Jun, B.-H. Mono-6-deoxy-6-aminopropylamino-β-cyclodextrin on Ag-embedded SiO_2 nanoparticle as a selectively capturing ligand to flavonoids. *Nanomaterials* **2019**, *9*, 1349. [CrossRef] [PubMed]
6. Das, S.; Gazdag, Z.; Szente, L.; Meggyes, M.; Horváth, G.; Lemli, B.; Kunsági-Máté, S.; Kuzma, M.; Kőszegi, T. Antioxidant and antimicrobial properties of randomly methylated β cyclodextrin – captured essential oils. *Food Chem.* **2019**, *278*, 305. [CrossRef]
7. Xing, M.; Yanli, Z. Biomedical applications of supramolecular systems based on host–guest interactions. *Chem. Rev.* **2015**, *115*, 7794. [CrossRef]
8. Ryu, J.H.; Hong, D.J.; Lee, M. Aqueous self-assembly of aromatic rod building blocks. *Chem. Commun.* **2008**, *1054*, 1043. [CrossRef]
9. Zang, W.; Jin, W.; Fukushima, T.; Saeki, A.; Seki, S.; Aida, T. Supramolecular linear heterojunction composed of graphite-like semiconducting nanotubular segments. *Science* **2011**, *334*, 340. [CrossRef]
10. Saha, S.; Roy, A.; Roy, K.; Roy, M.N. Study to explore the mechanism to form inclusion complexes of b-cyclodextrin with vitamin molecules. *Sci. Rep.* **2016**, *6*, 35764. [CrossRef]
11. Kfoury, M.; Landy, D.; Fourmentin, S. Characterization of cyclodextrin/volatile inclusion complexes: A Review. *Molecules* **2018**, *23*, 1204. [CrossRef] [PubMed]
12. Menezes, P.D.P.; Andrade, T.A.; Frank, L.A.; de Souza, E.P.B.S.S.; Trindade, G.D.G.G.; Trindade, I.A.S.; Serafini, M.R.; Guterres, S.S.; Araújo, A.A.S. Advances of nanosystems containing cyclodextrins and their applications in pharmaceuticals. *Int. J. Pharm.* **2019**, *559*, 312. [CrossRef] [PubMed]
13. Song, Y.; Jiang, J.; Ma, J.; Zhou, Y.; von Gunten, U. Enhanced transformation of sulfonamide antibiotics by manganese(IV) oxide in the presence of model humic constituents. *Water Res.* **2019**, *153*, 200. [CrossRef] [PubMed]
14. Le-Minh, N.; Stuetz, R.M.; Khan, S.J. Determination of six sulfonamide antibiotics, two metabolites and trimethoprim in wastewater by isotope dilution liquid chromatography/ tandem mass spectrometry. *Talanta* **2012**, *89*, 407. [CrossRef] [PubMed]
15. Jiménez, V.; Adrian, J.; Guiteras, J.; Marco, M.P.; Companyó, R. Validation of an enzyme-linked immunosorbent assay for detecting sulfonamides in feed resources. *J. Agric. Food Chem.* **2010**, *58*, 7526. [CrossRef]
16. Boreen, A.L.; Arnold, W.A.; McNeill, K. Photochemical fate of sulfa drugs in the aquatic environment: Sulfa drugs containing five-membered heterocyclic groups. *Env. Sci. Technol.* **2004**, *38*, 3933. [CrossRef]
17. Nesterenko, I.S.; Nokel, M.A.; Eremin, S.A. Immunochemical methods for the detection of sulfanylamide drugs. *J. Anal. Chem.* **2009**, *64*, 453. [CrossRef]
18. Bani-Yaseen, A.D.; Mo'ala, A. Spectral, thermal, and molecular modeling studies on the encapsulation of selected sulfonamide drugs in b-cyclodextrin nano-cavity. *Spectrochim. Acta A* **2014**, *131*, 424. [CrossRef]
19. Saha, S.; Roy, A.; Roy, M.N. Mechanistic investigation of inclusion complexes of a sulfa drug with a- and b-cyclodextrins. *Ind. Eng. Chem. Res.* **2017**, *56*, 11672. [CrossRef]
20. Prabhu, A.A.M.; Venkatesh, G.; Rajendiran, N. Spectral characteristics of sulfa drugs: Effect of solvents, pH and β-cyclodextrin. *J. Solution. Chem.* **2010**, *39*, 1061. [CrossRef]
21. Rajendiran, N.; Siva, S. Inclusion complex of sulfadimethoxine with cyclodextrins: Preparation and characterization. *Carbohydr. Polym.* **2014**, *101*, 828. [CrossRef] [PubMed]
22. Pal, A.; Gaba, R.; Soni, S. Absorption and fluorescence spectral studies of the interaction of sulpha drugs with α-cyclodextrin. *Phys. Chem. Liq.* **2019**, *57*, 362. [CrossRef]

23. Zoppi, A.; Delviro, A.; Aissa, V.; Longhi, M.R. Binding of sulfamethazine to b-cyclodextrin and methyl-b-cyclodextrin. AAPS Pharm. *Sci. Tech.* **2013**, *14*, 727. [CrossRef]

24. Zoppi, A.; Quevedo, M.A.; Delviro, A.; Longhi, M.R. Complexation of sulfonamides with b-cyclodextrin studied by experimental and theoretical methods. *J. Pharm. Sci.* **2010**, *99*, 3166. [CrossRef]

25. Morisue, M.; Ueno, I. Preferential solvation unveiled by anomalous conformational equilibration of porphyrin dimers: Nucleation growth of solvent–solvent segregation. *J. Phys. Chem. B* **2018**, *122*, 5251. [CrossRef]

26. Wintgens, V.; Lorthioir, C.; Miskolczy, Z.; Amiel, C.; Biczók, L. Substituent effects on the inclusion of 1-alkyl-6-alkoxy-quinolinium in 4-sulfonatocalix[n]arenes. *ACS Omega* **2018**, *3*, 8631. [CrossRef]

27. Nicolas, H.; Yuan, B.; Xu, J.; Zhang, X.; Schönhoff, M. pH-responsive host–guest complexation in pillar[6]arene-containing polyelectrolyte multilayer films. *Polymers* **2017**, *9*, 719. [CrossRef]

28. Dan, Z.; Cao, H.; He, X.; Zhang, Z.; Zou, L.; Zeng, L.; Xu, Y.; Yin, Q.; Xu, M.; Zhong, D.; et al. A pH-responsive host-guest nanosystem loading succinobucol suppresses lung metastasis of breast cancer. *Theranostics* **2016**, *6*, 435. [CrossRef]

29. Li, B.; Meng, Z.; Li, Q.; Huang, X.; Kang, Z.; Dong, H.; Chen, J.; Sun, J.; Dong, Y.; Li, J.; et al. A pH responsive complexation-based drug delivery system for oxaliplatin. *Chem. Sci.* **2017**, *8*, 4458. [CrossRef]

30. Mohamed Ameen, H.; Kunsági-Máté, S.; Szente, L.; Lemli, B. Encapsulation of sulfamethazine by native and randomly methylated β-cyclodextrins: The role of the dipole properties of guests. *Spectrochim. Acta A Spectrosc.* **2020**, *225*, 117475. [CrossRef]

31. Liu, M.; Guo, Q.; Shi, Y.; Cai, C.; Pei, W.; Yan, H.; Jia, H.; Han, J. Studies on pH and temperature dependence of inclusion complexes of bisdemethoxycurcumin with β-cyclodextrin derivatives. *J. Mol. Struct* **2019**, *1179*, 336. [CrossRef]

32. Poór, M.; Matisz, G.; Kunsági-Máté, S.; Derdák, D.; Szente, L.; Lemli, B. Fluorescence spectroscopic investigation of the interaction of citrinin with native and chemically modified cyclodextrins. *J. Lumin.* **2016**, *172*, 23. [CrossRef]

33. Texido, M.; Pignatello, J.J.; Beltran, J.L.; Granados, M.; Peccia, J. Speciation of the ionizable antibiotic sulfamethazine on black carbon (Biochar). *Environ. Sci. Technol.* **2011**, *45*, 100020. [CrossRef]

34. Lazar, P.; Lee, Y.; Kim, S.; Chandrasekaran, M.; Lee, K.W. Molecular dynamics simulation study for ionic strength dependence of RNA-host factor interaction in Staphylococcus aureus Hfq. *Bull. Korean Chem. Soc.* **2010**, *31*, 1519. [CrossRef]

35. Liu, L.; Gou, Q.-X. The driving forces in the inclusion complexation of cyclodextrins. *J. Incl. Phenom. Macro.* **2002**, *42*, 1. [CrossRef]

36. Terekhova, I.V.; Chislov, M.V.; Brusnikina, M.A.; Chibunova, E.S.; Volkova, T.V.; Zvereva, I.A.; Proshin, A.N. Thermodynamics and binding mode of novel structurally related 1,2,4-thiadiazole derivatives with native and modified cyclodextrins. *Chem. Phys. Lett.* **2017**, *671*, 28. [CrossRef]

37. Dragan, A.I.; Read, C.M.; Crane-Robinson, C. Enthalpy-entropy compensation: The role of solvation. *Eur Biophys. J.* **2017**, *46*, 301. [CrossRef]

38. Pan, A.; Biswas, T.; Rakshit, A.K.; Moulik, S.P. Enthalpy-entropy compensation (EEC) effect: A revisit. *J. Phys. Chem. B* **2015**, *119*, 15876. [CrossRef]

39. Pan, A.; Kar, T.; Rakshit, A.K.; Moulik, S.P. Enthalpy-entropy compensation (EEC) effect: Decisive role of free energy. *J. Phys. Chem. B* **2016**, *120*, 10531. [CrossRef]

40. Rekharsky, M.V.; Inoue, J. Complexation thermodynamics of cyclodextrins. *Chem. Rev.* **1998**, *98*, 1875. [CrossRef]

41. Kunsági-Máté, S.; Iwata, K. Effect of cluster formation of solvent molecules on the preferential solvatation of anthracene in binary alcoholic solutions. *Chem. Phys. Lett.* **2009**, *473*, 284. [CrossRef]

42. Kunsági-Máté, S.; Iwata, K. Electron density dependent composition of the solvation shell of phenol derivatives in binary solutions of water and ethanol. *J. Solut. Chem.* **2013**, *42*, 165. [CrossRef]

43. Szejtli, J. Introduction and general overview of cyclodextrin chemistry. *Chem. Rev.* **1998**, *98*, 1743. [CrossRef] [PubMed]

44. Poór, M.; Bálint, M.; Hetényi, C.; Gődér, B.; Kunsági-Máté, S.; Kőszegi, T.; Lemli, B. Investigation of non-covalent interactions of aflatoxins (B1, B2, G1, G2 and M1) with serum albumin. *Toxins* **2017**, *9*, 339. [CrossRef]

45. Poór, M.; Boda, G.; Kunsági-Máté, S.; Needs, P.W.; Kroon, P.A.; Lemli, B. Fluorescence spectroscopic evaluation of the interactions of quercetin, isorhamnetin, and quercetin-3'-sulfate with different albumins. *J. Lumin.* **2018**, *194*, 156. [CrossRef]
46. Gans, P.; Sabatini, A.; Vacca, A. Investigation of equilibria in solution. Determination of equilibrium constants with the HYPERQUAD suite of programs. *Talanta* **1996**, *43*, 1739. [CrossRef]
47. HyperChem, Hypercube Inc. 2007. Available online: http://www.hyper.com/ (accessed on 12 December 2019).
48. Faisal, Z.; Kunsági-Máté, S.; Lemli, B.; Szente, L.; Bergmann, D.; Humpf, H.-U.; Poór, M. Interaction of dihydrocitrinone with native and chemically modified cyclodextrins. *Molecules* **2019**, *24*, 1328. [CrossRef]
49. Fliszár-Nyúl, E.; Lemli, B.; Kunsági-Máté, S.; Szente, L.; Poór, M. Interactions of mycotoxin alternariol with cyclodextrins and its removal from aqueous solution by beta-cyclodextrin bead polymer. *Biomolecules* **2019**, *9*, 428. [CrossRef]

Sample Availability: Samples of the compounds BCD and RAMEB are available from CycloLab Ltd. (L.S.).

Review

Research Progress on Synthesis and Application of Cyclodextrin Polymers

Yuan Liu [†], Ting Lin [†], Cui Cheng *, Qiaowen Wang, Shujin Lin, Chun Liu * and Xiao Han *

College of Biological Science and Engineering, Fuzhou University, Fuzhou 350108, China;
N195720006@fzu.edu.cn (Y.L.); lintingtt33@163.com (T.L.); wendyeveryday@163.com (Q.W.);
linshujin32@163.com (S.L.)
* Correspondence: ibptcc@fzu.edu.cn (C.C.); ibptlc@fzu.edu.cn (C.L.); hanxiao@fzu.edu.cn (X.H.)
† These authors contributed equally to this work.

Abstract: Cyclodextrins (CDs) are a series of cyclic oligosaccharides formed by amylose under the action of CD glucosyltransferase that is produced by Bacillus. After being modified by polymerization, substitution and grafting, high molecular weight cyclodextrin polymers (pCDs) containing multiple CD units can be obtained. pCDs retain the internal hydrophobic-external hydrophilic cavity structure characteristic of CDs, while also possessing the stability of polymer. They are a class of functional polymer materials with strong development potential and have been applied in many fields. This review introduces the research progress of pCDs, including the synthesis of pCDs and their applications in analytical separation science, materials science, and biomedicine.

Keywords: cyclodextrin polymers; synthesis; separation science; materials science; biomedicine

Citation: Liu, Y.; Lin, T.; Cheng, C.; Wang, Q.; Lin, S.; Liu, C.; Han, X. Research Progress on Synthesis and Application of Cyclodextrin Polymers. *Molecules* **2021**, 26, 1090. https://doi.org/10.3390/molecules26041090

Academic Editors: Angelina Angelova, Marina Isidori, Margherita Lavorgna and Rosa Iacovino

Received: 17 December 2020
Accepted: 2 February 2021
Published: 19 February 2021

Publisher's Note: MDPI stays neutral with regard to jurisdictional claims in published maps and institutional affiliations.

1. Introduction

Cyclodextrins (CDs) [1] is the general term for a series of cyclic oligosaccharides produced by amylose under the action of enzymes that are produced by Bacillus. They usually contain 6 to 12 glucopyranose units, and natural CDs are divided into α-CD, β-CD, and γ-CD with cavity sizes of ~0.5, 0.6, and 0.8 nm, respectively [2]. While single CD molecules can no longer meet the present practical application needs [3–5], the development of polymers has continued because of their excellent properties. They have become an important field of materials research and have brought new opportunities for CDs [6–9]. Studies involving CDs have demonstrated that they can also be used to form living polymers [10–12]. CD polymers (pCDs) can effectively address issues related to the manipulation of CD molecules, and they can endow unique functions and physical and chemical properties that are absent in single CD molecules. The advantages of the cavity structure of dextrin [13] include the simple formation of inclusions with guest molecules, control of the direction and rate of release of the guest molecules, and modifiability of the groups at the edge of the cavity; additionally, they can combine the excellent properties of polymers, including the mechanical strength and hardness, high relative molecular weight, and good thermal stability. In some cases, pCDs are also called cyclodextrin nanosponges (CD NSs). They do not appear all at once. They have undergone a long development. Since they were proposed in 1990, they have overcome the limitations of CDs, especially in water solubility. Great breakthroughs have been made in synthesis and application, and the form of cyclodextrins has been continuously changed in subsequent developments, especially in the past 50 years. The development of current pCDs started from a relatively simple cross-linking network in the 1960s, which was later developed into a multifunctional polymer [14]. Therefore, pCDs are widely used in various fields, including pharmaceutical, food, chemistry, chromatographic, catalysis, biotechnology, agriculture, cosmetics, hygiene, medicine, textile, and environmental fields [15–18].

This article introduces research progress in pCDs, including the synthesis of pCDs and their applications in analytical separation science, materials science, and biomedicine. It focuses on applications in biomedicine, and in particular, the technological innovations for application as drug delivery vehicles. Finally, the trends related to the development of pCDs are summarized and directions for future research are discussed.

2. Synthesis of pCDs

pCDs are polymer compounds containing CD units. They include crosslinked pCDs, linear pCDs, fixed pCDs, pCD inclusion compounds, and hyperbranched pCDs (Figure 1).

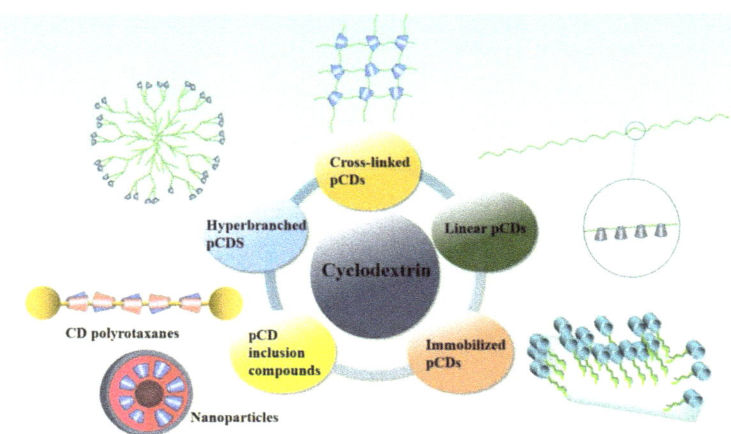

Figure 1. Five kinds of cyclodextrin polymers (pCDs) that containing cyclodextrin units.

pCDs have different synthesis methods that depend on the final form. Crosslinked pCDs are obtained by pCDs and their derivatives using crosslinking agents with bifunctional or multifunctional groups. Commonly used crosslinking agents are citric acid, peroxides, isocyanates, acid anhydrides, and *N,N*-methylene bisacrylamide [19]. As an example, citric acid is a non-toxic crosslinking agent that has been used to graft α-CD onto cellulose fibers [20,21]. Ghorpade et al. [22] prepared a β-cyclodextrin-carboxymethyl cellulose (β-CD-CMC) hydrogel film via the esterification crosslinking method with citric acid for controlled release of ketoconazole (model drug). β-CD helps minimize the sudden release of the drugs. In crosslinked pCDs, the CDs are polymerized through the special functional groups of the crosslinking agent. The synthesis method is relatively simple, has strong operability, and can produce polymers with a high relative molecular weight, but the products have poor mechanical properties.

Linear pCDs are polymer compounds prepared from modified CD through alkenyl copolymerization or condensation polymerization of other special functional groups. In this process, the CD is first modified and then polymerized with almost no side reactions. Linear pCDs will not destroy the cavity structure of the CDs, so they have high potential for applications including for ion exchange, drug loading, separation, and adsorption. For example, the supramolecular linear polyacrylamide (SL-PAM) synthesized by You et al. [23] is a combination of β-CD and adamantane-terminated polyacrylamide (AT-PAM). It was prepared by the interaction between the host and the guest. SL-PAM samples were investigated using 2D NOESY NMR and thermal analysis to verify the formation of the inclusion compounds.

Fixed pCDs are polymer compounds formed by bonding CDs and its derivatives to a carrier. The material properties vary with the carrier, which includes inorganic polymers (such as silica gel and graphene) [24], natural polymers (such as cellulose and chitosan) [25], synthetic polymers (such as polystyrene, polyacrylate) [26]. For example, Shang et al. [27]

synthesized an immobilized polyvinyl alcohol/CD ecological adsorbent and studied its application for removing ibuprofen from pharmaceutical wastewater. The adsorbent was prepared by solution blending (2-hydroxypropyl)-β-cyclodextrin (HPBCD) and polyvinyl alcohol (PVA), followed by glutaraldehyde treatment. The experimental process was simple and the product could be easily obtained. It can also be quickly reused via a simple soaking procedure. Immobilized pCDs have the advantages of good mechanical properties and a wide range of stable applications.

pCDs that form clathrates are produced by complexation between the polymer and the CDs. This inclusion compound has significantly better structure and properties than cyclodextrin and polymers. For example, CDs can improve the solubility of the guest [28–30]. Their synthesis methods include the saturated aqueous solution method, the ultrasonic method, grinding, colloid milling, freeze drying, and spray drying [31–33]. There are various synthetic methods that are widely used in the field of biomedicine and their applications will be explained in detail below. It is worth mentioning that the cyclodextrin polyrotaxane is a kind of pCD inclusion compound. Its structure includes a linear axis, multiple rings connected to the linear axis, and two end-capping groups connected to the linear axis. At each end, when the end of the linear shaft becomes larger than the inner diameter of the ring, or when molecules larger than the ring are bound to the end of the shaft, the ring on the polyrotaxane cannot be dissociated from the dumbbell-shaped shaft to make the polymer stably exist. The association constant K_a is a quantitative indicator reflecting the progress of the complex reaction. Angelina Angelova et al. [34] first reported a method for determining the association constant of amphiphilic water-soluble drugs. The amphiphilic peptide antibiotic polymyxin B (PMB) reacts with CD and assuming that CD and the compound drug are surface-inactive, and the two substances do not affect the surface properties of free PMB, the formula is calculated:

$$ka = \frac{C_D^T - [D]}{(C_{CD}^T - C_D^T + [D])[D]} \tag{1}$$

where D represents PMB, C_D^T and C_{CD}^T respectively represent the total concentration of CD and PMB, [D] needs to use the concentration dependence of surface tension to be evaluated.

Further extended to water-insoluble drugs, the authors of [35] inferred the formula of the complexation of retinol (RL) and CD:

$$ka = \frac{[RL]_T - [RL]}{[RL][CD]} \tag{2}$$

Among them, $[RL]$ and $[CD]$ represent the remaining interface concentration and the concentration of free CD molecules after RL molecules are exhausted. Here $[RL]_T$ represents the RL concentration before depletion. From this we can quantify the output of pCD, and this method is widely used in applications.

Similar to CDs, hyperbranched polymers also have a certain cavity structure and hydrophilic and hydrophobic properties. Some hyperbranched polymers have been used in the field of self-assembly [36]. The so-called hyperbranched polymer, i.e., a macromolecule with a highly branched structure, has the advantages of low viscosity and non-crystallinity. It has a highly branched structure with cavities and a large number of terminal functional groups. These characteristics give hyperbranched polymers the advantages of high solubility and reactivity [37]. Hyperbranched pCDs [38] have been developed on the basis of previous polymers, and are divided into three categories: (1) Bonding CDs to hyperbranched polymer; (2) complexing CDs to hyperbranched polymer, where the inclusion compound is formed on the polymer; (3) hyperbranched polymers synthesized with CDs as the core [38]. These methods combine the advantages of hyperbranched polymers and CDs, and have advantages such as good reactivity, high solubility, and broad application prospects.

Research needs to rethink traditional craftsmanship and request the latest synthesis methods. The new method is not mature enough, but with the development of the process, the yield and degree of polymerization will also become mature. It is worth noting that pCDs most commonly react with a suitable cross-linking agent in an organic polar aprotic solvent such as N, N-dimethylformamide (DMF) solution, but there will be some pollution. In recent years, people have been exploring solvent-free/green synthesis methods. Rubin Pedrazzo et al. proposed a green synthesis method through mechanochemical methods [39]. The test method is simple and the product is no different from the traditional organic solvent method. It is obtained by rotating anhydrous cyclodextrin and carbonyl diimidazole in a ball mill, washing with deionized water and acetone, and finally extracting. Max Petitjean et al. [40] cross-linked β-cyclodextrin-functionalized chitosan, xanthan gum, and locust bean gum to form a polymer under solvent-free conditions. The polymer has high stability, a large degree of crosslinking, and the method is simple, but homogenization of the solid mixture may occur. The article mentions that a small amount of water can be used to knead the mixture to prepare a paste solution, which has potential in the treatment of biologically active phenolic compounds, the purification of wastewater or the reuse of agricultural waste. Giancarlo Cravotto et al. [41] used low-boiling epoxy reagents in high-energy ball mills (HEBM) to simplify the preparation and purification of low-substitution (2-hydroxy) propylated β-and γ-cyclodextrins (β/γ-CDs). Compared with traditional methods, the properties of mechanically synthesized pCD, such as the degree of complexation, are different, and most of them are better. There are many such examples, which shows that the solvent-free/green synthesis method of pCDs, as a new direction, has attracted more and more attention and has great development potential.

The above text introduced the basic attributes and synthesis techniques for pCDs. So, what applications does the brand have in reality? The following mainly introduces research progress on pCDs from the latest applications in the fields of analysis and separation science, materials science, and biomedicine. It is worth noting that applications in the field of biomedicine, especially as a drug delivery system, have become a topic of intense research interest in recent years, and continued technological progress has also promoted the continuous development of pCDs and realization of their potential (Figure 2).

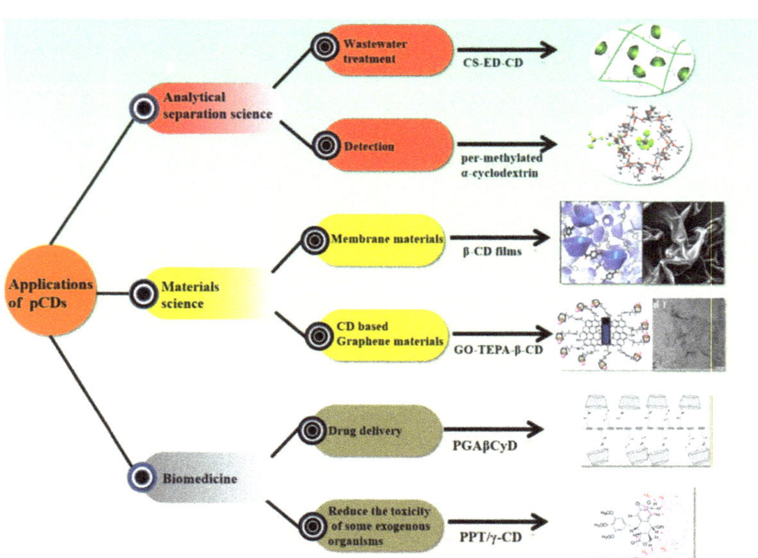

Figure 2. Multi-domain applications of CDs [42–47]. Adapted with permission from ref. [42]. Copyright 2017 Springer Nature; ref. [43]. Copyright 2018 John Wiley and Sons; ref. [44]. Copyright 2020 John Wiley and Sons; ref. [45]. Copyright 2018 Royal Society of Chemistry; ref. [46]. Copyright 2016 John Wiley and Sons; ref. [47]. Copyright 2017 Elsevier.

3. Application of pCDs in Analytical Separation Science

3.1. Application of pCDs in Wastewater Treatment and Water Purification

Over the past 30 years, water-related inorganic and organic micropollutants (such as heavy metals, drugs, and endocrine-damaging chemicals) are increasingly present in global water resources, and environmental issues have become a primary concern for society, public institutions, and industries [48–52]. The adsorption method has been extensively studied because it is economical, highly efficient, recyclable, and has good selectivity [53–55]. New pCDs can simultaneously adsorb and encapsulate a variety of organic and inorganic impurities, such as polycyclic aromatic hydrocarbons [56], pesticides [57], heavy metals [58], dyes [59], phenol compounds [60,61], phthalates [62], and pharmaceutically active compounds [63,64]. Hence, they have become topics of intense research interest because of their low cost and reusability. For example, β-CD/chitosan polymer was prepared by using glutaraldehyde as a cross-linking agent through solution polymerization, which could be developed as a new type of wastewater treatment purification material. Compared with traditional activated carbon, it has many advantages, such as renewability, low energy consumption, and low cost [65].

Laura et al. [66] studied the organic matter adsorption capacity of β-pCDs beads (BCPB) with different chemical compositions and thicknesses in solution. BCPB is a macromolecule produced by crosslinking β-CD with epichlorohydrin. They used a model solution containing ibuprofen and a total organic carbon (TOC) analyzer to determine the adsorption capacity. The results show that BCPB has excellent adsorption capacity for active organic drugs. Alsbaiee et al. [67] developed a porous CD derivative that crosslinked β-CD with rigid aromatics to form a high surface area mesoporous polymer. It was capable of quickly adsorbing a variety of organic impurities and it had 15 to 200 times the adsorption rate constant of traditional activated carbon. [68] Additionally, the polymer could be reused. Zhao et al. [42] reported a chitosan-EDTA-β-cyclodextrin (CS-ED-CD) multifunctional adsorbent prepared using EDTA as a crosslinking agent for the adsorption of toxic metals and organic trace pollutants in wastewater. The fixed CD cavity captured organic compounds while the EDTA-group complexed with the metal. This multifunctional adsorbent had improved potential for complex practical applications. The work provided new insights for the future design and preparation of sustainable materials for water purification. With continued development of CD and its derivatives, they are likely to become increasingly important in wastewater treatment and water purification, especially with the development of β-CD, which makes efficient water treatment possible (Figure 3). The above-mentioned relatively single treatment method has become increasingly unable to meet people's needs. Scientists are exploring a non-polluting, efficient, and recyclable material to deal with pollution problems, and solar energy can be used to treat water pollution. Xuejiao Hu et al. [69] synthesized a new type of magnetic carboxymethylated β-CD-based porous polymer with fast adsorption performance and superparamagnetism in the water phase. The polymer has large pores and is adsorbed in the printing and dyeing wastewater through positive and negative electric attraction. The widespread anionic dyes are renewable materials with great potential. Garcia-Diaz et al. [70] developed a ROS-resistant fluorinated pCDs, which uses its adsorption capacity to adsorb pollutants near the catalyst, improves the utilization rate of photo-living oxygen, and optimizes the coating thickness on TiO_2 microspheres. To improve the efficiency of pollutant degradation, the two microspheres combined to form CDP-TiO_2 are expected to be used in photocatalytic water treatment. Sanaz Khammar et al. [71] grafted carboxymethyl-β-cyclodextrin (CM-β-CD) to the surface of core-shell titanium dioxide magnetic nanoparticles and successfully prepared recyclable CMCD-Fe_3O_4@TiO_2, which is conducive to the adsorption of pollutants, protects nanoparticles, and promotes the photocatalytic activity of TiO_2. Its cost-reduction, simple material, non-volatile and non-toxic properties have excellent application value in reducing the toxicity of polluted oil.

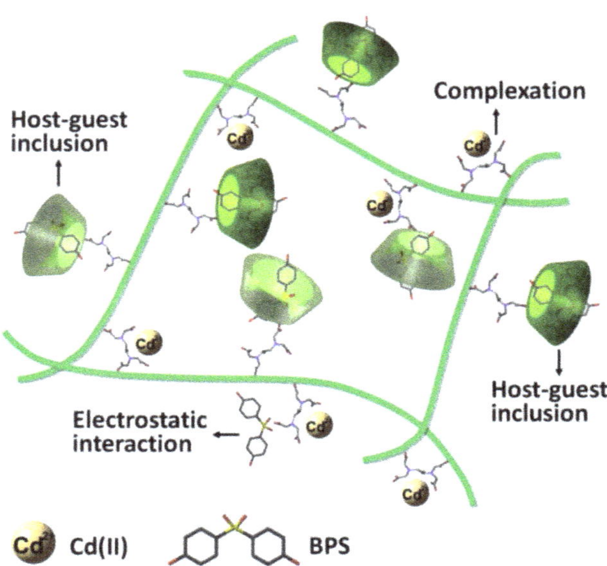

Figure 3. The schematic illustration of the related adsorption mechanisms of CS-ED-CD toward Cd(II) and BPS [42]. Adapted with permission from ref. [42]. Copyright 2017 Springer Nature.

3.2. Application of pCDs in Analysis and Detection

Chirality is ubiquitous in nature [72,73] and it is very important in scientific research. For example, complex biological activity in the human body requires chirality [74,75]. However, the analysis and detection of chiral compounds is difficult [76]. Generally, achiral separation analysis of PMA and BMA [77–82] is not satisfactory. The unique cavity structure of pCDs and the molecule itself has multiple chiral centers, so it has good chiral resolution capabilities and can be used for analysis and detection. For example, by means of the different affinities between the β-CD unit and the two configurations of the racemate, enantioselective separation and detection of the racemate can be achieved.

Immohra et al. [83] applied the chiral characteristics of CD using the CD derivatives (hydroxypropyl-β-CD, CD oligomer, sulfobutyl ethyl ether)-β-CD, triacetyl-β-CD and hepta(2,6-di-*O*-methyl)-β-CD) identified as D- and L-glutamic acid, for the chiral recognition of t-asparagine, L-praziquantel, and its racemates. Ryvlin et al. [43] used the cavity structure of CDs and permethylated pCDs to detect and remove trichlorofluoromethane, which is harmful to the environment. The reaction produced a stable supramolecular molecule, which is a color transparent crystal complex that can be used repeatedly. Girschikofsky and Maiko [84] used permethyl, perethyl, and allallyl which substituted α-CDs, β-CDs, and γ-CDs as the sensitive sensor materials. Liang et al. [85] used the CD molecule itself to obtain multiple chiral centers, and they synthesized benzylureido-β via the reaction of 6-amino-β-cyclodextrin and reactive benzyl isocyanate. The -CD was bonded to silica gel through an addition reaction to obtain a new chiral stationary phase (BzCDP) based on benzylureido-β-cyclodextrin; it was successfully used to separate phenylthioglycolic acid (PMA) and benzyl mercapto acid (BMA) enantiomers, which have been shown to be biomarkers in human urine for benzene and toluene exposure. The separation of enantiomers has also been optimized through the study of their related factors. BzCDP is of great significance for the in-depth study of the presence and content of chiral markers in human urine, and for better understanding and evaluating of the harmful effects of benzenes on humans. Poor and Miklos [86] studied CDs and certain mycotoxins to form host–guest complexes, and removed alternan from aqueous solutions using insoluble β-CD bead polymers (BBP). Carcu-Dobrin et al. [87] studied the use of CD derivatives as chiral selectors to identify an optimized method for the chiral identification of amlodipine (AML) enantiomers. Carboxymethylethyl-β-CD (Cm-β-CD) was selected for enantiomeric identification. Through analysis and research, several factors were modified simultaneously to obtain an optimized separation method. Zheng et al. [88] used β-CD-gel and D- or L-tryptophan (homotype D-or L-Trp-gel) modified polyacrylamide-based gel for visible chiral recognition. In the NaCl aqueous solution, due to the obvious changes, the β-CD gel successfully distinguished the D- and L-Trp gels macroscopically, and the chirality difference becomes obvious, which will be very conducive to more in-depth research. It will also be more conducive to the understanding of chirality in the general public. The pCDs have shown their unique advantages in the field of analysis and testing in recent years, and they are a promising research direction.

4. Application of pCDs in Materials Science

4.1. Application of pCD Films

In recent years, membrane technology has been widely used in many fields of production because of its high separation efficiency, easy operation, and the absence of secondary pollution [89–93]. Compared with pCDs membranes, molecular sieves are expensive and have high energy consumption, and these issues are difficult to resolve. Because CD is produced at a large scale from starch, it has the advantages of being sustainable [94], non-toxic [95], and inexpensive; furthermore, it has been proven to be suitable for a variety of separations [96] and the production process is well established [97]. CD has been used as filler in films, a part of film-forming polymers, and as surface modifiers. pCDs membranes have excellent potential for applications, including in isomer separation and metal ion transport.

Villalobos et al. [44] studied the molecular level design of a new type of crosslinked CD filter membrane, which forms a CD film through interfacial polymerization. The filter membrane is cheap macrocyclic glucose with a shape similar to that of a hollow truncated cone (Figure 4). The channel-shaped cavity of the CD creates many pathways with a defined pore size in the separation layer, which can effectively distinguish molecules. The transport of molecules through these membranes is highly shape-sensitive. In addition, the cavities are hydrophobic and the ester-crosslinked outer part is hydrophilic, resulting in the high permeability of these membranes for polar and non-polar solvents (Figure 4).

Pangeniet al. [98] used titanium glycinate-N,N-dimethylphosphonate to prepare cross-linked sulfonated polyvinyl alcohol membranes; then, they modified them by incorporating sulfonated β-CD. The ion exchange capacity of the membrane was found to be in the range of 1.40 to 2.55 meq/g. A high-precision impedance analyzer was used to evaluate the proton conductivity of the membrane at different temperatures and 100% relative humidity. The membranes containing 16% by mass and 20% by mass of sulfonated β-CD exhibited excellent proton conductivity of 0.121 and 0.143 S/cm at 80 °C, respectively. Wang, Yunze, and Lin et al. [99] proposed a new strategy to improve the flux and antifouling performance of ethylene vinyl alcohol (EVAL) membranes by blending with macrocyclic hyperamphiphiles (CD). A CD-rich layer was formed on the membrane surface. During the phase inversion process, the synergistic interaction between the hydrophobic and hydrophilic segments of the amphiphilic pCDs increases the membrane flux and increases the surface roughness and hydrophilicity of the membrane. In addition, the macrocyclic super amphiphilic hybrid membrane exhibited improved antifouling performance compared with the original EVAL membrane. The introduction of pCDs enabled the formation of a hydrophilic membrane surface, which has high application potential for practical membrane applications.

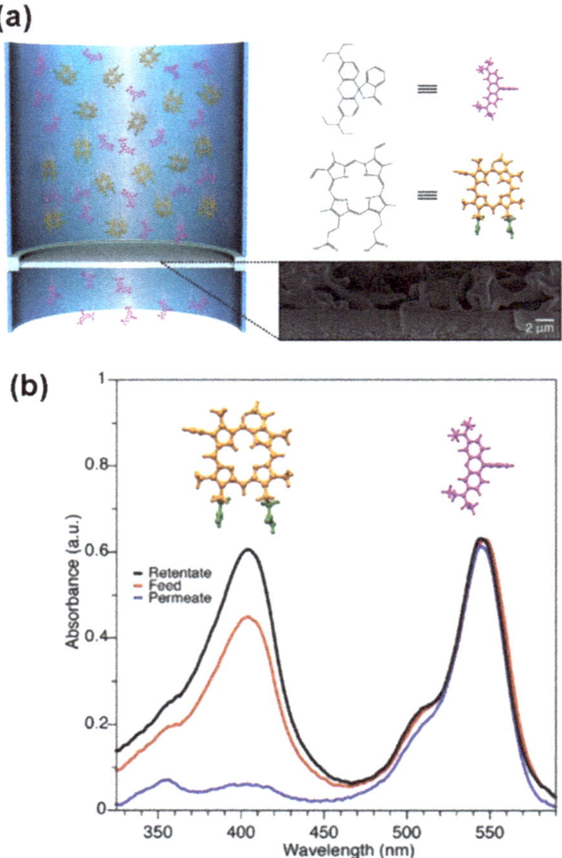

Figure 4. (**a**) Schematic showing how a β-CD membrane separates molecules based on their shape. Cross-section SEM image corresponds to a β-CD membrane prepared using 2 m NaOH aqueous solution. (**b**) UV–vis absorption spectra of a methanol solution with PPIX (orange molecule) and RB (pink molecule) to evidence the separation performance of the β-CD membrane [44]. Adapted with permission from ref. [44]. Copyright 2020 John Wiley and Sons.

4.2. Application of CDs Functionalized Graphene Materials

The introduction of CD into the graphene family of materials is an important direction for graphene research. Graphene-based materials are widely used in macro/microstructures, sensors, oil/water separation membranes, and biomimetic interfaces [100–104]. CDs can improve their water solubility, biocompatibility, and supramolecular screening ability; hence, it may introduce new and interesting properties for these materials. CD-functionalized graphene materials have the properties of graphite, the inherent properties of olefins (high surface area, easy functionality [105], biocompatibility [106]), and the inherent properties of CDs.

Liu et al. [45] synthesized an excellent water-soluble nanosensor based on CD derivatives and graphene oxide; it was a supramolecular system in which the CD was loaded on the graphene oxide. The supramolecular system was very sensitive to Al^{3+}, and the large specific surface area of the graphene oxide could capture Al^{3+}. Simultaneously, the introduction of CDs could enhance the water solubility of graphene oxide, and this was the first self-assembled nanosensor composed of graphene oxide and CD derivatives. Its water solubility and excellent sensing activity effectively improved the application value of fluorescent nanosensors for tracking and detecting Al^{3+} in the environment and organisms. Chen et al. [107] developed a new type of β-CD with a large adsorption capacity and high throughput to effectively remove bisphenol A (BPA) (an environmental endocrine disruptor that can affect human health), i.e., a fine (β-CD) modified graphite oxide (CDGO) film. CDGO nanosheets are made by chemically grafting β-CD molecules to both sides of the GO nanosheets. The β-CD molecules can recognize and form stable complexes with BPA molecules to achieve efficient BPA removal, and the β-CD molecules on both sides of the CDGO nanosheets have a high grafting density, large surface area, and an interception efficiency of ~100%; its adsorption capacity is several times higher than the traditional method. Further, it could go through multiple operation cycles and is very promising for water treatment and molecular separation applications. Hu et al. [108] developed a host–guest recognition method using β-CD and Azo to prepare a new sandwich-type graphene/CD/C_{60} nanohybrid, which loaded β-CD through a one-pot reduction reaction. On graphene, it can be used to control the release of C_{60} and has better nitric oxide (NO) quenching ability than other graphene/C_{60} nanohybrids, which can serve as an effective nanoplatform against oxidative damage. The hybridization of rGO, β-CD, and Azo-C_{60} enhanced cell uptake and limited the aggregation of C_{60} and showed enhanced protection against NO-induced cytotoxicity. The rGO/β-CD platform can also be reused. Because host–guest chemistry and diazo chemistry are universal and generally applicable, this strategy can also be used to prepare other light-responsive nanohybrids, which should be valuable in the life science and materials science fields. Wang and Zhe [109] successfully synthesized β-CD functionalized three-dimensional graphene foam (CDGF) using a simple, one-step hydrothermal method. The effect of pH on the material was studied. Because the anion species of Cr(VI) are partially located on the positively charged surface of CDGF, when the pH of the Cr(VI) solution = 3, the CDGF has good selectivity for Cr(VI). As the pH increases, the adsorption capacity gradually decreases and the hydroxyl groups on CDGF play a major role in the adsorption process, which is a simple separation strategy. After adsorption of Cr(VI), CDGF maintains a fixed form and the separation process is simplified. This work provides a novel material for the adsorption of hexavalent chromium from water, and it provides direction for easy and fast solid–liquid separation strategies for adsorption and other applications (Figure 5).

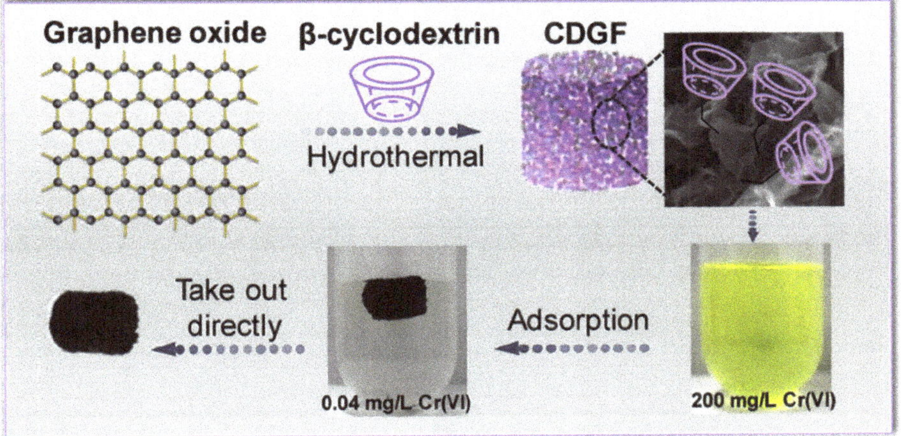

Figure 5. The β-CD functionalized three-dimensional structured graphene foam (CDGF) was applied for the adsorption of Cr(VI) with the easy and rapid separation strategy [109]. Adapted with permission from ref. [109]. Copyright 2019 Elsevier.

5. Application of pCDs in Biomedicine

5.1. pCDs Reduce the Toxicity of Some Exogenous Organisms

The amorphous cavity of pCDs can capture a variety of drugs, thereby adjusting the physical and chemical properties of the guest molecule. Forming highly stable host–guest complexes with pCDs will reduce the biological effects of the guest molecules, and the improved bioavailability would reduce the toxicity of some exogenous organisms [110–113].

Guo et al. [46] prepared an inclusion compound of podophyllotoxin (PPT) and -CD, which could greatly reduce the toxicity of PPT. The behavior, characterization, and water solubility of the inclusion compound were carefully studied using multiple techniques. The inclusion compound was formed with a ratio of 1:1 and had a considerable stability constant K-s (4245.5 Lmol^{-1}). The anti-cancer activity of the inclusion compound was better than that of cisplatin (DDP, positive control). Faisal et al. [114] studied the protective effect of -CD on the toxicity of HeLa cells and zebrafish embryos induced by zelaketone. Under certain conditions, sulfobutyl, methyl, and succinyl substituted CDs formed stable complexes, which significantly reduced or even eliminated the toxicity of azilenone. In addition, co-treatment with -CD also significantly reduced the sublethal effect of zaraketone. Studies have also shown that the formation of a stable zelenone-CD complex can significantly reduce or even eliminate the toxicity of zelenone in vivo and in vitro. Therefore, CD is expected to become a new mycotoxin binder.

5.2. Application of pCDs in Drug Delivery

The bioavailability of drugs is often limited by poor water solubility [115–118]. pCDs have been used for drug encapsulation, which improves the stability of drugs and effectively regulates their release. pCDs can greatly improve the solubility of poorly water-soluble drugs; additionally, they can also be used to prepare carrier systems that control drug release over a long time. CD has a special structure, which can enhance the biocompatibility and degradability of the drug by forming an inclusion compound, as well as a high loading rate of drug molecules and improvement of the controlled release performance. CDs can control the release of drugs and reduce their toxicity [119,120].

Oliveria et al. [47] used a high-yield reaction route based on polyglutamic acid, and used β-CD or γ-CD for the synthesis. The novel polymer had an average of ~17 CD cavities and was characterized using nuclear magnetic resonance, MALDI-MS, and DLS. It was determined to be a carrier of doxorubicin in human tumor cells, and this inclusion compound has antiproliferative activity in the tumor cells. Bisphosphonate is a mature drug with a wide range of applications in medicine. However, the side chain and nature of the phosphorus group may cause poor water solubility, thereby affecting its bioavailability. Mallard et al. [121] proved that CDs can be used as a bisphosphonate carrier. Through bisphosphonate functionalized CD, a cyclodextrin/bisphosphonate polymer (CD/BP) was synthesized and characterized. The formation of CD/BP was characterized by one-dimensional and two-dimensional nuclear magnetic resonance spectroscopy, isothermal titration, calorimetry, and ultraviolet-visible spectrophotometry, which showed that cyclodextrin is an effective carrier for bisphosphonates. CD/BP can be used to treat parasitic diseases, in particular, to prevent Chagas disease [122,123]. This provides a better treatment plan for the treatment of sleeping sickness caused by parasites. Ho et al. [124] proposed a submicrocarrier with an average hydrodynamic size of 400 to 900 nm through electrostatic gelation of anionic β-CD and chitosan (SMCs). This could address the issues of poor solubility of drugs and the limited bioavailability and pharmacological effects. It was used to improve the solubility and clinically relevant anti-infective controlled release properties of ciprofloxacin. In the study, it was found that when the encapsulation efficiency (~90%) and load capacity (~9%) of sMC were maximized, the molar ratio of ciprofloxacin to β-CD was 1:1. The results showed that regardless of their size, after 24 h of incubation, the cells absorbed sMC well without pathological changes. The sMC was non-toxic and had very good biocompatibility, and it is a suitable system with promising prospects for the treatment of extracellular lung infections. Khelghati et al. [125] focused on reducing the side effects of adriamycin by designing a pH-sensitive magnetic hyperbranched β-CD as a nano-level drug carrier. Nanoparticles were released more under acidic conditions and were released less under neutral alkaline conditions. The results of hemolysis analysis showed that the synthesized nanocarriers were completely biocompatible. In vitro studies have shown that free doxorubicin has a higher cytotoxicity than doxorubicin-loaded nanocarriers, demonstrating its high potential to deliver doxorubicin to tumor tissues. Li et al. [126] reported the design and synthesis of a new multifunctional nuclear initiator based on octanorbornene functionalized γ-CD, which is an eight-arm star polymer [127,128] prepared by a nuclear-preferred ring-opening metathesis polymerization reaction. Hexaethylene glycol functionalized with norbornene was used to graft from the initiator. Using norbornene-functionalized polyethylene glycol (PEG) to extend the chain of the omega functional group to produce water-soluble diblock brush arm star copolymer (DBASC), the size of DBASC was between 10–11 nm, with very good thermal stability, long-range order, and crystallinity. Because of the introduction of γ-CD, DBASC has excellent solubility, enhanced drug binding ability, shows low toxicity, and has a strong inhibitory effect on MCF-7 breast cancer cells. Star polymers represent a new type of modular polymer platform with potential applications in nanostructure self-assembly and drug delivery. Das et al. [129] used β-CD as a drug delivery vehicle for drug cells with amino acid-based ionic liquid (AAIL) substitutes, which could improve the bioavailability of the therapeutic agent. Therefore, the use of β-CD is preferred to stabilize AAIL in our work, and AAIL can be further separated in a controlled manner. Because of its importance in the field of biotechnology, electrodialysis separation, and drug delivery, AAILs, i.e., proline nitrate PN and β-CD are proposed as a model system. In this work, the experimental measurements were theoretically related to quantum chemistry methods. Two-dimensional correlation experiments show the characteristics of PN and β-CD binding. These fascinating results were clarified with the help of molecular docking simulation studies. The confirmation with antibacterial research was consistent with the experimental results. The new AILs-based inclusion compounds have little toxicity and can be used as potential carriers. The results reported were encouraging for the practical application of AAIL and β-CD. Viale et al. [130] synthesized fibrin gel (FBG) and an amino-pCDs inclusion complex (oCD-NH$_2$/Dox) in 2019, demonstrating that the FBG can be used in clinical or experimental applications to release different doxorubicin (Dox) nanoparticles. The fibrinogen (FG) and Ca^{2+} concentrations may change this activity. In vivo data support that the overall and local toxicity of FBG loaded with oCD-NH$_2$/Dox is lower than that of FBG loaded with Dox. The results indicate that when administered locally via FBG loaded with oCD-NH$_2$/Dox, the therapeutic index of Dox may increase, providing the possibility for using these delivery systems to treat neuroblastoma. Haley et al. [131] reported that a drug delivery system made of pCDs allowed for local administration of amphotericin B (AmB) (the leading drug for the treatment of clinical fungal infections), which can reduce toxicity to the host cells while maintaining the ability to eliminate fungal activity. By exploiting the molecular interaction between the CD cavity and the drug, a slow and sustained delivery rate of AmB was achieved. Lin [132] reported that β-cyclodextrin-{poly(lactide)-poly(2-(dimethylamino)ethylmethacrylate)-poly[oligo(2-ethyl-2-oxazoline)methacrylate]}$_{21}$ [β-CD-(PLA-PDMAEMA-PEtOxMA)$_{21}$] monomolecular micelles act as gold nanoparticles (GNP); the in situ formation and subsequent Dox encapsulated template were applied for the development of anti-cancer drug delivery and computer tomography (CT) imaging to form a pH-responsive therapeutically reactive nanocomposites in situ. Through a combination of experiments and dissipative particle dynamics (DPD) simulations, the formation, microstructure, and distribution of GNP and Dox were studied. Under acidic tumor conditions, Dox-loaded micelles had an encapsulation efficiency of 41–61%, showing rapid release (88% after 102 h). Both in vitro and in vivo experiments have shown excellent anti-cancer efficacy and effective CT imaging performance for β-CD-(PLA-PDMAEMA-PEtOxMA)$_{21}$/Au/DOX. Single-molecule micelles represent a class of multifunctional nanocarriers for therapeutic diagnostics. Nanoparticle carriers are now a hot frontier field, and the intelligentization of drug loading can greatly increase chances of curing diseases such as cancer and NPC. In recent years, research on biodegradable nanocarriers has gradually increased, and pCDs have been used in new forms of drug delivery. Liu et al. [133] synthesized a new type of star-shaped nanocarrier $(C_{12}H_{25})_{14}$-β-CD-(SS-mPEG)$_7$ (CCSP) for anti-tumor activity in 2019. The DOX-loaded nanocarrier CCSP has good biocompatibility, high drug loading, good stimulus response release performance, and low leakage, which have potential in anti-cancer intelligence. It is worth noting that Ran Namgung et al. [134] designed a new type of nano-assembled drug delivery system formed by the interaction between polymer-cyclodextrin conjugate and polymer-paclitaxel conjugate, which is the most popular CD polymer. This is one of the successful drug delivery systems. Nano-components have high stability, which can effectively deliver paclitaxel to targeted cancer cells through passive and active targeting mechanisms and effectively release them. Ying et al. [135] introduced the cationic CD loop into the multi-arm PEG backbone in a sterically selective manner, and developed a multi arm pCD polyrotaxane nanocarrier platform that protectively encapsulates the interleukin 12 (IL-12) encoding plasmid for immune gene therapy of colon cancer. Compared with the linear pCD polyrotaxane, the multi-arm polymer design significantly improves the circulating half-life, and the reported tumor suppressor effect is excellent and non-toxic. Nowadays, pCD as an API is a hot topic in the field of biomedicine. For example, Atsushi Tamura et al. [136] have developed an acid-labile β-CD/Pluronic P123-based polyrotaxane for the treatment of a fatal metabolic disease, Niemann-Pick Type C (NPC), compared to the general 2-hydroxypropyl β-cyclodextrin (HP-β-CD) drug treatment, it can not only promote cholesterol excretion and prolong the life of the animal in animal models, and the required dose is greatly reduced, which has huge advantages compared with traditional medicine. In addition, in order to enhance the pharmacokinetics and biodistribution char acteristics, and thereby improve the efficacy at lower doses, Aditya Kulkarn et al. [137] designed a β-CD-based polymer prodrug ORX-301, which is composed of β-CD two. It is formed by the polymerization of azide monomer and bifunctional ketal. It overcomes the main limitations of current β-CD-based NPC therapy, and is a potential alternative to existing treatment methods (Table 1).

Table 1. Abstracts of papers for pCDs used as drug delivery vehicles.

Formulations	Drugs Carried	Application Effect	System Size	Ref.
PGAβCyD	Doxorubicin	Inhibit tumor tissue growth, Reduce the side effects of doxorubicin	5.5 nm	[47]
PGAγCyD.	Doxorubicin	Inhibit tumor tissue growth, Reduce the side effects of doxorubicin	5.5 nm	[47]
CD/BP	Bisphosphonate	Treat parasitic diseases	ND	[121]
sMC	Ciprofloxacin	Improve the solubility and release of the drug	400–900 nm	[124]
HPG-β-CD	Doxorubicin	Reduce the side effects of doxorubicin	30 nm	[125]
DBASC	Doxorubicin	Inhibit tumor tissue growth, Reduce the side effects of doxorubicin	10.0–11.0 nm	[126]
PN-β-CD	Real drug cell drugs	Non-toxic, drug-acceptable, low-cost, and environmentally friendly carrier	ND	[129]
CD-NH$_2$/Dox	Doxorubicin	Treatment of neuroblastoma	3.2 nm	[130]
pCDs	AmB	Treatment of clinical fungus	2–15 kDa	[131]
β-CD-(PLA-PDMAEMA-PEtOxMA)$_{21}$	Doxorubicin	Inhibit tumor tissue growth	27–28 nm	[132]
CCSP	Doxorubicin	Reduce drug leakage and increase drug load content	40–50 nm	[133]
Polymer—cyclodextrin conjugate	Polymer—paclitaxel conjugate	Confer high stability to the nano-assembly	54.6 ± 11.6 nm	[134]
pCD polyrotaxane	interleukin 12	Protective packaging	193 ± 6.2 nm	[135]
β-CD/Pluronic P123-based polyrotaxane	β-CD	Treating NPC	29,000 Da	[136]

6. Conclusions

pCDs have many advantages, including a sustainable source of raw materials, low cost, multiple synthetic methods, and a simple synthesis procedure. They not only have the characteristics of cyclodextrin molecules, such as an internal hydrophobic-external hydrophilic cavity structure and hand-shaped features, but they also have the excellent mechanical strength, hardness, and good thermal stability of polymers.

This article introduced the research progress of pCDs, including their synthesis and applications in analytical separation science, materials science, and biomedicine. The field of analysis and separation science mainly uses the cavity structure and hand shape characteristics of pCDs to realize separation and detection. In the field of materials science, CDs have been incorporated into various materials. Compared with traditional materials, pCDs have the characteristics of a high flux and solubility, resulting in practical materials with superior properties. In the field of biomedicine, compared with traditional materials, pCDs have the advantages of specific identification, non-toxicity, and good biocompatibility. In short, pCDs have been increasingly applied in the fields of analysis and separation science, material science, and biomedicine because of their unique characteristics.

While emphasizing the multi-field applications of pCDs, this article focused on its scientific and technological innovation as drug delivery vehicles, using the latest research as examples to illustrate their potential for cancer treatment and the elimination of fungi. A summary of the development trend in pCDs from multiple fields and perspectives illustrates the large growth in the amount of pCD research and the large potential for development, and this work provides direction for future research.

Author Contributions: Literature search, Y.L. and T.L.; figures, Q.W., Y.L. and S.L.; study design, C.C. and X.H.; data collection, C.L. and Q.W.; data analysis, X.H. and C.L.; data interpretation, T.L. and S.L.; writing, C.C. and Y.L. All authors have read and agreed to the published version of the manuscript.

Funding: This work was supported by National Natural Science Foundation of China (81901896); and Scientific Research Foundation of Fuzhou University (GXRC-19025).

Data Availability Statement: Data are available in a publicly accessible repository.

Conflicts of Interest: The authors declare no conflict of interest.

References

1. Eastburn, S.D.; Tao, B.Y. Applications of modified cyclodextrins. *Biotechnol. Adv.* **1994**, *12*, 325–339. [CrossRef]
2. Gharib, R.; Greige-Gerges, H.; Fourmentin, S.; Charcosset, C.; Auezova, L. Liposomes incorporating cyclodextrin-drug inclusion complexes: Current state of knowledge. *Carbohyd. Polym.* **2015**, *129*, 175–186. [CrossRef]
3. Storsberg, J.; Hartenstein, M.; Muller, A.H.E.; Ritter, H. Cyclodextrins in polymer synthesis: Polymerization of methyl methacrylate under atom-transfer conditions (ATRP) in aqueous solution. *Macromol. Rapid Commun.* **2000**, *21*, 1342–1346. [CrossRef]
4. Yamamoto, K.; Nakai, Y. Inclusion Compound Formation by Co-Grinding of Cyclodextrin and Host Drugs. *Sib. Khimicheskii Zhurnal* **1991**, *1991*, 51–55.
5. Buschmann, H.J.; Schollmeyer, E. Applications of cyclodextrins in cosmetic products: A review. *J. Cosmet. Sci.* **2002**, *53*, 185–191. [PubMed]
6. Gerard, M.; Chaubey, A.; Malhotra, B.D. Application of conducting polymers to biosensors. *Biosens. Bioelectron.* **2002**, *17*, 345–359. [CrossRef]
7. Xu, Y.H.; Jin, S.B.; Xu, H.; Nagai, A.; Jiang, D.L. Conjugated microporous polymers: Design, synthesis and application. *Chem. Soc. Rev.* **2013**, *42*, 8012–8031. [CrossRef] [PubMed]
8. Borzenkov, M.; Mitina, N.; Lobaz, V.; Hevus, O. Synthesis and Properties of Novel Surface Active Monomers Based on Derivatives of 4-Hydroxybutyric Acid and 6-Hydroxyhexanoic Acid. *J. Surfactants Deterg.* **2015**, *18*, 133–144. [CrossRef]
9. Hajipour, A.R.; Zahmatkesh, S.; Ruoho, A.E. Optically active polymer: Synthesis and characterization of new optically active poly (hydrazide-imide)s incorporating L-leucine. *E-Polymers* **2007**, *7*, 1018–1030. [CrossRef]
10. Rusa, C.C.; Luca, C.; Tonelli, A.E. Polymer-cyclodextrin inclusion compounds: Toward new aspects of their inclusion mechanism. *Macromolecules* **2001**, *34*, 1318–1322. [CrossRef]
11. Wenz, G.; Keller, B. Threading Cyclodextrin Rings on Polymer-Chains. *Angew. Chem. Int. Ed. Engl.* **1992**, *31*, 197–199. [CrossRef]
12. Harada, A.; Takashima, Y.; Yamaguchi, H. Cyclodextrin-based supramolecular polymers. *Chem. Soc. Rev.* **2009**, *38*, 875–882. [CrossRef] [PubMed]

13. Wang, Z.; Zou, W.; Liu, L.Y.; Wang, M.; Li, F.; Shen, W.Y. Characterization and bacteriostatic effects of beta-cyclodextrin/quercetin inclusion compound nanofilms prepared by electrospinning. *Food Chem.* **2021**, *338*, 127980. [CrossRef]
14. Krabicova, I.; Appleton, S.L.; Tannous, M.; Hoti, G.; Caldera, F.; Pedrazzo, A.R.; Cecone, C.; Cavalli, R.; Trotta, F. History of Cyclodextrin Nanosponges. *Polymers (Basel)* **2020**, *12*, 1122. [CrossRef] [PubMed]
15. Crini, G. Review: A History of Cyclodextrins. *Chem. Rev.* **2014**, *114*, 10940–10975. [CrossRef] [PubMed]
16. Yu, G.C.; Jie, K.C.; Huang, F.H. Supramolecular Amphiphiles Based on Host-Guest Molecular Recognition Motifs. *Chem. Rev.* **2015**, *115*, 7240–7303. [CrossRef]
17. Miao, C.Y.; Zhang, L.M.; Yuan, W.J.; Su, D.F. Angiotensin II and AT(1) receptor in hypertrophied ventricles and aortas of sinoaortic-denervated rats. *Acta Pharmacol. Sin.* **2003**, *24*, 812–818.
18. Yao, X.K.; Huang, P.; Nie, Z.H. Cyclodextrin-based polymer materials: From controlled synthesis to applications. *Prog. Polym. Sci.* **2019**, *93*, 1–35. [CrossRef]
19. Gidwani, B.; Vyas, A. Synthesis, characterization and application of Epichlorohydrin-β-cyclodextrin polymer. *Colloids Surf. B Biointerfaces* **2014**, *114*, 130–137. [CrossRef]
20. Gawish, S.M.; Ramadan, A.M.; Abo El-Ola, S.M.; Abou El-Kheir, A.A. Citric Acid Used as a Cross-Linking Agent for Grafting β-Cyclodextrin onto Wool Fabric. *Polym. Plast. Technol. Eng.* **2009**, *48*, 701–710. [CrossRef]
21. Rukmani, A.; Sundrarajan, M. Inclusion of antibacterial agent thymol on β-cyclodextrin-grafted organic cotton. *J. Ind. Text.* **2012**, *42*, 132–144. [CrossRef]
22. Ghorpade, V.S.; Yadav, A.V.; Dias, R.J. Citric acid crosslinked β-cyclodextrin/carboxymethylcellulose hydrogel films for controlled delivery of poorly soluble drugs. *Carbohydr. Polym.* **2017**, *164*, 339–348. [CrossRef] [PubMed]
23. You, Q.; Zhang, P.; Bai, S.; Huang, W.; Jia, Z.; Zhou, C.; Li, D. Supramolecular linear polymer formed by host–guest interactions of β-cyclodextrin dimers and polyacrylamide end-capped with adamantane. *Colloids Surf. A Physicochem. Eng. Asp.* **2015**, *484*, 130–135. [CrossRef]
24. Noomen, A.; Penciu, A.; Hbaieb, S.; Parrot-Lopez, H.; Amdouni, N.; Chevalier, Y.; Kalfat, R. Silicon Based Polymers. In *Grafting β-Cyclodextrins to Silicone, Formulation of Emulsions and Encapsulation of Antifungal Drug*; Ganachaud, F., Boileau, S., Boury, B., Eds.; Springer: Dordrecht, The Netherlands, 2008; pp. 163–179.
25. Wu, J.; Shen, Q.; Fang, L. Sulfobutylether-β-cyclodextrin/chitosan nanoparticles enhance the oral permeability and bioavailability of docetaxel. *Drug Dev. Ind. Pharm.* **2013**, *39*, 1010–1019. [CrossRef] [PubMed]
26. Han, F.; Liu, Q.; Lai, X.; Li, H.; Zeng, X. Compatibilizing effect of β-cyclodextrin in RDP/phosphorus-containing polyacrylate composite emulsion and its synergism on the flame retardancy of the latex film. *Prog. Org. Coat.* **2014**, *77*, 975–980. [CrossRef]
27. Shang, S.; Chiu, K.-L.; Jiang, S. Synthesis of immobilized poly(vinyl alcohol)/cyclodextrin eco-adsorbent and its application for the removal of ibuprofen from pharmaceutical sewage. *J. Appl. Polym. Sci.* **2017**, *134*, 21. [CrossRef]
28. Jiang, L.X.; Yan, Y.; Huang, J.B. Versatility of cyclodextrins in self-assembly systems of amphiphiles. *Adv. Colloid Interface* **2011**, *169*, 13–25. [CrossRef]
29. Van der Veen, B.A.; Uitdehaag, J.C.M.; Dijkstra, B.W.; Dijkhuizen, L. Engineering of cyclodextrin glycosyltransferase reaction and product specificity. *BBA-Protein Struct. Mol. Enzymol.* **2000**, *1543*, 336–360. [CrossRef]
30. Messner, M.; Kurkov, S.V.; Flavia-Piera, R.; Brewster, M.E.; Loftsson, T. Self-assembly of cyclodextrins: The effect of the guest molecule. *Int. J. Pharm.* **2011**, *408*, 235–247. [CrossRef]
31. Abou-Okeil, A.; Rehan, M.; El-Sawy, S.M.; El-bisi, M.K.; Ahmed-Farid, O.A.; Abdel-Mohdy, F.A. Lidocaine/β-cyclodextrin inclusion complex as drug delivery system. *Eur. Polym. J.* **2018**, *108*, 304–310. [CrossRef]
32. Ol'khovich, M.; Sharapova, A.; Blokhina, S.; Perlovich, G.; Skachilova, S.; Shilova, E. A study of the inclusion complex of bioactive thiadiazole derivative with 2-hydroxypropyl-β-cyclodextrin: Preparation, characterization and physicochemical properties. *J. Mol. Liq.* **2019**, *273*, 653–662. [CrossRef]
33. Aytac, Z.; Sen, H.S.; Durgun, E.; Uyar, T. Sulfisoxazole/cyclodextrin inclusion complex incorporated in electrospun hydroxypropyl cellulose nanofibers as drug delivery system. *Colloid Surf. B* **2015**, *128*, 331–338. [CrossRef] [PubMed]
34. Angelova, A.; Ringard-Lefebvre, C.; Baszkin, A. Drug–Cyclodextrin Association Constants Determined by Surface Tension and Surface Pressure Measurements: I. Host–Guest Complexation of Water Soluble Drugs by Cyclodextrins: Polymyxin B–β Cyclodextrin System. *J. Colloid Interface Sci.* **1999**, *212*, 275–279. [CrossRef]
35. Angelova, A.; Ringard-Lefebvre, C.; Baszkin, A. Drug–Cyclodextrin Association Constants Determined by Surface Tension and Surface Pressure Measurements: II. Sequestration of Water Insoluble Drugs from the Air–Water Interface: Retinol–β Cyclodextrin System. *J. Colloid Interface Sci.* **1999**, *212*, 280–285. [CrossRef]
36. Frieler, L.; Ho, T.M.; Anthony, A.; Hidefumi, Y.; Yago, A.J.E.; Bhandari, B.R. Crystallisation properties of amorphous cyclodextrin powders and their complexation with fish oil. *J. Food Sci. Technol.* **2019**, *56*, 1519–1529. [CrossRef] [PubMed]
37. Gao, C.; Yan, D. Hyperbranched polymers: From synthesis to applications. *Prog. Polym. Sci.* **2004**, *29*, 183–275. [CrossRef]
38. Tang, B.; Liu, X.B.; Zhao, X.L.; Zhang, J.H. Highly Efficient In Situ Toughening of Epoxy Thermosets with Reactive Hyperbranched Polyurethane. *J. Appl. Polym. Sci.* **2014**, *131*. [CrossRef]
39. Rubin Pedrazzo, A.; Caldera, F.; Zanetti, M.; Appleton, S.L.; Dhakar, N.K.; Trotta, F. Mechanochemical green synthesis of hyper-crosslinked cyclodextrin polymers. *Beilstein J. Org. Chem.* **2020**, *16*, 1554–1563. [CrossRef] [PubMed]
40. Petitjean, M.; Aussant, F.; Vergara, A.; Isasi, J.R. Solventless Crosslinking of Chitosan, Xanthan, and Locust Bean Gum Networks Functionalized with beta-Cyclodextrin. *Gels (Basel)* **2020**, *6*, 51. [CrossRef]

41. Jicsinszky, L.; Calsolaro, F.; Martina, K.; Bucciol, F.; Manzoli, M.; Cravotto, G. Reaction of oxiranes with cyclodextrins under high-energy ball-milling conditions. *Beilstein J. Org. Chem.* **2019**, *15*, 1448–1459. [CrossRef]

42. Zhao, F.; Repo, E.; Yin, D.; Chen, L.; Kalliola, S.; Tang, J.; Iakovleva, E.; Tam, K.C.; Sillanpää, M. One-pot synthesis of trifunctional chitosan-EDTA-β-cyclodextrin polymer for simultaneous removal of metals and organic micropollutants. *Sci. Rep.* **2017**, *7*, 15811. [CrossRef]

43. Ryvlin, D.; Girschikofsky, M.; Schollmeyer, D.; Hellmann, R.; Waldvogel, S.R. Pollutant Adsorbtion and Detection: Methyl-Substituted α-Cyclodextrin as Affinity Material for Storage, Separation, and Detection of Trichlorofluoromethane (Global Challenges 8/2018). *Glob. Chall.* **2018**, *2*, 1870184. [CrossRef]

44. Villalobos, L.F.; Huang, T.F.; Peinemann, K.V. Cyclodextrin Films with Fast Solvent Transport and Shape-Selective Permeability. *Adv. Mater.* **2017**, *29*, 1606641. [CrossRef]

45. Liu, Z.C.; Yang, W.J.; Tian, B.S.; Liu, J.; Zhu, W.P.; Ge, G.W.; Xiao, L.N.; Meng, Y.N. Fabrication of a self-assembled supramolecular fluorescent nanosensor from functional graphene oxide and its application for the detection of Al^{3+}. *New J. Chem.* **2018**, *42*, 17665–17670. [CrossRef]

46. Guo, Y.; Zhang, Y.; Li, J.; Zhao, F.; Liu, Y.; Su, M.; Jiang, Y.; Liu, Y.; Zhang, J.; Yang, B.; et al. Inclusion Complex of Podophyllotoxin with γ-Cyclodextrin: Preparation, Characterization, Anticancer Activity, Water-Solubility and Toxicity. *Chin. J. Chem.* **2016**, *34*, 425–431. [CrossRef]

47. Oliveri, V.; Bellia, F.; Viale, M.; Maric, I.; Vecchio, G. Linear polymers of β and γ cyclodextrins with a polyglutamic acid backbone as carriers for doxorubicin. *Carbohydr. Polym.* **2017**, *177*, 355–360. [CrossRef]

48. Zheng, Y.C.; Li, S.P.; Weng, Z.L.; Gao, C. Hyperbranched polymers: Advances from synthesis to applications. *Chem. Soc. Rev.* **2015**, *44*, 4091–4130. [CrossRef]

49. Crini, G.; Lichtfouse, E. Advantages and disadvantages of techniques used for wastewater treatment. *Environ. Chem. Lett.* **2019**, *17*, 145–155. [CrossRef]

50. Schwarzenbach, R.P.; Escher, B.I.; Fenner, K.; Hofstetter, T.B.; Johnson, C.A.; von Gunten, U.; Wehrli, B. The challenge of micropollutants in aquatic systems. *Science* **2006**, *313*, 1072–1077. [CrossRef] [PubMed]

51. Qu, J.H.; Dong, M.; Wei, S.Q.; Meng, Q.J.; Hu, L.M.; Hu, Q.; Wang, L.; Han, W.; Zhang, Y. Microwave-assisted one pot synthesis of beta-cyclodextrin modified biochar for concurrent removal of Pb(II) and bisphenol a in water. *Carbohydr. Polym.* **2020**, *250*, 117003. [CrossRef] [PubMed]

52. An, T.C.; Yang, H.; Li, G.Y.; Song, W.H.; Cooper, W.J.; Nie, X.P. Kinetics and mechanism of advanced oxidation processes (AOPs) in degradation of ciprofloxacin in water. *Appl. Catal. B-Environ.* **2010**, *94*, 288–294. [CrossRef]

53. Li, M.; Li, M.Y.; Feng, C.G.; Zeng, Q.X. Preparation and characterization of multi-carboxyl-functionalized silica gel for removal of Cu (II), Cd (II), Ni (II) and Zn (II) from aqueous solution. *Appl. Surf. Sci.* **2014**, *314*, 1063–1069. [CrossRef]

54. Modwi, A.; Khezami, L.; Taha, K.; Al-Duaij, O.K.; Houas, A. Fast and high efficiency adsorption of Pb(II) ions by Cu/ZnO composite. *Mater. Lett.* **2017**, *195*, 41–44. [CrossRef]

55. Sani, H.A.; Ahmad, M.B.; Hussein, M.Z.; Ibrahim, N.A.; Musa, A.; Saleh, T.A. Nanocomposite of ZnO with montmorillonite for removal of lead and copper ions from aqueous solutions. *Process. Saf. Environ.* **2017**, *109*, 97–105. [CrossRef]

56. Allabashi, R.; Arkas, M.; Hormann, G.; Tsiourvas, D. Removal of some organic pollutants in water employing ceramic membranes impregnated with cross-linked silylated dendritic and cyclodextrin polymers. *Water Res.* **2007**, *41*, 476–486. [CrossRef] [PubMed]

57. Liu, H.H.; Cai, X.Y.; Wang, Y.; Chen, J.W. Adsorption mechanism-based screening of cyclodextrin polymers for adsorption and separation of pesticides from water. *Water Res.* **2011**, *45*, 3499–3511. [CrossRef] [PubMed]

58. Sancey, B.; Trunfio, G.; Charles, J.; Badot, P.M.; Crini, G. Sorption onto crosslinked cyclodextrin polymers for industrial pollutants removal: An interesting environmental approach. *J. Incl. Phenom. Macrocycl.* **2011**, *70*, 315–320. [CrossRef]

59. Yilmaz, E.; Memon, S.; Yilmaz, M. Removal of direct azo dyes and aromatic amines from aqueous solutions using two beta-cyclodextrin-based polymers. *J. Hazard Mater.* **2010**, *174*, 592–597. [CrossRef] [PubMed]

60. Aoki, N.; Kinoshita, K.; Mikuni, K.; Nakanishi, K.; Hattori, K. Adsorption of 4-nonylphenol ethoxylates onto insoluble chitosan beads bearing cyclodextrin moieties. *J. Incl. Phenom. Macrocycl.* **2007**, *57*, 237–241. [CrossRef]

61. Romo, A.; Peñas, F.J.; Isasi, J.R.; García-Zubiri, I.X.; González-Gaitano, G. Extraction of phenols from aqueous solutions by β-cyclodextrin polymers. Comparison of sorptive capacities with other sorbents. *React. Funct. Polym.* **2008**, *68*, 406–413. [CrossRef]

62. Chen, C.Y.; Chung, Y.C. Removal of phthalate esters from aqueous solutions by chitosan bead. *J. Environ. Sci. Health Part A* **2006**, *41*, 235–248. [CrossRef] [PubMed]

63. Kitaoka, M.; Hayashi, K. Adsorption of bisphenol A by cross-linked beta-cyclodextrin polymer. *J. Incl. Phenom. Macrocycl.* **2002**, *44*, 429–431. [CrossRef]

64. Li, J.; Chen, C.; Zhao, Y.; Hu, J.; Shao, D.; Wang, X. Synthesis of water-dispersible Fe_3O_4@β-cyclodextrin by plasma-induced grafting technique for pollutant treatment. *Chem. Eng. J.* **2013**, *229*, 296–303. [CrossRef]

65. Repo, E.; Warchol, J.K.; Bhatnagar, A.; Mudhoo, A.; Sillanpaa, M. Aminopolycarboxylic acid functionalized adsorbents for heavy metals removal from water. *Water Res.* **2013**, *47*, 4812–4832. [CrossRef] [PubMed]

66. Jurecska, L.; Dobosy, P.; Barkacs, K.; Fenyvesi, E.; Zaray, G. Characterization of cyclodextrin containing nanofilters for removal of pharmaceutical residues (Reprinted from International Journal of Pharmaceutical and Biomedical Analysis, vol 98, pg 90–93, 2014). *J. Pharm. Biomed.* **2015**, *106*, 124–128. [CrossRef] [PubMed]

67. Alsbaiee, A.; Smith, B.J.; Xiao, L.; Ling, Y.; Helbling, D.E.; Dichtel, W.R. Rapid removal of organic micropollutants from water by a porous β-cyclodextrin polymer. *Nature* **2016**, *529*, 190–194. [CrossRef]
68. Lo Meo, P.; Lazzara, G.; Liotta, L.; Riela, S.; Noto, R. Cyclodextrin-calixarene co-polymers as a new class of nanosponges. *Polym Chem.* **2014**, *5*, 4499–4510. [CrossRef]
69. Hu, X.; Hu, Y.; Xu, G.; Li, M.; Zhu, Y.; Jiang, L.; Tu, Y.; Zhu, X.; Xie, X.; Li, A. Green synthesis of a magnetic β-cyclodextrin polymer for rapid removal of organic micro-pollutants and heavy metals from dyeing wastewater. *Environ. Res.* **2020**, *180*, 108796 [CrossRef]
70. García-Díaz, E.; Zhang, D.; Li, Y.; Verduzco, R.; Alvarez, P.J.J. TiO$_2$ microspheres with cross-linked cyclodextrin coating exhibit improved stability and sustained photocatalytic degradation of bisphenol A in secondary effluent. *Water Res.* **2020**, *183*, 116095 [CrossRef]
71. Khammar, S.; Bahramifar, N.; Younesi, H. Preparation and surface engineering of CM-β-CD functionalized Fe$_3$O$_4$@TiO$_2$ nanoparticles for photocatalytic degradation of polychlorinated biphenyls (PCBs) from transformer oil. *J. Hazard Mater.* **2020**, *394*, 122422. [CrossRef]
72. Wagner, A.J.; Zubarev, D.Y.; Aspuru-Guzik, A.; Blackmond, D.G. Chiral Sugars Drive Enantioenrichment in Prebiotic Amino Acid Synthesis. *Acs Central Sci.* **2017**, *3*, 322–328. [CrossRef] [PubMed]
73. Knight, B.; Stache, E.; Ferreira, E. An analysis of the complementary stereoselective alkylations of imidazolidinone derivatives toward α-quaternary proline-based amino amides. *Tetrahedron* **2015**, *71*, 5814–5823. [CrossRef] [PubMed]
74. Camacho-Munoz, D.; Kasprzyk-Hordern, B. Multi-residue enantiomeric analysis of human and veterinary pharmaceuticals and their metabolites in environmental samples by chiral liquid chromatography coupled with tandem mass spectrometry detection. *Anal. Bioanal. Chem.* **2015**, *407*, 9085–9104. [CrossRef] [PubMed]
75. Wang, X.Y.; Li, Z.; Zhang, H.; Xu, J.F.; Qi, P.P.; Xu, H.; Wang, Q.; Wang, X.Q. Environmental Behavior of the Chiral Organophosphorus Insecticide Acephate and Its Chiral Metabolite Methamidophos: Enantioselective Transformation and Degradation in Soils. *Environ. Sci. Technol.* **2013**, *47*, 9233–9240. [CrossRef] [PubMed]
76. Subramanian, G.; Stalcup, A.M. Chiral Separation Techniques: A Practical Approach. *J. Am. Chem. Soc.* **2007**, *129*, 8922–8923.
77. Pieri, M.; Miraglia, N.; Acampora, A.; Genovese, G.; Soleo, L.; Sannolo, N. Determination of urinary S-phenylmercapturic acid by liquid chromatography-tandem mass spectrometry. *J. Chromatogr. B* **2003**, *795*, 347–354. [CrossRef]
78. Sabatini, L.; Barbieri, A.; Indiveri, P.; Mattioli, S.; Violante, F.S. Validation of an HPLC–MS/MS method for the simultaneous determination of phenylmercapturic acid, benzylmercapturic acid and o-methylbenzyl mercapturic acid in urine as biomarkers of exposure to benzene, toluene and xylenes. *J. Chromatogr. B* **2008**, *863*, 115–122. [CrossRef] [PubMed]
79. B'Hymer, C. Validation of an HPLC-MS-MS Method for the Determination of Urinary S-Benzylmercapturic Acid and S-Phenylmercapturic Acid. *J. Chromatogr. Sci.* **2011**, *49*, 547–553. [CrossRef]
80. Schettgen, T.; Musiol, A.; Alt, A.; Kraus, T. Fast determination of urinary S-phenylmercapturic acid (S-PMA) and S-benzylmercapturic acid (S-BMA) by column-switching liquid chromatography-tandem mass spectrometry. *J. Chromatogr. B* **2008**, *863*, 283–292. [CrossRef] [PubMed]
81. Barbieri, A.; Sabatini, L.; Accorsi, A.; Roda, A.; Violante, F.S. Simultaneous determination of t,t-muconic, S-phenylmercapturic and S-benzylmercapturic acids in urine by a rapid and sensitive liquid chromatography/electrospray tandem mass spectrometry method. *Rapid Commun. Mass Spectrom.* **2004**, *18*, 1983–1988. [CrossRef]
82. Chou, J.S.; Lin, Y.C.; Ma, Y.C.; Sheen, J.F.; Shih, T.S. Measurement of benzylmercapturic acid in human urine by liquid chromatography-electrospray ionization-tandem quadrupole mass spectrometry. *J. Anal. Toxicol.* **2006**, *30*, 306–312. [CrossRef]
83. Immohra, L.I.; Pein-Hackelbusch, M. Development of stereoselective e-tongue sensors considering the sensor performance using specific quality attributes-A bottom up approach. *Sensor Actuat. B Chem.* **2017**, *253*, 868–878. [CrossRef]
84. Girschikofsky, M.; Ryvlin, D.; Waldvogel, S.R.; Hellmann, R. Optical Sensor for Real-Time Detection of Trichlorofluoromethane. *Sensors* **2019**, *19*, 632. [CrossRef] [PubMed]
85. Li, L.; Wang, H.; Jin, Y.; Shuang, Y.; Li, L. Preparation of a new benzylureido-β-cyclodextrin-based column and its application for the determination of phenylmercapturic acid and benzylmercapturic acid enantiomers in human urine by LC/MS/MS. *Anal. Bioanal. Chem.* **2019**, *411*, 5465–5479. [CrossRef]
86. Fliszar-Nyul, E.; Lemli, B.; Kunsagi-Mate, S.; Szente, L.; Poor, M. Interactions of Mycotoxin Alternariol with Cyclodextrins and Its Removal from Aqueous Solution by Beta-Cyclodextrin Bead Polymer. *Biomolecules* **2019**, *9*, 428. [CrossRef]
87. Carcu-Dobrin, M.; Sabau, A.G.; Hancu, G.; Arpad, G.; Rusu, A.; Kelemen, H.; Papp, L.A.; Carje, A. Chiral discrimination of amlodipine from pharmaceutical products using capillary electrophoresis. *Braz. J. Pharm. Sci.* **2020**, *56*, 56. [CrossRef]
88. Zheng, Y.; Kobayashi, Y.; Sekine, T.; Takashima, Y.; Hashidzume, A.; Yamaguchi, H.; Harada, A. Visible chiral discrimination via macroscopic selective assembly. *Commun. Chem.* **2018**, *1*, 4. [CrossRef]
89. Fane, A.G.; Wang, R.; Hu, M.X. Synthetic Membranes for Water Purification: Status and Future. *Angew. Chem. Int. Ed.* **2015**, *54*, 3368–3386. [CrossRef]
90. Saufi, S.M.; Fee, C.J. Mixed matrix membrane chromatography based on hydrophobic interaction for whey protein fractionation. *J. Membr. Sci.* **2013**, *444*, 157–163. [CrossRef]
91. Moron-Lopez, J.; Nieto-Reyes, L.; Senan-Salinas, J.; Molina, S.; El-Shehawy, R. Recycled desalination membranes as a support material for biofilm development: A new approach for microcystin removal during water treatment. *Sci. Total. Environ.* **2019**, *647*, 785–793. [CrossRef]

92. Eren, E.; Sarihan, A.; Eren, B.; Gumus, H.; Kocak, F.O. Preparation, characterization and performance enhancement of polysulfone ultrafiltration membrane using PBI as hydrophilic modifier. *J. Membr. Sci.* **2015**, *475*, 1–8. [CrossRef]
93. Zhang, Y.Q.; Wang, L.L.; Xu, Y. ZrO$_2$ solid superacid porous shell/void/TiO2 core particles (ZVT)/polyvinylidene fluoride (PVDF) composite membranes with anti-fouling performance for sewage treatment. *Chem. Eng. J.* **2015**, *260*, 258–268. [CrossRef]
94. Dhiman, P.; Bhatia, M. Pharmaceutical applications of cyclodextrins and their derivatives. *J. Incl. Phenom. Macrocycl.* **2020**, *98*, 171–186. [CrossRef]
95. Martin del Valle, E. Cyclodextrins and their uses: A review. *Process. Biochem.* **2004**, *39*, 1033–1046. [CrossRef]
96. Schneiderman, E.; Stalcup, A.M. Cyclodextrins: A versatile tool in separation science. *J. Chromatogr. B* **2000**, *745*, 83–102. [CrossRef]
97. Morin-Crini, N.; Crini, G. Environmental applications of water-insoluble β-cyclodextrin–epichlorohydrin polymers. *Prog. Polym. Sci.* **2013**, *38*, 344–368. [CrossRef]
98. Pangeni, R.; Choi, J.U.; Panthi, V.K.; Byun, Y.; Park, J.W. Enhanced oral absorption of pemetrexed by ion-pairing complex formation with deoxycholic acid derivative and multiple nanoemulsion formulations: Preparation, characterization, and in vivo oral bioavailability and anticancer effect. *Int. J. Nanomed.* **2018**, *13*, 3329–3351. [CrossRef]
99. Xiong, Z.; Lin, H.B.; Liu, F.; Yu, X.M.; Wang, Y.Z.; Wang, Y. A new strategy to simultaneously improve the permeability, heat-deformation resistance and antifouling properties of polylactide membrane via bio-based beta-cyclodextrin and surface crosslinking. *J. Membr. Sci.* **2016**, *513*, 166–176. [CrossRef]
100. Peng, X.; Peng, L.L.; Wu, C.Z.; Xie, Y. Two dimensional nanomaterials for flexible supercapacitors. *Chem. Soc. Rev.* **2014**, *43*, 3303–3323. [CrossRef]
101. Akbarzade, S.; Chamsaz, M.; Rounaghi, G.H. Highly selective preconcentration of ultra-trace amounts of lead ions in real water and food samples by dispersive solid phase extraction using modified magnetic graphene oxide as a novel sorbent. *Anal. Methods* **2018**, *10*, 2081–2087. [CrossRef]
102. Yang, K.; Chen, B.; Zhu, L. Graphene-coated materials using silica particles as a framework for highly efficient removal of aromatic pollutants in water. *Sci. Rep.* **2015**, *5*, 11641. [CrossRef]
103. Gao, P.; Liu, Z.Y.; Sun, D.D.; Ng, W.J. The efficient separation of surfactant-stabilized oil-water emulsions with a flexible and superhydrophilic graphene-TiO$_2$ composite membrane. *J. Mater. Chem. A* **2014**, *2*, 14082–14088. [CrossRef]
104. Peng, Y.M.; Nie, J.P.; Cheng, W.; Liu, G.; Zhu, D.W.; Zhang, L.H.; Liang, C.Y.; Mei, L.; Huang, L.Q.; Zeng, X.W. A multifunctional nanoplatform for cancer chemo-photothermal synergistic therapy and overcoming multidrug resistance. *Biomater. Sci.* **2018**, *6*, 1084–1098. [CrossRef] [PubMed]
105. Dreyer, D.R.; Park, S.; Bielawski, C.W.; Ruoff, R.S. The chemistry of graphene oxide. *Chem. Soc. Rev.* **2010**, *39*, 228–240. [CrossRef] [PubMed]
106. Mao, H.Y.; Laurent, S.; Chen, W.; Akhavan, O.; Imani, M.; Ashkarran, A.A.; Mahmoudi, M. Graphene: Promises, Facts, Opportunities, and Challenges in Nanomedicine. *Chem. Rev.* **2013**, *113*, 3407–3424. [CrossRef] [PubMed]
107. Chen, Z.-H.; Liu, Z.; Hu, J.-Q.; Cai, Q.-W.; Li, X.-Y.; Wang, W.; Faraj, Y.; Ju, X.-J.; Xie, R.; Chu, L.-Y. β-Cyclodextrin-modified graphene oxide membranes with large adsorption capacity and high flux for efficient removal of bisphenol A from water. *J. Membr. Sci.* **2020**, *595*, 117510. [CrossRef]
108. Hu, Z.; Zhang, D.; Yu, L.; Huang, Y. Light-triggered C60 release from a graphene/cyclodextrin nanoplatform for the protection of cytotoxicity induced by nitric oxide. *J. Mater. Chem. B* **2018**, *6*, 518–526. [CrossRef]
109. Wang, Z.; Lin, F.Y.; Huang, L.Q.; Lu, Y.X.; Chen, J. Cyclodextrin functionalized 3D-graphene for the removal of Cr(VI) with the easy and rapid separation strategy. *Abstr. Pap. Am. Chem. S* **2019**, *258*, 112854. [CrossRef]
110. Connors, K.A. The stability of cyclodextrin complexes in solution. *Chem. Rev.* **1997**, *97*, 1325–1357. [CrossRef]
111. Szejtli, J. Introduction and general overview of cyclodextrin chemistry. *Chem. Rev.* **1998**, *98*, 1743–1753. [CrossRef] [PubMed]
112. Szejtli, J. Past, present, and future of cyclodextrin research. *Pure Appl. Chem.* **2004**, *76*, 1825–1845. [CrossRef]
113. Del Castillo, T.; Marales-Sanfrutos, J.; Santoyo-González, F.; Magez, S.; Lopez-Jaramillo, F.J.; Garcia-Salcedo, J.A. Monovinyl Sulfone β-Cyclodextrin. A Flexible Drug Carrier System. *Chemmedchem* **2014**, *9*, 383–389. [CrossRef]
114. Faisal, Z.; Garai, E.; Csepregi, R.; Bakos, K.; Fliszar-Nyul, E.; Szente, L.; Balazs, A.; Cserhati, M.; Koszegi, T.; Urbanyi, B.; et al. Protective effects of beta-cyclodextrins vs. zearalenone-induced toxicity in HeLa cells and Tg(vtg1:mCherry) zebrafish embryos. *Chemosphere* **2020**, *240*, 124948. [CrossRef]
115. Hu, Z.M.; Li, S.N.; Wang, S.K.; Zhang, B.; Huang, Q. Encapsulation of menthol into cyclodextrin metal-organic frameworks: Preparation, structure characterization and evaluation of complexing capacity. *Food Chem.* **2021**, *338*, 127839. [CrossRef]
116. Chen, L.; Okuda, T.; Lu, X.Y.; Chan, H.K. Amorphous powders for inhalation drug delivery. *Adv. Drug Deliv. Rev.* **2016**, *100*, 102–115. [CrossRef] [PubMed]
117. Stegemann, S.; Leveiller, F.; Franchi, D.; de Jong, H.; Linden, H. When poor solubility becomes an issue: From early stage to proof of concept. *Eur. J. Pharm. Sci.* **2007**, *31*, 249–261. [CrossRef] [PubMed]
118. Wauthoz, N.; Amighi, K. *Formulation Strategies for Pulmonary Delivery of Poorly Soluble Drugs*; John Wiley & Sons, Ltd.: Hoboken, NJ, USA, 2015; pp. 87–122.
119. Nishimura, K.; Hidaka, R.; Hirayama, F.; Arima, H.; Uekama, K. Improvement of Dispersion and Release Properties of Nifedipine in Suppositoriesby Complexation with 2-Hydroxypropyl-β-cyclodextrin. *J. Incl. Phenom. Macrocycl.* **2006**, *56*, 85–88. [CrossRef]
120. Wang, Y.; Han, B. ChemInform Abstract: Cyclodextrin-Based Porous Nanocapsules. *ChemInform* **2013**, *44*, 38. [CrossRef]

121. Monteil, M.; Lecouvey, M.; Landy, D.; Ruellan, S.; Mallard, I. Cyclodextrins: A promising drug delivery vehicle for bisphosphonate. *Carbohyd. Polym.* **2017**, *156*, 285–293. [CrossRef]
122. Hudock, M.P.; Sanz-Rodriguez, C.E.; Song, Y.C.; Chan, J.M.W.; Zhang, Y.H.; Odeh, S.; Kosztowski, T.; Leon-Rossell, A.; Concepcion, J.L.; Yardley, V.; et al. Inhibition of Trypanosoma cruzi hexokinase by bisphosphonates. *J. Med. Chem.* **2006**, *49*, 215–223. [CrossRef] [PubMed]
123. Szajnman, S.H.; Rosso, V.S.; Malayil, L.; Smith, A.; Moreno, S.N.J.; Docampo, R.; Rodriguez, J.B. 1-(Fluoroalkylidene)-1,1-bisphosphonic acids are potent and selective inhibitors of the enzymatic activity of Toxoplasma gondii farnesyl pyrophosphate synthase. *Org. Biomol. Chem.* **2012**, *10*, 1424–1433. [CrossRef]
124. Ho, D.K.; Costa, A.; De Rossi, C.; Carvalho-Wodarz, C.D.; Loretz, B.; Lehr, C.M. Polysaccharide Submicrocarrier for Improved Pulmonary Delivery of Poorly Soluble Anti-infective Ciprofloxacin: Preparation, Characterization, and Influence of Size on Cellular Uptake. *Mol. Pharm.* **2018**, *15*, 1081–1096. [CrossRef]
125. Khelghati, N.; Rasmi, Y.; Farahmandan, N.; Sadeghpour, A.; Mir, S.M.; Karimian, A.; Yousefi, B. Hyperbranched polyglycerol β-cyclodextrin as magnetic platform for optimization of doxorubicin cytotoxic effects on Saos-2 bone cancerous cell line. *J. Drug Deliv. Sci. Technol.* **2020**, *57*, 101741. [CrossRef]
126. Li, R.; Li, X.; Zhang, Y.; Delawder, A.O.; Colley, N.D.; Whiting, E.A.; Barnes, J.C. Diblock brush-arm star copolymers via a core-first/graft-from approach using γ-cyclodextrin and ROMP: A modular platform for drug delivery. *Polym. Chem.* **2020**, *11*, 541–550. [CrossRef]
127. Blencowe, A.; Tan, J.F.; Goh, T.K.; Qiao, G.G. Core cross-linked star polymers via controlled radical polymerisation. *Polymer* **2009**, *50*, 5–32. [CrossRef]
128. Ren, J.M.; McKenzie, T.G.; Fu, Q.; Wong, E.H.; Xu, J.; An, Z.; Shanmugam, S.; Davis, T.P.; Boyer, C.; Qiao, G.G. Star Polymers. *Chem. Rev.* **2016**, *116*, 6743–6836. [CrossRef] [PubMed]
129. Das, K.; Sarkar, B.; Roy, P.; Basak, C.; Chakraborty, R.; Gardas, R.L. Physicochemical investigations of amino acid ionic liquid based inclusion complex probed by spectral and molecular docking techniques. *J. Mol. Liq.* **2019**, *291*, 111255. [CrossRef]
130. Viale, M.; Vecchio, G.; Monticone, M.; Bertone, V.; Giglio, V.; Maric, I.; Cilli, M.; Bocchini, V.; Profumo, A.; Ponzoni, M.; et al. Fibrin Gels Entrapment of a Poly-Cyclodextrin Nanocarrier as a Doxorubicin Delivery System in an Orthotopic Model of Neuroblastoma: Evaluation of In Vitro Activity and In Vivo Toxicity. *Pharm. Res.* **2019**, *36*, 115. [CrossRef] [PubMed]
131. Haley, R.M.; Zuckerman, S.T.; Gormley, C.A.; Korley, J.N.; von Recum, H.A. Local delivery polymer provides sustained antifungal activity of amphotericin B with reduced cytotoxicity. *Exp. Biol. Med.* **2019**, *244*, 526–533. [CrossRef]
132. Lin, W.J.; Yao, N.; Qian, L.; Zhang, X.F.; Chen, Q.; Wang, J.F.; Zhang, L.J. pH-responsive unimolecular micelle-gold nanoparticles-drug nanohybrid system for cancer theranostics. *Acta Biomater.* **2017**, *58*, 455–465. [CrossRef]
133. Liu, H.; Chen, J.; Li, X.; Deng, Z.; Gao, P.; Li, J.; Ren, T.; Huang, L.; Yang, Y.; Zhong, S. Amphipathic β-cyclodextrin nanocarriers serve as intelligent delivery platform for anticancer drug. *Colloids Surf. B Biointerfaces* **2019**, *180*, 429–440. [CrossRef]
134. Namgung, R.; Lee, Y.; Kim, J.; Jang, Y.; Lee, B.-H.; Kim, I.-S.; Sokkar, P.; Rhee, Y.; Hoffman, A.; Kim, W. Poly-cyclodextrin and poly-paclitaxel nano-assembly for anticancer therapy. *Nat. Commun.* **2014**, *5*, 3702. [CrossRef] [PubMed]
135. Ji, Y.; Liu, X.; Huang, M.; Jiang, J.; Liao, Y.-P.; Liu, Q.; Chang, C.H.; Liao, H.; Lu, J.; Wang, X.; et al. Development of self-assembled multi-arm polyrotaxanes nanocarriers for systemic plasmid delivery in vivo. *Biomaterials* **2019**, *192*, 416–428. [CrossRef] [PubMed]
136. Tamura, A.; Yui, N. Polyrotaxane-based systemic delivery of β-cyclodextrins for potentiating therapeutic efficacy in a mouse model of Niemann-Pick type C disease. *J. Control. Release* **2018**, *269*, 148–158. [CrossRef] [PubMed]
137. Kulkarni, A.; Caporali, P.; Dolas, A.; Johny, S.; Goyal, S.; Dragotto, J.; Macone, A.; Jayaraman, R.; Fiorenza, M.T. Linear Cyclodextrin Polymer Prodrugs as Novel Therapeutics for Niemann-Pick Type C1 Disorder. *Sci. Rep.* **2018**, *8*, 9547. [CrossRef]

 *molecules*

Review

Cyclodextrin-Based Contrast Agents for Medical Imaging

Yurii Shepelytskyi [1,2], Camryn J. Newman [3], Vira Grynko [1,2], Lauren E. Seveney [4], Brenton DeBoef [4], Francis T. Hane [2,5] and Mitchell S. Albert [2,5,6,*]

[1] Chemistry and Materials Science Program, Lakehead University, Thunder Bay, ON P7B 5E1, Canada; yshepely@lakeheadu.ca (Y.S.); vgrynko@lakeheadu.ca (V.G.)

[2] Thunder Bay Regional Health Research Institute, Thunder Bay, ON P7B 6V4, Canada; francishane@gmail.com

[3] Biology Department, Lakehead University, Thunder Bay, ON P7B 5E1, Canada; cjnewma1@lakeheadu.ca

[4] Chemistry Department, University of Rhode Island, Kingston, RI 02881, USA; laurenseveney@uri.edu (L.E.S.); bdeboef@uri.edu (B.D.)

[5] Chemistry Department, Lakehead University, Thunder Bay, ON P7B 5E1, Canada

[6] Northern Ontario School of Medicine, Thunder Bay, ON P7B 5E1, Canada

* Correspondence: malbert1@lakeheadu.ca; Tel.: +1-807-355-9191

Academic Editors: Marina Isidori, Margherita Lavorgna and Rosa Iacovino

Received: 30 October 2020; Accepted: 26 November 2020; Published: 27 November 2020

Abstract: Cyclodextrins (CDs) are naturally occurring cyclic oligosaccharides consisting of multiple glucose subunits. CDs are widely used in host–guest chemistry and biochemistry due to their structural advantages, biocompatibility, and ability to form inclusion complexes. Recently, CDs have become of high interest in the field of medical imaging as a potential scaffold for the development of a large variety of the contrast agents suitable for magnetic resonance imaging, ultrasound imaging, photoacoustic imaging, positron emission tomography, single photon emission computed tomography, and computed tomography. The aim of this review is to summarize and highlight the achievements in the field of cyclodextrin-based contrast agents for medical imaging.

Keywords: medical imaging; contrast agents; α-cyclodextrin; β-cyclodextrin; γ-cyclodextrin; MRI; PET; CT; SPECT; PAI

1. Introduction

Cyclodextrins (CDs) are chemically stable naturally occurring cyclic oligosaccharides consisting of multiple glucose subunits connected by α-1,4 glycosidic bonds [1,2]. There are three main types of CDs that contain six (α-CD), seven (β-CD), and eight (γ-CD) glucose subunits in a ring (Figure 1). These cyclodextrin macromolecules are cone-shaped with a hydrophobic interior cavity and polar exterior surface [3]. Due to their non-toxic nature [1,4–6] and water solubility [7–9], CDs became widely used in various biomedical fields such as drug solubilization [7,10–12], drug delivery [8,13–17], and nucleic acid transfer [18–20].

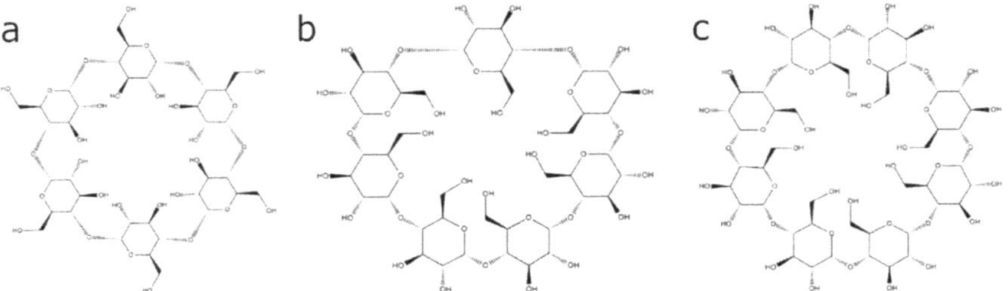

Figure 1. Chemical structure of (**a**) α-cyclodextrin, (**b**) β-cyclodextrin, and (**c**) γ-cyclodextrin.

Compared to other macrocyclic hosts, cyclodextrins are by far the most extensively used in host–guest chemistry applications and medical imaging [21–25]. They tend to be the macrocycle of choice due to their structural advantages and robust ability to form inclusion complexes [21,26]. An inclusion complex is formed when a guest molecule, commonly a small drug, is partially or fully encapsulated inside the host's interior cavity [1,3]. In the case of cyclodextrins, their preferred guest molecules tend to be hydrophobic, making them suitable for binding in the hydrophobic interior. Therefore, cyclodextrins possess the ability to form inclusion complexes with a wide variety of hydrophobic guest molecules [26–28]. Formation of inclusion complexes, or molecular encapsulation, can affect the physiochemical properties of the drug or molecule itself, such as solubility and rate of dissolution [3]. CDs are often exploited because of this property in addition to enhancing water solubility of water-insoluble molecules [3]. The exterior of cyclodextrin is predominantly hydrophilic due to the extensive hydrogen bonding network, making it a biocompatible agent for a wide range of applications [1–3,6–17]. These structural factors are largely why CDs are favored when synthesizing inclusion complexes. Electronics and thermodynamics both play a role in determining if a CD will form an inclusion complex with a guest molecule [1]. The driving force for inclusion complexation involves various noncovalent interactions such as desolvation, or removal, of water molecules from the interior cavity and formation of Van der Waals, hydrophobic, and hydrogen bonding interactions [29]. The major driving forces for cyclodextrin complexation are van der Waals interaction and hydrophobic interaction, whereas hydrogen bonding and electrostatic interaction mostly affect the conformation of a particular inclusion complex [29]. CD's ability to form inclusion complexes with small organic molecules has pioneered the field of supramolecular chemistry. More recently, CDs have successfully been used to build various molecular architectures such as catenanes, rotaxanes, pseudorotaxanes, polyrotaxanes, and other molecular machines [30,31].

During the past few decades, cyclodextrins have become of interest as contrast agents and potential biosensors for different medical imaging modalities [32]. In the context of imaging, CDs have been used primarily as a scaffold to support and/or solubilize smaller molecules that produce or quench a signal for enhanced imaging. However, in a few cases, the unique supramolecular nature of the CD is essential in producing the signal for imaging. Examples of CD-containing constructs that can be imaged by a wide variety of modern imaging technologies are discussed herein.

Magnetic resonance imaging (MRI) was the first imaging modality to utilize cyclodextrins as contrast agents [33]. CD-based MRI contrast agents produced contrast through the reduction in spin-lattice relaxation (T_1) time of the water protons. There are two different established methods of synthesis of CD-based MRI contrast agents: (1) host–guest interactions between CD cavity and metal–organic complexes [33,34] and (2) by direct conjugation of CDs to the metal–organic complexes through the external hydroxyl groups of CD molecule [35,36]. The reduction in T_1 relaxation for CD-based contrast agents, and therefore their contrast, is substantially stronger compared to the metal–organic complexes on their own [34,37].

Secondly, positron emission computed tomography (PET) utilizes CD-based molecular imaging probes [25]. The PET probes emit positron, which annihilates with a stationary electron from the surroundings producing two gamma-photons, which are detected [38]. The PET tracers based on CD can be divided in two classes. The first class contains CD-based nanoparticles (NPs) radiolabeled with either ^{64}Cu [25] or ^{18}F [39]. Another recently developed type of CD-based PET imaging agents contains CD molecules conjugated to the p-NCS-benzyl-NODA-GA (NODAGA) chelator labeled with ^{68}Ga [40,41]. Followed by PET, the CD-based contrast agents were developed for single photon emission computed tomography (SPECT). SPECT contrast agents were created by radiolabeling of CD-based NPs either with ^{99m}Tc [42] or ^{125}I [43].

Recently, CDs were applied as contrast agents for ultrasound (US) and photo-acoustic imaging (PAI). The mechanism of contrast creation for US imaging relies on the substantial differences in the acoustic impedances between the biological tissue and the CD-based agents [44]. The mechanism of PA imaging is more complicated. The PA tracer absorbs the light with subsequent heating. Due to the temperature increase, the contrast agent undergoes thermoelastic expansion resulting in emission of the ultrasonic acoustic waves that can be detected by US receiver [45,46]. The developed CD photoacoustic contrast agents absorbed the light in the infrared range [47–49].

Lastly, multiple studies were conducted to evaluate the performance of the CD-based contrast agents for computed tomography (CT) [50–52]. All of the developments were focused on CD-based NPs that contained metal atoms (Au, Yb, Dy) [50–53]. The presence of the element with high atomic number rises up the X-ray absorption coefficient yielding to the contrast increase. These CD-based contrast agents demonstrated better performance compared to conventional iodine-based CT agents [50,51].

The purpose of this review is to provide an update on recent developments in CD-based contrast agents. A comparison of the developed contrast agents to the clinically available are presented as well as a comparison between different CD-based agents.

2. CD-Based Contrast Agents for MRI

MRI was the first medical imaging modality that began to develop CD-based contrast agents. The vast majority of the developed CD-based contrast agents reduce T_1 relaxation of the surrounding protons and were used to produce substantial contrast on T_1-weighted images. Nevertheless, several studies demonstrated the potential of T_2 CD-based contrast agents as well. An overall summary of the achievements and developments in CD-based MRI contrast agents and molecular imaging probes is presented in detail below.

Due to the wide disfavor of traditional MRI probe modalities, mainly consisting of diethylenetriamine pentaacetic acid (DTPA) and dodecane tetraacetic acid (DOTA)-derived small molecule Gd(III) complexes, the continuous development and synthesis of innovative contrast agents is needed [54]. Cyclodextrin-based MRI contrast agents have gained notoriety in the last two decades and prove to be viable, robust candidates as potential MRI sensors for biomedical imaging [54]. In addition to their high biocompatibility and ability to form inclusion complexes, their large molecular weights, ease of functionalization and conjugation, and multivalent loading capacity make them well-suited as a new class of paramagnetic macromolecules [35,55,56]. CDs have proved to enhance contrast, sensitivity, and diagnostic imaging time [35]. This is attributed mainly due to their large molecular weights, which allow for longer and tunable rotational times [35,54].

2.1. Contrast Agents Based on Host–Guest Complexation between CDs and Metal–Organic Chelates

2.1.1. In Vitro Studies of the Host–Guest CD-Based MRI Contrast Agents

The genesis of CD-based MRI contrast agents occurred in 1991 when Aime et al. [33] applied Freed's theory [57] to predict that the relaxation rates of the solvent protons increase when the paramagnetic ion or complex is bound to a macrocycle. This principle became a fundamental base for the future development of CD-based MRI contrast agents. Aime et al. reported numerous results for the inclusion complexes formation between CD and gadolinium chelates such as DOTA and DTPA. Host–guest interaction is dependent on internal cavity size: α-CD did not demonstrate host–guest interaction, while β-CD non-covalently bonded with DOTA/DTPA with significant increase in proton relaxivity in vitro (reciprocal spin-lattice relaxation time ($1/T_1$)) observed for both β-CD-DOTA and β-CD-DTPA contrast agents [33]. Subsequent studies utilizing Gd(III)-bis(benzylamide)diethylenetriaminepentaacetic acid (BBA-DTPA) [34], gadolinium (III) 3,6,9,15-tetraazabi-cyclo[1,3,9]pentadeca-1(15),11,13-triene-3,6,9-triacetic acid (PCTA) complexed with poly-β-CD [58], and DOTA-benzyloxymethyl (DOTA(bom)$_3$) [59] demonstrated increasing water proton relaxivity due to the larger quantity of paramagnetic complexes accumulating in the region of interest due to interactions with a β-CD polymers [59]. The relaxivity of the developed poly- β-CD contrast agents exceeded up to six times the clinical analogues ($r_1 = 61$ mM^{-1}s^{-1}) [59]. Furthermore, the results of modified Gd(III)-PCTA in blood serum suggested the possibility of application of this developed CD-based contrast agent in MRI angiography [58]. Despite such promising results, the relaxation measurements were conducted at 20 MHz NMR spectrometer at 0.5 T magnetic field far below the clinical MRI magnetic fields (1.5 or 3.0 T).

Larger CD-based NPs have also been the subject of recent research. These NPs have the potential for lower toxicity [60] compared to Gd-CD-based contrast agents. Examples of a proposed β-CD-based Gd-loaded NPs include a supramolecular assembly between DO3A-based gadolinium chelate (conjugated to adamantane through an acetamide spacer), poly-β-CD, and modified dextran [60] and cleavable β-CD-based Gd(III)-loaded nanocapsules, a promising agent as a redox-sensitive MRI contrast agent [61]. Besides toxicity reduction in Gd chelates, the proposed NPs are utilized the concept of further increasing of the number of Gd(III) atoms per unit of contrast agent resulting in increasing relaxivities.

2.1.2. In Vivo Imaging of the Host–Guest CD-Based Contrast Agents

The performance of some CD-based MRI imaging agents has been evaluated in vivo. Lahrech et al. demonstrated imaging of C6 glioma rats using a Gd-α-CD complex to quantify cerebral blood volume [62]. Although performance of Gd-α-CD contrast agents was significantly better compared to Gd-DOTA in terms of relaxation rates ($r_1 = 7.3$ mM^{-1}s^{-1} at 9.4 T), the developed supramolecular agent did not accumulate in tumors and almost no enhancement was observed on T_1-weighted images during the first hour after Gd-α-CD injection. On the contrary, the CBV fraction was successfully measured using rapid steady state T_1 method and Gd-α-CD contrast agent.

More advanced CD-based contrast agents were demonstrated by Sun et al. who created a supramolecular complex between bridged bis(permethyl-β-cyclodextrin)s with Mn-porphyrin bearing polyethylene glycol side chains (Mn-TPP) [63]. Mice injected with this supramolecular polymer demonstrated strong contrast observed in the blood, kidneys, and bladder (Figure 2) [63]. A supramolecular polymer built using the non-covalent interaction between Mn(II)-TPP and bridged tris(permethyl-β-CD)s resulted in a longitudinal relaxivity only 7% higher compared to previously developed Mn(II)-containing linear polymer [63,64].

Work by Feng et al. conducted on CD-based NPs as MRI contrast agents demonstrated neodymium doped NaHoF$_4$ NPs as T_2 imaging agents cultured with human mesenchymal stem cells injected into the brain hemisphere of nude mice. This combination of CD-based contrast agents and stem cells [23] supports the idea of using stem cells as an MRI contrast agent carrier. Furthermore, due to the high relaxivity of the developed probe ($r_2 = 143.7$ mM^{-1}s^{-1} at 11.7 T) [23], use of the developed contrast agent will be beneficial for ultrahigh field MRI imaging, since the transverse relaxivity increases highly with the magnetic field strength [23,65,66].

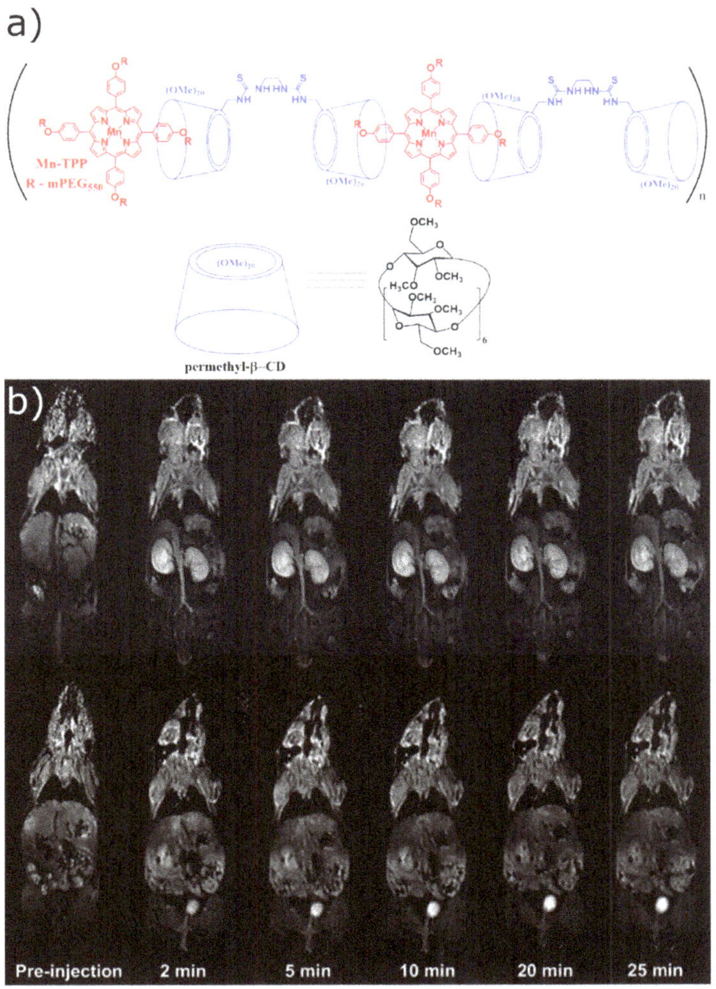

Figure 2. (**a**) Molecular structure of Mn(II)-TPP and bridged bis(permethyl-β-CD)s polymer. (**b**) Representative 2D coronal T1-weighted MR images of the mice at 2, 5, 10, 20, and 25 min after intravenous injection of Mn(II)-TPP/bridged β-CD magnetic resonance imaging (MRI) contrast agents at 0.03 mmol of Mn/kg [63]. The images are reprinted with permission from publisher [63].

2.1.3. In Vivo Tumor Imaging

Imaging of cancer is one of the hot topics in modern medical imaging field. Despite the numerous developed contrast agents discussed above, only Zhou et al. used multiple β-CDs attached to a polyhedral oligomeric silsesquioxane nano globule at a targeted nano globular contrast agent from host–guest assembly for magnetic resonance cancer molecular imaging [67]. The host–guest contrast agent bonds to $\alpha_v\beta_3$ integrinin 4T1 malignant breast tumor through cyclic RGDfK peptide and gives greater contrast enhancement, due to the $\alpha_v\beta_3$ that is overexpressed in tumors (Figure 3a–c). This designed contrast agent produced superior contrast and signal enhancement compared to the clinically used Gd-based ProHance and non-targeted control cRAD-POSS-bCD-(DOTA-Gd)-Cy5 contrast agent. Molecular structure of cRGD-POSS-βCD-(DOTA-Gd)-Cy5 imaging agent is shown on Figure 3d.

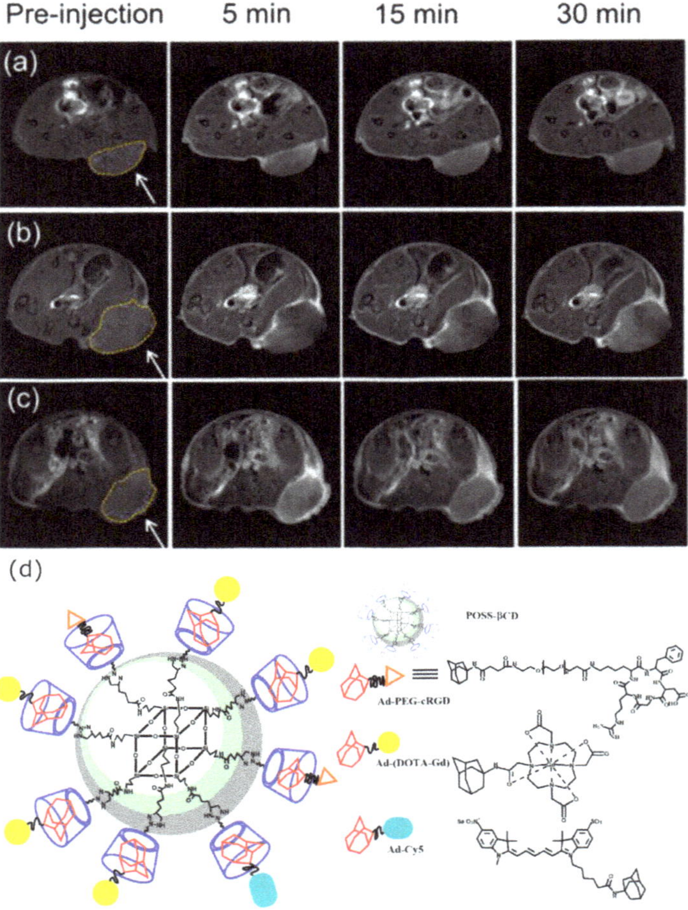

Figure 3. Magnetic resonance molecular imaging with cRGD-POSS-βCD-(DOTA-Gd)-Cy5 in mice bearing 4T1-Luc2-CFP tumor xenografts. The representative 2D axial fat-suppressed T1-weighted spin-echo MRI images before and at 5, 15, and 30 min post-injection of ProHance (**a**), cRAD-POSS-bCD-(DOTA-Gd)-Cy5 (**b**), and cRGD-POSS-bCD-(DOTA-Gd)-Cy5 (**c**) at 0.1 mmol-Gd/kg. The injection of cRGD-POSS-βCD-(DOTA-Gd)-Cy5 creates superior signal enhancement in tumor region. The images are reprinted with permission from publisher [67]. (**d**) Molecular structure of the developed cRGD-POSS-βCD-(DOTA-Gd)-Cy5 contrast agent.

2.2. Direct Labeling of the CD Molecules

Another approach of synthesis of CD-based contrast agents for MRI imaging purposes is to conjugate metal–organic Gd(III)-containing complexes to the CD molecule through covalent bonds. We refer to this method as direct labeling of CD molecules. The key advantage of this approach is the availability of CD cavity for host–guest interaction with other molecules that can be effectively used for drug delivery study and imaging of molecular interactions.

Synthetically, modifying CDs in order to meet the criteria of an ideal contrast agent can be summarized by three key components: having a point of functionalization to attach the chelating group, designing a rigid linker in order to slow local movements, and lastly, conjugation of the macrocyclic complex in order to encapsulate multiple lanthanide chelates—thus, overall enhancing MRI signal, while improving overall stability and relaxivity profiles [35,54–56,68].

Various synthetic approaches have been explored, which demonstrates CD's ability to be functionalized and conjugated in a robust fashion dependent on the desired application(s) [35,54–56,68]. Bryson et al. synthesized a monodisperse β-CD Click cluster containing seven paramagnetic chelates encompassing two water exchange sites [35]. Using Click chemistry, an alkyne-functionalized dendron was reacted with the per-azido-β-CD to yield the desired product (Figure 4). Using similar methods, Champagne et al. recently reported the synthesis of a different β-CD MRI probe containing seven iminodiacetate arms connected at the C6-position of β-CD by a triazole-based linker following a copper(I)-mediated 1,3-dipolar cycloaddition [54]. In addition, CDs can be differentially conjugated to produce multifunctional probes. Kotková et al. synthesized a novel bimodal fluorescence/MRI probe using a β-CD scaffold [55]. β-CD was labeled first using fluorescein isothiocyanate (FITC) and subsequentially with an isothiocyanate derivative containing a DOTA-based ligand [55]. The rigidity of the linker between the CD and the Gd-containing ligand plays an important role in increasing T_R, thus enhancing the overall MRI signal [35,55]. Additionally, multifunctional NPs have been modified using an asymmetrically functionalized β-CD-based star copolymer by conjugating β-CD using doxorubicin (DOX), folic acid (FA), and DOTA-Gd moieties [54]. Similar to the work of Brysor et al. [35] and Champagne et al. [54], the key conjugation method used was azide-alkyne Huigsen cycloaddition, creating rigid triazole linkers.

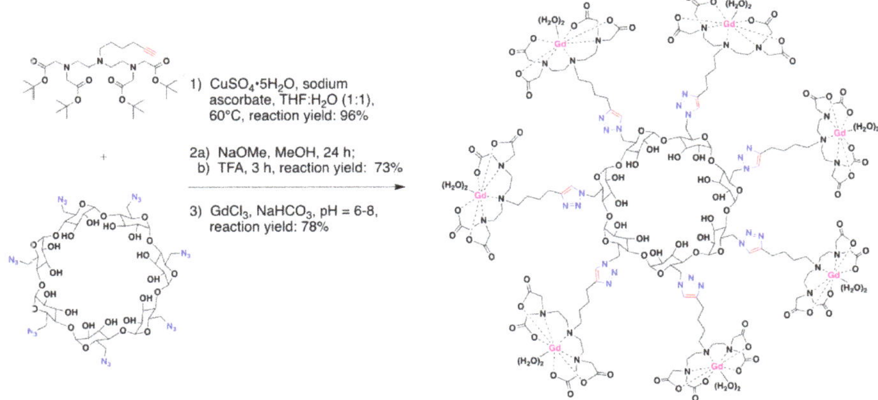

Figure 4. Multivalent β-CD "Click cluster", containing seven paramagnetic chelating groups, each with two water exchange sites linked via triazole-based linkers. The β-CD "Click cluster" was synthesized from per-azido-β-CD precursor and conjugated using the well-established Huigsen cycloaddition reaction. Figure adapted from Bryson et al. with permission from publisher [35].

2.2.1. In Vitro Development

Skinner et al. demonstrated the first labeling of CD macrocycle with a Gd(III) chelate albeit the CD cavity was still used for non-covalent binding to another Gd(III) chelate in order to increase the proton relaxivity. [69]. Relaxivity of this complex increased when it was bound noncovalently to another gadolinium complex with the addition of two phenyl moieties.

Bryson et al. created a contrast agent with a ten-fold increase in relaxivity at 9.4 T compared to clinically available Magnevist by labeling of per-azido-β-cyclodextrin core with seven diethylenetriaminetetraacetic acid (DTTA) Gd(III) chelates [35]. This development can be attributed to the increased Gd(III) ions per molecule and further increase in relaxivity due to conjugation to the macrocycle. In addition to the high relaxivity, unoccupied cavity of β-CD makes the developed contrast agent an excellent host scaffold to functionalize through noncovalent assembly with biological receptor-specific targets.

Bimodal MRI-fluorescence probes were demonstrated by Kotkova et al. [55] who combined a DOTA-based ligand with fluorescein functionality to simultaneously obtain fluorescence and MR images. Although the developed CD-based agent was studied in vitro, the benefit of this scaffold for MRI visualization under in vivo conditions was assumed due to its low cytotoxicity and high cell uptake. Fredy et al. developed cyclodextrin polyrotaxanes as a highly modular platform for an imaging agent [70]. Selectively functionalized cyclodextrins with a Gd(III) complex or BODIPY fluorescent tag were put on to a polyammonium chain to form polyrotaxanes. From this, polyrotaxanes could be assembled with fluorescent CDs and CDs with dia- or paramagnetic lanthanide complexes. Each threaded cyclodextrin was molecularly defined, which is an advantage over statistical post-functionalization of CD-polyrotaxanes. In vitro studies demonstrated that the Gd-bearing polyrotaxanes have relaxivities that are five times higher than Gd-DOTA, which makes them effective contrast agents for MRI applications [70].

NPs fabricated using biological macromolecules have been demonstrated by both Liu et al. [68] and Su et al. [71]. Liu designed pH disintegrable β-CD-based micellar NPs, while Su reported star-like dextran wrapped superparamagnetic iron oxide NPs. Both groups reported dual effects: an imaging contrast agent and cytotoxicity to HeLa cells at high concentrations, making these molecules both imaging probes and potential chemotherapeutics agents. Later, the synthesis of the contrast agent that affects the spin-spin (T_2) relaxation was suggested by conjugating β-CD to magnetic NPs [72].

2.2.2. In Vivo Imaging of CD-Based MRI Contrast Agents Based on the Direct Labeling of CD Cavity

In vitro studies have paved the way for a number of groups to demonstrate CD-based MRI contrast agents in vivo. Vascular imaging has long been a molecular imaging goal allowing investigators the ability to image the vasculature with high contrast. The G2/MOP–DTPA–Gd contrast agents synthesized from polyester dendrimers with β-CD core have been demonstrated with high yield and fast synthesis while providing for 2.7 times the relaxivity of Magnevist (DTPA-Gd) [24]. This contrast agent is rapidly hydrolyzed at the pH of 7.4 in the presence of esterase and slowly hydrolyzed at an acidic pH. The superior signal enhancement in vivo was observed following 0.1 mM Gd/kg injection (Figure 5a) and was significantly higher compared to Magnevist (Figure 5b) Furthermore, the G2/MOP–DTPA–Gd contrast agent (Figure 5c) did not show tissue retention, making the ideal blood pool and kidney imaging agent.

Zhou et al. synthesized Gd(III)-1,4,7,10-Tetraazacyclododecane-1,4,7-triacetic-2-hydroxypropyl-β-CD/Pluronic polyrotaxane contrast agent [73]. Interestingly, Gd-DO3A-HPCD/Pluronic polyrotaxane construct circulated for more than 30 min in the living mouse and caused about 100-fold vascular enhancement when compared to the monomeric form (Figure 6a). Furthermore, the polyrotaxane derivative showed a much higher signal enhancement after 5 min in the heart than the monomeric form but underwent rapid elimination by renal filtration, preventing blood enhancement. Thus, the Gd(III)-DO3A-HPCD/Pluronic polyrotaxane (Figure 6b) is a promising contrast agent, which enables higher anatomic detail of blood vessel organization.

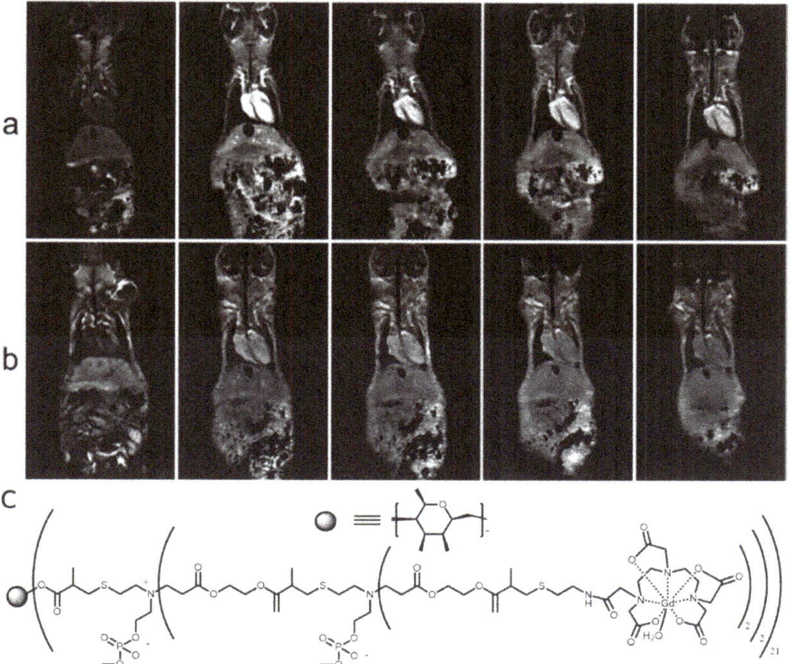

Figure 5. Representative T1 weighted multislice image of the mice heart after 0.1 mM Gd/kg injection of G2/MOP–DTPA–Gd contrast (**a**) and Magnevist (**b**). The superior contrast-to-noise ratio was observed from the heart after the G2/MOP–DTPA–Gd injection. The images (**a**) and (**b**) are reprinted with permission from publisher [24]. (**c**) The molecular structure of the developed G2/MOP-DTPA-Gd CD-based contrast agent.

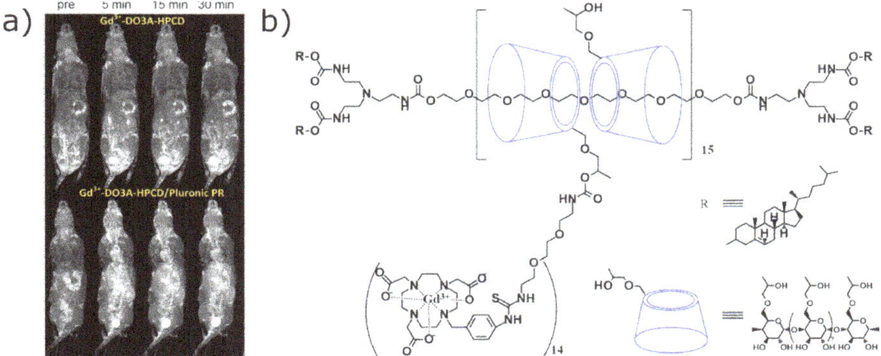

Figure 6. (**a**) T1 weighted 3D maximum intensity projection images of Balb/c mice. Mice were injected with Gd(III)-DO3A-HPCD (top row) or Gd(III) -DO3A-HPCD/Pluronic polirotaxane (bottom row) at a 0.03 mM-Gd/kg dose. Contrast agent distribution is shown in the images for pre injection and 5, 15, and 30 min after injection in to the tail vein with Gd3+ complexes. Images have been re-printed with permission from publisher [73]. (**b**) Molecular structure of the Gd(III)-DO3A-HPCD/Pluronic polirotaxane developed by Zhou et al.

2.2.3. In Vivo Imaging of Cancer

Cancer metastasis is the final insult leading to patient mortality. The early detection of metastasis is therefore a primary concern in the field of oncology. CD-based MRI contrast agents hold promise for the imaging of metastasis. Although CD-based contrast agents have yet been clinically tested, a number of groups have demonstrated these contrast agents in mammalian models.

Zhou et al. [74] reported the enhancement of MRI of liver metastases with a zwitterionized biodegradable dendritic CD-based contrast agent. Traditionally, the sensitivity in the liver for MR imaging of metastases is low due to the accumulation of the contrast agent in the Kupffer cells and hepatocytes instead of cancer cells. Zhou et al. used a novel dendritic contrast agent with β-CD core and the net size of 9 nm. The developed dendritic contrast agent reduces background signals in the liver significantly by avoiding being uptakes by hepatocytes and Kupffer cells through the zwitterionization, while increasing the signal in tumors through the enhanced permeability and retention effect. This CD-based zwitterionized dendritic contrast agent also showed shorter Gd(III) retention in all organs and tissues, because it could degrade into small fragments.

Zhang et al. developed polyethyleneimine- β-CD (PEI-β-CD) as a novel vector for carrying ferritin gene modified by alpha-fetoprotein promoter [75] to create a highly specific endogenous T_2 contrast agent for hepatocellular carcinoma. In vitro T_2-weighted and $T_2{}^*$-weighted MRI was used to examine the effect of ferritin heavy gene transfection. Zhang et al. observed the significant $T_2/T_2{}^*$-induced MRI signal decay (up to 40%) from the BEL-7402 hepatocellular carcinoma cells treated with the developed PEI-β-CD/ferritin gene. Therefore, it was proposed that the ferritin gene carried by PEI-β-CD has a high potential to be used for early-stage MRI detection as an endogenous contrast agent for hepatocellular carcinoma imaging.

Gd (III) oxide NPs coated with folic acid functionalized poly (β-CD-co-pentetic acid) (Gd$_2$O$_3$@PCD-FA) as a biocompatible targeted nano-contrast agent was proposed by Mortezazadeh et al. [76]. Mortezazadeh et al. observed that Gd$_2$O$_3$@PCD-FA demonstrated significantly higher r_1 and r_2 relaxivities at 3T (r_1 = 3.95 mM^{-1}s^{-1}; r_2 = 4.6 mM^{-1}s^{-1}) than Gd(III)-DOTA. On the other hand, the measured relaxivities were lower compared to the pure Gd$_2$O$_3$ (r_1 = 4.86 mM^{-1}s^{-1}; r_2 = 5.97 mM^{-1}s^{-1}) due to the reduced water accessibility to Gd$_2$O$_3$ core in Gd$_2$O$_3$@PCD-FA. In order to study the performance of the developed NPs in vivo, the Gd$_2$O$_3$@PCD-FA contrast agent has been evaluated in the animal tumor model. Maximization of CNR was observed in 1h post-injection of the Gd$_2$O$_3$@PCD-FA contrast agent. Interestingly, Gd$_2$O$_3$@PCD-FA NPs demonstrated almost no cytotoxicity after 12 and 24 h administering to MCF-10A human normal breast cell lines.

Han et al. developed a hypoxia-targeting dendritic MRI contrast agent based on internally hydroxy dendrimer (IHD) with β-CD core [77]. The disturbance of the zwitterionic surface reduces unspecific cellular uptake by normal cells. In vivo imaging of an orthotopic breast tumor in mice injected with the developed contrast agent showed the maximization of CNR in 1 h post-injection. Han et al. observed CNR reaching the level of 10, remaining constant during the second hour, and slightly decaying 3 h post-injection.

2.3. Cyclodextrin-Based Contrast Agents for Hyperpolarized MRI

Although molecular imaging using MRI is challenging due to the lack of sensitivity of this imaging modality [78,79], progress has been made with the development of hyperpolarized (HP) MRI [79,80]. HP MRI utilizes the advantage of a metastable state with spin population excess significantly larger compared to the thermal equilibrium state [81]. Hyperpolarization of noble gases (such as ^{3}He and ^{129}Xe) is conducted via spin exchange optical pumping (SEOP), whereas polarization of ^{13}C or ^{1}H containing molecules is created through dynamic nuclear polarization (DNP) [82–85].

2.3.1. Hyperpolarized ^{13}C CD-Based Contrast Agents

Due to the high signal enhancement, molecular imaging using HP ^{13}C-containing molecule becomes possible. β-CD was used to create a contrast for HP ^{13}C MRI [86]. Keshari et al. showed that the host–guest interaction between a β-CD cavity and HP benzoic acid substantially decrease the T_1 relaxation of both HP ^{13}C nuclei resulting in negative image contrast (decrease in signal intensity) induced by the presence of supramolecular cage (Figure 7). Based on this result, the authors proposed that similar mechanism of the negative contrast can be used to study the interaction of ligand–receptor pairs in vivo.

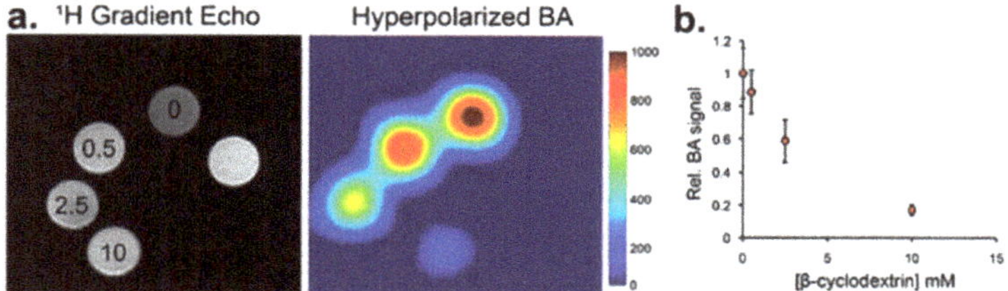

Figure 7. In vitro experiment at 14T demonstrating the potential application of β-CD as a contrast agent for hyperpolarized (HP) ^{13}C MRI. (**a**) Proton gradient echo image demonstrating the position of phantoms with a different concentration of β-CD (0–10 mM). The HP ^{13}C imaging was performed after the administration of 2.5 mM HP [1-13C] benzoic acid. It can be seen that the MRI signal decreases with β-CD concentration. (**b**) Relative MRI ^{13}C signal dependence on β-CD concentration [86]. The images are reprinted with permission from publisher [86].

DNP was used to create a CD contrast agent for DNP HP MRI [87]. Caracciolo et al. observed the polarization level of 10%. Unfortunately, the T_1 relaxation of β-CD protons was equal to 1s at 300 K, which made HP β-CD inapplicable for molecular imaging purposes. On the other hand, HP β-CD can be of interest in the fields, which require the production of strong ¹H NMR signal from CD molecules [87]. Following this work was the hyperpolarization of methylated β-CD [88]. The methylation has been conducted using [13]CH$_3$I, which enriched the potential contrast agent with ^{13}C and ¹H nuclei. Methylated β-CDs underwent DNP and polarization levels of 7.5 and 7% were achieved for ¹H and ^{13}C, respectively. The proton T_1 relaxation times were found to be similar to those published in [87]; however, the T_1 relaxation time of ^{13}C nuclei was equal to 3.3 and 4.9 s for fully methylated β-CDs and partially methylated β-CDs, respectively. These relaxation times allow further application of HP β-CD as contrast agents in the molecular imaging field. In addition, authors demonstrated the method of further increasing relaxation times [88].

2.3.2. CD-Based Molecular Probes for Hyperpolarized [129]Xe MRI

Current studies of molecular imaging with HP [129]Xe MRI utilizes hyperpolarized chemical exchange saturation transfer (HyperCEST) [89,90]. The HyperCEST effect relies on a constant chemical exchange between the dissolved HP [129]Xe nuclei in the solution and supramolecular host that can effectively encapsulate [129]Xe [79]. Following selective depolarization of the [129]Xe nuclei encapsulated in the supramolecular cage, the decrease in the dissolved phase [129]Xe MRI signal can be observed as a result of the exchange dynamic [79,89].

The interaction between HP [129]Xe and the α-CD cavity was studied for the first time in 1997 [91]. Authors observed that spin polarization induced nuclear Overhauser effect (SPINOE) and related transfer of nuclear polarization to the α-CD protons [91]. However, this approach did not become widely used in the field of molecular imaging using HP [129]Xe.

CD-based contrast agents for HyperCEST molecular imaging became of interest recently. The first detection of HyperCEST effect using α-CD-based molecules was achieved from the α-CD pseudorotaxane complex with five carbon diethylimidazolium bar in aqueous solution [92]. The observed HyperCEST depletion was equal to 30%, which is significantly smaller compared to other molecular imaging probes [89,93]. The first β-CD-based molecular imaging probe for HyperCEST detection has been developed recently [94]. The HyperCEST contrast agent was realized as cucurbit[6]uril-based rotaxane in which β-CDs played the role of a stopper. Although rotaxane interaction with ^{129}Xe was through cucurbit[6]uril cavity and no HyperCEST effect was observed from β-CDs, the presence of β-CDs was required in order to increase the solubility of the end groups [94].

Finally, the HyperCEST effect from γ-CD-based pseudorotaxane was observed after the complexation of γ-CD with bisimidazolium guest [95]. Although γ-CD cavity is too large to sufficiently interact with HP ^{129}Xe, the cavity size can be decreased by threading it with a long alkyl chain. Threading these guest molecules through the cavity of γ-CD reduces the cavity size in order for adequate HP ^{129}Xe binding to occur, thus making it suitable for HyperCEST detection (Figure 8a). Potentially, this same concept can be applied for any supramolecular cage with a reasonably large cavity. The HyperCEST effect detected from γ-CD-based pseudorotaxanes was equal to 47.5% on average, which makes them interesting candidates for further application in vivo [95]. The main advantage of the HyperCEST contrast agents based on pseudorotaxane architecture over the other studied HP ^{129}Xe hosts is the ease synthesis and of functionalization of the pseudorotaxanes.

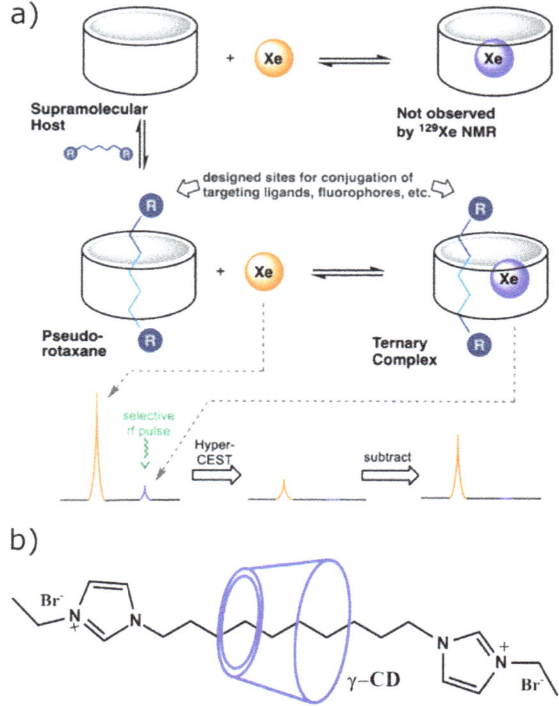

Figure 8. (a) Schematic representation of how CD-based ternary complexes are formed in the presence of HP ^{129}Xe. The guest is threaded through the hydrophobic cavity of cyclodextrin and HP ^{129}Xe is introduced. Detection via HyperCEST is obtained in order to determine if HP ^{129}Xe is bound in the cavity of CD. The images are reprinted with permission from publisher [95]. (b) The developed cyclodextrin-based pseudorotaxane used for in vitro HyperCEST detection [95].

In addition, γ-CD pseudorotaxane complexes prove to be a sufficient paradigm for HP ^{129}Xe MRI. The synthesis of the threads proceeds in one step and can be functionalized with different terminal end groups as well as different chain lengths. The 8- and 10-carbon alkyl chains were functionalized with ethylimidazoliums (Figure 8b), strategically used to enhance water solubility of the inherently hydrophobic alkyl chain [95]. The γ-CD pseudorotaxane was comparable to that of CB6, a known xenon cage that has been responsive to in vivo HP ^{129}Xe MRI [80]. Similar to CB6, the maximum HyperCEST depletion for γ-CD pseudorotaxanes was present when the samples were irradiated at a frequency of +128 ppm [95]. γ-CD pseudorotaxanes exhibited binding on the order of 10^3 at 1:1 host–guest complexation, proving to be sufficient association for HP ^{129}Xe. In addition, binding proved to be two magnitudes of order lower in fetal bovine serum, indicating that this system would work sufficiently as a potential biosensor in vivo [95].

3. CDs-Based Contrast Agent for Ultrasound Imaging and Photoacoustic Imaging

In addition to MRI contrast agents, CDs have been used for ultrasound (US) imaging. The first approach of synthetizing the potential contrast agent was developed by Cavalieri et al., in 2006 [96]. Air-filled polymer microbubbles functionalized with β-CDs were used as a source for contrast. The contrast originates from significant difference between acoustic impedance of tissue and air encapsulated inside of a microbubble. Cavalieri et al. found that conjugating microbubbles to β-CD preserves them from random coil to α- helix conformation transition. In addition, due to the presence of β-CD, these contrasts allow hosting molecules with hydrophobic features [96].

After nearly a decade of inactivity, the second attempt of using a β-CD-based contrast agent for US imaging was made in 2015 [44]. The authors demonstrated use of a perfluorinated FC-77/ β-CD complex. A visible blurring of signal from FC-77/β-CD caused by a disruption of the inclusion complexes under US was detected [44].

Another US medical imaging modality that utilized CD-based contrast agents was PAI. The first CD-based contrast agent was synthetized by surface modification of oleic-acid (OA) stabilized upconversional NPs (UCNPs) NaYF$_4$:Yb^{3+}, Er^{3+} with α-CD [49]. α-CD formed an inclusion complexes with an OA yielding to luminescence quenching of UCNPs and production of a strong PA signal instead (Figure 9a). Cytotoxicity studies demonstrated no toxicity of α-CD/UCNPs contrast agents. Following in vitro imaging (Figure 9b), the first PAI in a living mouse was conducted using 980 nm excitation laser. These results demonstrated the ability of α-CD/UCNPs to be an efficient PAI contrast agent for diagnostic purposes [49].

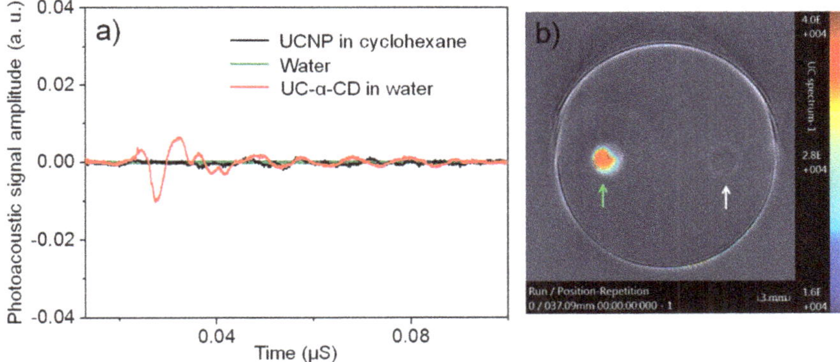

Figure 9. (**a**) High-resolution PA signal originated from upconversional NPs (UCNPs) in cyclohexane (black curve), distilled water (green curve), and α-CD/UCNPs (red curve) in water. The excitation was conducted using 980 nm nanosecond pulsed laser. (**b**) Photo-acoustic imaging (PAI) of tissue-mimicking phantom containing chambers filled with α-CD/UCN in water (green arrow) and distilled water (white arrow) [49]. The images are reprinted with permission from the publisher [49].

An effective PAI agent for imaging prosthetic joint infection (PJI) [48] was demonstrated by conjugation of β-CD to indocyanine green (ICG). The β-CD-ICG PAI agent was demonstrated in the mice model of PJI. Wang et al. found that conjugation of ICG to β-CD improves its PA signal generation. Although the PAI signal increase was not significant with β-CD-ICG contrast, it still demonstrated the ability to serve as a contrast agent for non-invasive diagnostic of PJI [48].

Yu et al. recently developed a CD-based PAI contrast agent sensitive to tumor environment [47]. Gold NPs (AuNP) were modified initially with pyridine-2-imine-terminated single stand DNA via gold-thiol bonds, and α-CDs were capped on the end of DNA through hydrophobic interaction with CD's cavity. The α-CD-AuNP agent produced no PA signal under neutral pH conditions, but upon entering the tumor, α-CDs separate from the DNA ends due to reduction in non-covalent forces. This study demonstrated that a developed α-CD-AuNP contrast agent can be successfully used as a tumor-selective theranostic agent [47].

4. Radiolabeled CD-Based Contrast Agents

PET and SPECT are imaging modalities that require probe radiolabeling to produce tomographic images. These modalities have superior sensitivity and deeper tissue penetration compared with luminescence-based imaging and MRI [97]; they require lower dose of imaging agent for functioning. However, the application of PET/SPECT tracers is limited to the half-life of radiolabeled isotopes and biodistribution in the living organism.

The first demonstration of the CD-based PET molecular imaging probe was done by Bartlett et al., in 2007 [25]. CD-containing NPs were studied as delivery agents for transferrin (Tf)-targeted delivery to tumors of siRNA molecules. Nontargeted and Tf-targeted siRNA NPs were synthetized by using cyclodextrin-containing polycation. One percent of the targeted NPs containing adamantane-PEG molecules on the surface were modified with Tf. Si-RNA were conjugated with DOTA to the 5′ end with further ^{64}Cu labeling [25]. MicroPET revealed negligible impact of the attachment of the Tf targeting ligand to the NPs on biodistribution. Unfortunately, nearly identical tumor localization kinetics of both targeted and nontargeted ^{64}Cu-DOTA-siRNA NPs were observed; tumor accumulation was also similar at 1 day after injection ($\approx$1% ID/cm^3). However, bioluminescent imaging showed the intracellular localization and functional activity of siRNA delivered by Tf-targeted NPs in the tumor cells 1 day after injection.

Interestingly, the presence of CD in therapeutic compounds allows radio isotopic labelling for PET imaging of the drug circulation. In vivo biodistribution of the IT-101, clinically developed drug for cancer treatment, in mice with Neuro2A tumors was studied by Schluep et al. [98]. IT-101 contains molecule drug camptothecin (CPT) conjugated with β-CD based polymers (CDP), which acts as a carrier system for active molecule. The CDP site was labeled with ^{64}Cu through attaching of DOTA complex for microPET imaging; the obtained nanoparticle was similar to the IT-101 structure. Plasma pharmacokinetics of ^{64}Cu-IT-101 ere studied at 1, 4, and 24 h after injection with microPET/CT. It was shown that a low-molecular-weight fraction cleared rapidly through kidneys due to biphasic elimination profile, whereas remaining NPs circulated with a terminal half-life of $\approx$13 h. Biodistribution of IT-101 was examined at 24 h after administration; the highest tissue concentration was found in the tumor followed by the liver.

Another study conducted to investigate the effect of CD inclusion in NPs for drug delivery revealed the possibility of CD-based SPECT imaging agent [42]. Areses et al. compared the adhesive abilities and biodistribution of orally administered poly(anhydride) NPs and CD containing NP (CD-NP) in rats utilizing labeling with ^{99m}Tc for SPECT imaging. ^{99m}Tc-NP showed activity only in the gastrointestinal tract on SPECT images, whereas ^{99m}Tc-CD-NP revealed extended residence time in stomach: about 13% of ^{99m}Tc-CD-NP administered dose and 3% of ^{99m}Tc-NP given dose were found in the stomach after 8 h.

Liu et al., in 2011 [39], studied the improvement of the biodistribution of NPs using CD. Rare-earth UCNPs were modified by α-CD and OA for increasing of the water-solubility. UCNP-OA-CD complexes with Tm inclusion were labeled with ^{18}F (^{18}F-UCNP(Tm)-OA-CD) for microPET imaging of ex vivo and in vivo biodistribution in mice at 5 min and 2 h. This was the first labeling of CD-based probe with an ^{18}F. Ex vivo imaging displayed rapid accumulation of NPs in the liver (~90.8% injected dose(ID)/g) and spleen (~62.5% ID/g) at 5 min and further decreasing of liver uptake to ~57.6% ID/g, while increased spleen accumulation to 118.9% ID/g after 2 h post-injection. In vivo microPET images were consistent with ex vivo biodistribution results and showed intense radioactive signals in the liver and spleen at 5 min after injection.

In addition to the previously discussed reports, a pre-targeted approach for molecular imaging probe development was presented by Hou et al. [99]. This work was based on biorthogonal conjugation chemistry between NPs, which have tendency to accumulate in tumors to enhanced permeability and retention (EPR) effect and radiolabeled imaging agents for PET imaging. One of NPs components was synthesized from CD-grafted polyethyleneimine (CD-PEI) and trans-cyclooctene N-hydroxysuccinimide (TCO-NHS), resulted in TCO/CD-PEI building block. Actual tumor-targeting NP (TCO⊂SNPs) was prepared via self-assembly from four different blocks TCO/CD-PEI, CD-PEI, adamantane-grafted polyamidoamine (Ad-PAMAM), and Ad-grafted polyethylene glycol (Ad-PEG) and injected to the tail vein of the mice with U87 glioblastoma cells. After EPR-driven accumulation in tumor, TCOresulted in TCO/CD-PEI building block. Actual tumor-targeting NP (TCO⊂SNPs can dynamically disassemble to release TCO/CD-PEI. After 24 h post-injection of TCO⊂SNPs, freshly prepared tetrazine compound radiolabeled with ^{64}Cu through DOTA (^{64}Cu-Tz) was injected. Subsequently, distributed ^{64}Cu-Tz can undergo biorthogonal reaction with TCO/CD-PEI parts left after TCO⊂SNPs disassembling in vivo, yielding the dihydropyrazine conjugation adduct ^{64}Cu-DHP/CD-PEI, which acts as a contrast agent for tumors. Multiple microPET and anatomical CT images were acquired following the injection of pre-targeted NPs and ^{64}Cu-Tz compound with in vivo reaction (Figure 10a), along with two series of control PET imaging of a fully ex vivo prepared ^{64}Cu-DHP/CD-PEI adduct (Figure 10b) and free radiolabeled reporter ^{64}Cu-Tz (Figure 10c). Pre-targeted studies showed the accumulation and retention of radioactivity mainly in the glioblastoma tumor and liver and some nonspecific uptake by tissues. Supramolecular nanoparticles (SNP) control and probe (^{64}Cu-Tz) imaging did not present highly distinguishable tumor uptake. Although, high radioactivity was observed in the liver in all three cases, which can be explained by ^{64}Cu^{2+} dissociation from DOTA ligand. This issue can be eliminated by using different radioisotopes for labelling.

Modification of CD by grafting alkyl chains (C6-C14) can lead to self-organization of obtained derivatives into NPs potentially useful for drug delivery [100]. Further co-nanoprecipitation of bio-esterified alkylated cyclodextrins with PEGylated phospholipids (PEG) can lead to surface-modified NPs [101]. Perret et al. researched the effect of the PEG chain length on the plasma protein absorptivity and blood kinetics of NPs [43]. β-CD derivatives with C10 alkyl chain (β-CD-C10) were co-nanoprecipitated with PEG with chain length of 2000 (^{125}I-βCD-C10-PEG$_{2000}$-NP) and 5000 Da (^{125}I-βCD-C10-PEG$_{5000}$-NP) and radiolabeled with ^{125}I for SPECT ex vivo and in vivo biodistribution studies. In vivo SPECT/CT images were acquired at 10 min, 1, 3, 6, and 24h following the injection of NPs without PEG (^{125}I-βCD-C10-NP) and with PEG (^{125}I-βCD-C10-PEG$_{2000}$-NP, ^{125}I-βCD-C10-PEG$_{5000}$-NP). Hepatic activity was observed with all NPs; however, splanchnic activity was observed only with ^{125}I-βCD-C10-NP. Additionally, ^{125}I-βCD-C10-PEG$_{5000}$-NP systems showed reduced elimination and increased circulating concentration following in vivo intravenous injection in comparison with other NPs.

β-CD-based rotaxane were used in developing theranostic shell-crosslinked NPs (SCNPs) by Yu et al. for improving drug delivery and controllable release in supramolecular medicine [102]. The core-shell-structured self-assembling NPs were obtained from polyrotaxanes consisted of amphiphilic diblock copolymer and the primary-amino-containing β-CD (β-CD-NH$_2$), which undergoes complexation with poly(ε-caprolactone) (PCL) segment. In the gained structure, amphiphilic deblock copolymer acts as the axle, and β-CD-NH$_2$ acts as a wheel in complex with PCL, whose chains can experience hydrophobic interactions along with the perylene diimide (PDI) stoppers, which has a tendency to π-π stacking interactions. Obtained SCNPs were labeled with radioactive ^{64}Cu through DOTA attachment to SCNPs (^{64}Cu SCNPs@DOTA) for PET imaging of the dynamic biodistributions and accumulations of SCNPs in the main organs. HeLa tumor-bearing mice were imaged at various time-points after intra-venous injection 150 μCi of ^{64}Cu SCNPs@DOTA. Images revealed the high liver uptake of ^{64}Cu SCNPs@DOTA along with increasing of tumor uptake from the point of injection to 12 h post-injection with further start of clearance at 48 h.

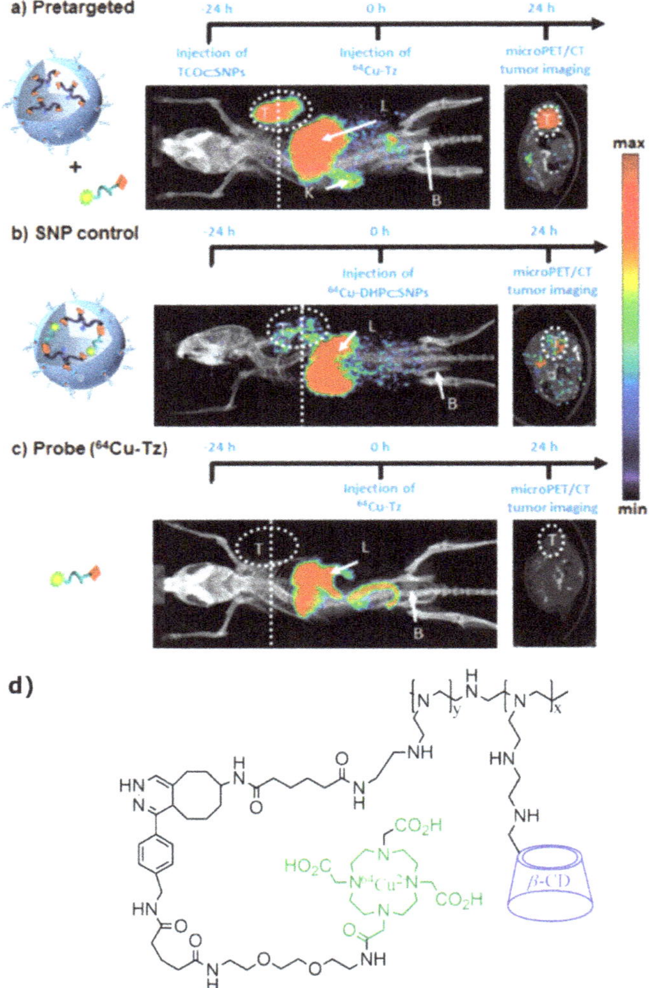

Figure 10. Timeline of the injection protocol employed for (**a**) pre-targeted, (**b**) SNP control (64Cu-DHP⊂SNPs), and (**c**) free radiolabeled reporter (64Cu-Tz) studies. Representative in vivo microPET/CT images of the mice (n = 4/group) subjected to the three studies at 24 h p.i. Labels T, L, K, and B refer to the tumor, liver, kidney, and bladder, respectively. Dashed lines correspond to the transverse cross-section through the center of each tumor mass, whose image is shown in the right panel [99]. The images are reprinted with permission from publisher [99]. (**d**) The chemical structure of the tumor targeting imaging probe developed by Hou et al. [99].

Another study utilized PET for investigation in vivo distribution of 2-Hydroxypropyl-β-CD (HPBCD), β-cyclodextrin derivative, and an orphan drug for the Niemann–Pick disease treatment [40]. Six-deoxy-6-monoamino-(2-Hydroxypropyl)-β-CD (NH2-HPBCD) was conjugated with p-NCS-benzyl-NODA-GA (NODAGA) and radiolabeled with ^{68}Ga for PET/CT imaging. Ex vivo and in vivo studies on healthy mice showed that ^{68}Ga-NODAGA-HPBCD was mainly excreted through the urinary system with low uptake of the abdominal and thoracic organs and tissues at 30 and 90 min post-injection.

The most recent study in the area of CD-based PET molecular imaging probes containing tumor targeting compounds was done by Trencsenyi et al., in 2019 [41]. Their aims were to develop novel radiolabeled compound specific to the prostaglandin E2 (PGE2), which plays an important role in tumor progress and formation of metastases. The high affinity of PGE2 to the randomly methylated β-CD (RAMEB) was reported by Sauer et al. [103]. Trencsenyi et al., in their research, aimed to synthesize PGA-specific RAMEB labeled with ^{68}Ga through NODAGA (^{68}Ga-NODAGA-RAMEB) for investigation of its tumor-targeting properties and in vivo biodistribution using PET. PancTu-1 and BxPC3 tumor-bearing SCID mice were intravenously injected with ^{68}Ga-NODAGA-RAMEB. The injection was followed with dynamic and static microPET imaging at 0–90 min. The accumulation of ^{68}Ga-NODAGA-RAMEB was significantly higher in BxPC3 tumors than in the PancTu-1; the highest post-injection tumor-background ratio (T/M) was obtained at 80–90 min post-injection. The T/M standardized uptake values (SUVs) were 10-fold lower in the PancTu-1 than those of BxPC3 tumors confirming the high PGE2 selectivity of ^{68}Ga-lebeled cyclodextrin.

5. CD-Based CT Contrast Agents

The initial development of CD-based contrast agents for CT was used to investigate multimodality imaging potential of phosphorescent-modified NaDyF$_4$ NPs (DyNPs) [53]. OA-coated DyNPs underwent complexation with α-CD followed by conjugation with Gd(III)-DTPA complex. The last stage of contrast agent formation was the loading of phosphorescent Ir(Dbz)$_2$(Pbi) complex with in the hydrophobic layer of Gd-α-CD-DyNPs [53]. Besides the contrast creation potential for fluorescence and MRI, Zhou et al. considered the developed Ir-Gd-α-CD-DyNPs as a potential contrast agent for CT due to the presence of heavy atoms of Dy, Gd, and Ir. The measured Hounsfield units (HU) value of 10 mg/mL Ir-Gd-α-CD-DyNPs aqueous solution was equal to 158 at 80 kV X-ray energies. Following in vitro experiments, Ir-Gd-α-CD-DyNPs were used for CT scanning of tumor-bearing mice (Figure 11). The original HU value of 109 for the tumor increased to 212 HU after intratumoral injection of 100 ul of 3 mg/mL Ir-Gd-α-CD-DyNPs (Figure 11G) solution. As a result, approximately 100% signal enhancement was observed on the CT scans.

Another multimodal CD-based contrast agent that was developed for CT scanning was a core-shell-structured alkali ion-doped CaF$_2$:Yb,Er UCNP [50]. Similar to the NPs discussed above, the core of NPs was coated with OA with subsequent complexation with α-CD. The final α-CD/UCNPs contrast agent was compared to well-known iopromide 300 CT clinical contrast agent. Yin et al. found that developed α-CD/UCNPs have 45% higher HU value at concentration 80 mM/L. This high X-ray absorption originates from the present high atomic number elements. In addition, investigated α-CD/UCNPs were loaded with doxorubicin, and the ability of simultaneous cancer imaging and treatment using this contrast agent was suggested [50].

The next achievement in the CT contrast agent development was obtained in 2017 by conjugating β-CD to poly(methyl vinyl ether-alt-maleic anhydride) (PVME-alt-MAH) with a subsequent reaction using modified dextran to create gel microspheres [104]. To make dextran microspheres visible for CT, they were loaded with iodine in *n*-hexane. Through the micro-CT phantom imaging, Zhu et al. observed the subsequent improvement of contrast after iodine loading.

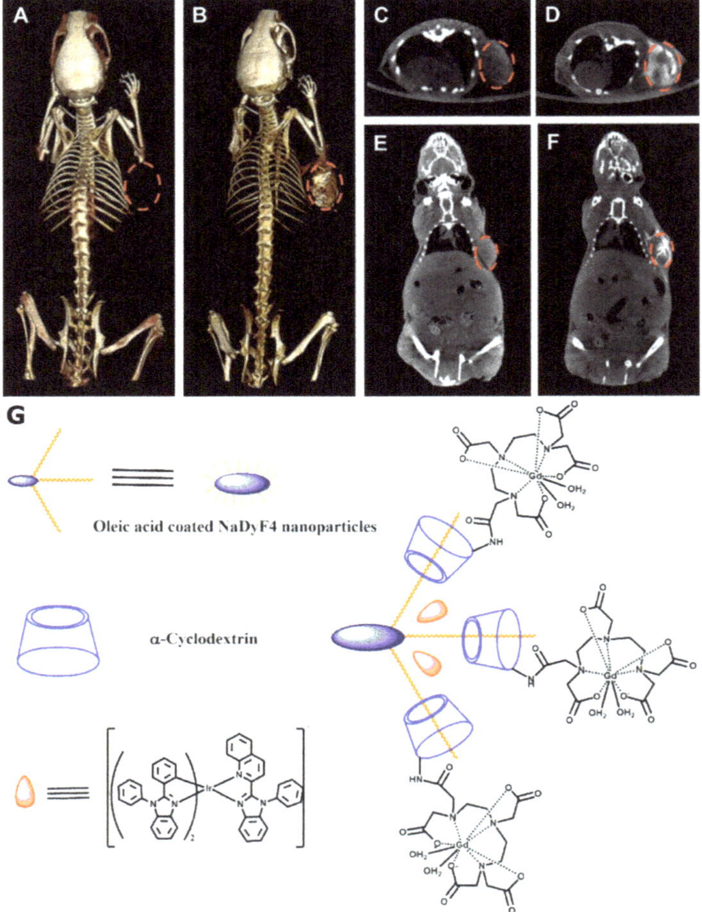

Figure 11. In vivo 3D volume-rendered (**A,B**) and maximum intensity projections in axial (**C,D**) and coronal (**E,F**) view CT images of the tumor-bearing mouse obtained pre- (**A,C,E**) and post- (**B,D,F**) injection of Ir-Gd-α-CD-DyNPs contrast agent [53]. The images are reprinted with permission from the publisher [53]. The position of the tumor was marked by red circles. The chemical structure of the developed CT contrast agent is shown in (**G**).

Another study, conducted in 2017, demonstrated synthesis of β-CD-{poly(ε-caprolactone)-poly(2-aminoethyl methacrylate)-poly[poly(ethylene glycol) methyl ether methacrylate]}$_{21}$ (β-CD-(PCL-PAEMA-PPEGMA)$_{21}$) with stable unimolecular micelles formed in aqueous solution [51]. β-CD-(PCL-PAEMA-PPEGMA)$_{21}$ had used a template for the creation of gold NPs (AuNPs) with uniform sizes, followed by the encapsulation of doxorubicin (DOX). The CT performance of the final β-CD-(PCL-PAEMA-PPEGMA)$_{21}$/AuNPs/DOX contrast agent was compared to the clinically available Omnipaque both in vitro and in vivo. At a concentration of 800 uM, Lin et al. found the X-ray absorption of β-CD-(PCL-PAEMA-PPEGMA)$_{21}$/AuNPs/DOX solution to be approximately 23% higher compared to Omnipaque. In vivo imaging of β-CD-(PCL-PAEMA-PPEGMA)$_{21}$/AuNPs/DOX contrast agent in HepG2 mice tumor model demonstrated significant enhancement of CT signal compared to the clinical iodine analogue [51].

Another step in the development of CD-based CT contrast agents was done by creation an unimolecular micelle system synthesized from 21-arm star-like polymer β-CD-{poly(lactide)-poly(2-(dimethylamino) ethyl methacry late)-poly[oligo(2-ethyl-2-oxazoline) methacrylate]}$_{21}$ (β-CD-(PLA-PDMAEMA-PEtOxMA)$_{21}$) followed by production of β-CD-(PLA-PDMAEMA-PEtOxMA)$_{21}$/AuNPs/DOX [105]. The developed Au-loaded β-CD-based contrast agent was tested in the animal HepG2 tumor model. Lin et al. found that intravenously injected β-CD-(PLA-PDMAEMA-PEtOxMA)$_{21}$/AuNPs/DOX produces substantially higher CT contrast compared to iodine-based Omnipaque, which was used for control scans.

Following the previous studies, the most recent advance in the field of CD-based CT contrast agents was achieved by the same group [106]. The authors synthetized 21-arm star-like polymers β-CD-g-{poly(2-(dimethylamino)ethyl methacrylate)-poly(2-hydroxyethyl methacrylate)-poly[poly(ethylene glycol) methyl ether methacrylate]} (β-CD-g-(PDMA-b-PHEMA-b-PPEGMA)). By adding HAuCl$_4$ solution into β-CD-g-(PDMA-b-PHEMA-b-PPEGMA) aqueous solution and triggering subsequent reduction with DMA, the AuNPs at the core of unimolecular micelles were formed. The CT contrast of β-CD-g-(PDMA-b-PHEMA-b-PPEGMA)/AuNPs agent was compared to contrast created by Omnipaque in vitro. Lin et al. found that, at a concentration of 800 uM, the X-ray attenuation of β-CD-g-(PDMA-b-PHEMA-b-PPEGMA)/AuNPs was approximately 37% bigger than that of Omnipaque. The developed β-CD-g-(PDMA-b-PHEMA-b-PPEGMA)/AuNPs demonstrated slightly higher X-ray absorption compared to previously synthetized β-CD-(PCL-PAEMA-PPEGMA)$_{21}$/AuNPs [51,106]. However, the average HU values of β-CD-g-(PDMA-b-PHEMA-b-PPEGMA)/AuNPs at each concentration were slightly lower than previously developed β-CD-g-(PLA-b-PDMA-b-PEtOxMA)$_{21}$/AuNPs [105,106].

6. Discussion

Starting in the early 1990s, CDs became of large interest for developing contrast agents and molecular probes for medical imaging. Although the primary application of CDs in medical imaging was the basis for developing novel Gd-based MRI contrast agents, CD-based contrast agents commence to be utilized by other imaging modalities such as PET, CT, US, and PAI. This growing interest is caused by excellent biocompatibility, low toxicity, and relative ease of modifying of CD molecules. Currently, the most frequently used CD molecule for contrast agent development is β-CD. Despite the extensive development of β-CD-based contrast agents, some of the medical imaging areas started utilizing α- and γ- CD-based contrast agents and molecular probes as well.

The working mechanism of the vast majority of CD-based contrast agents for conventional MRI relies on the Freed's theory [57] and the decreasing of the relaxation times of the solvent protons once either paramagnetic ion or metal–organic complex containing the paramagnetic ion binds to a supramolecular cage. Initially, this binding was done through hydrophobic interaction between CDs cavity and metal–organic complexes. Although contrast agents synthesized using this approach demonstrated significantly better performance compared to the clinically available analogs, the main disadvantage of this technique is blocking the supramolecular cage cavity, which makes the developed contrast agents and molecular probes uncapable to image some drug transport, biodistribution, and treatment monitoring. In addition, the number of paramagnetic ions per molecule is also limited, since usually only one metal–organic complex enters the cavity. This issue was overcome by creating different CD-based polymers and dendrimers, and the number of macromolecules determined the maximum number of paramagnetic ions per molecule. Starting from the early 2000s, the direct labeling of the CD macrocycles with chelates containing paramagnetic ions has been developed intensively. This method allows keeping the CD cavity free for interactions with different molecules. Therefore, the contrast agents based on the direct labeling of the CD macromolecules have a significantly larger field of applications compared to agents developed using non-covalent interactions.

Currently, CD-based MRI contrast agents contain mostly Gd(III) ions and are designed as T_1 contrast agents. On the other hand, the development of CD-based T_2 and T_2^* contrast agents have been demonstrated in vitro and in vivo [23,71,75]. Although some of the developed CD-based contrast agents for proton MRI demonstrated an overall low level of cytotoxicity, dedicated toxicology studies are needed prior to the further translation of this imaging agents into clinics. In addition, the vast majority of the relaxivity measurements were conducted at low magnetic fields, which are not for clinical imaging purposes, such as 0.47 and 0.5 T. The studies that focused on animal imaging were mostly conducted at high magnetic fields, such as 7 and 9.4 T. In order to facilitate the potential clinical transition of the developed CD-based contrast agents, the accurate measurement of relaxivities and comparisons to clinical analogs should be performed at 1.5 and 3 T, which are currently used for clinical MRI. During the last eight years, CDs drew attention from researchers working with hyperpolarized MRI. Although only a few studies were conducted, the obtained results demonstrated the high potential of CD macromolecules to become a basis for further development of hyperpolarized probes. Recently, Hane et al. demonstrated that γ-CD-based pseudorotaxanes can be a valuable platform for developing the molecular imaging probes that utilizes the HyperCEST effect [95]. The alkyne chains have a high affinity to the γ-CD cavity and can be easily designed to serve as a high-affinity probe that has a selective binding to the disease site. Functionalizing the alkyne chain should be done carefully such that there is no effect on the interaction with hyperpolarized ^{129}Xe. The HyperCEST depletion observed was only around 50% indicating partial depolarization of hyperpolarized ^{129}Xe encapsulated by γ-CD-based pseudorotaxanes. An accurate study for radiofrequency saturation pulses is required to maximize the HyperCEST depletion and to translate γ-CD-based pseudorotaxanes to in vivo imaging applications. Although the demonstrated results of hyperpolarized MRI molecular imaging probes look promising, further in vivo imaging studies are needed, as well as biodistribution and toxicity evaluations of the proposed biosensors.

Following the success of the CD-based MRI contrast agents, several attempts have been made to use CDs as a contrast agent for US imaging [44,96]. The developed US CD-based contrast agents caused significant acoustic impedance difference between the tissue and the contrast agents. Despite successful in vitro demonstration of this proof-of-principle [44], there was no further development of the CD-based contrast agent for US imaging purposes during the last five years. On the contrary, CD-coated NPs became of interest for the PAI. It was found that coating of UCNPs with α-CD cause photoacoustic signal enhancement [49] and, therefore, can serve as efficient PAI contrast. The most recent achievement in the field demonstrated PAI detection of prosthetic joint infection using β-CD-conjugated indocyanine green in mice [48]. All studies with CD-based PAI tracers were conducted in small animals. Since the PAI tracers required irradiation with infrared light to emit ultrasound, further studies should move to the larger animals in order to evaluate penetration depth of the excitation light and to estimate the performance of PAI contrast agents prior to clinical translation. Furthermore, the development of larger pallet of PAI contrast dedicated for specific diseases would be beneficial to facilitate the future translation into the clinic.

The vast majority of the radiolabeled CDs derivatives were used for PET and SPECT imaging of the drug delivery and biodistribution in vivo. To be suitable for PET, multiple CD-based tracers containing ^{64}Cu, ^{68}Ga, and ^{18}F were developed. For SPECT imaging purposes, CD macrocycles were radiolabeled with either ^{99m}Tc or ^{125}I. The further widening of the radiolabeled CD derivatives might become a useful tool for pharmaceutics and drug development.

The most recent CD-based NPs were studied as contrast agents for CT. To be able to produce a sufficient contrast, the CD-based NPs must contain atoms with a high atomic number. The heavy atoms absorb X-rays with higher efficiency, increasing the X-ray attenuation coefficient of the tissue in which the contrast agent is present [107–109]. With modern advances in X-ray detection allowing lower dose image acquisition [110–113], the utilization of the novel CD-based NPs containing heavy elements will be highly beneficial for accurate anatomical imaging purposes. In addition, the novel dual energy CT approach [114,115] will benefit even more from the implementation of the novel CD-based NPs. Implementation and further development of the demonstrated NPs with substantially higher X-ray attenuation compared to currently available iodine and barium contrasts could potentially allow a superior improvement of CNR of the dual energy CT image. CT contrast of CD-based NPs was initially observed from contrast agents developed for multimodal imaging [50,53]. Only during the last three years has the research in synthesis of dedicated CT CD-based contrast agents been conducted extensively. Currently, the vast majority of the CD-based CT contrast agents were developed specifically for cancer imaging, whereas only one agent for angiography was developed. Therefore, it might be potentially useful to develop CD-based contrast agents suitable for different types of clinical CT imaging. In addition, the potential CD-based CT contrast agents must undergo an accurate toxicity study prior to further translation to in vivo imaging. Furthermore, special attention should be given to the investigation of extraction pathways of the contrast agents from the living organism during in vivo studies.

One of the potential approaches for further advances in the field of CD-based contrast agents is the development of a contrast agent that can serve for dual imaging modalities such as PET/MRI and PET/CT. Usually, the PET-active component of the tracer is small, and therefore, the development of the PET/MRI and PET/CT contrast agents could be built around radiolabeling of the existing CD-based agents for MRI and CT imaging modalities, respectively.

Overall, CDs are of high interest in the medical imaging field and are currently a very promising basis for developing various contrast agents. The successful clinical translation of CD-based contrast agents for proton MRI can help significantly improve the quality of clinical MRI scans. Further development of functionalized CD-based imaging agents for MRI has the potential to make molecular imaging using clinical proton MRI possible. Despite the enormous development level of CD-based contrast agents for conventional proton MRI, other imaging modalities started utilizing CD-based contrast agents recently and further developments and investigations are needed prior to successful clinical translation. Nevertheless, CD-based contrast agents demonstrated exceptional performance in the areas of CT, PET, and PAI.

Author Contributions: Y.S. performed literature search, wrote the manuscript, and revised and edited the paper. C.J.N. performed literature search and wrote the manuscript. V.G. wrote the manuscript and revised and edited the paper. L.E.S. wrote the manuscript and edited the paper. B.D. wrote the manuscript and revised and edited the paper. F.T.H. revised and edited the paper. M.S.A. revised and edited the paper. All authors have read and agreed to the published version of the manuscript.

Funding: This work was partially supported by the Natural Science and Engineering Research Council of Canada (NSERC) Discovery grant. Y.S. was supported by an Ontario Graduate Fellowship. C.J.N. was supported by the NSERC Undergraduate Student Research Award (USRA). V.G. was supported by an Ontario Trillium Scholarship.

References

1. Stella, V.J.; He, Q. Cyclodextrins. *Toxicol. Pathol.* **2008**, *36*, 30–42. [CrossRef] [PubMed]
2. Jansook, P.; Ogawa, N.; Loftsson, T. Cyclodextrins: Structure, physicochemical properties and pharmaceutical applications. *Int. J. Pharm.* **2017**, *535*, 272–284. [CrossRef] [PubMed]
3. Del Valle, E.M.M. Cyclodextrins and their uses: A review. *Process. Biochem.* **2004**, *39*, 1033–1046. [CrossRef]
4. Lai, W.-F. Cyclodextrins in non-viral gene delivery. *Biomaterials* **2014**, *35*, 401–411. [CrossRef]
5. Irie. T.; Uekama, K. Pharmaceutical Applications of Cyclodextrins. III. Toxicological Issues and Safety Evaluation. *J. Pharm. Sci.* **1997**, *86*, 147–162. [CrossRef]
6. Bellringer, M.; Smith, T.; Read, R.; Gopinath, C.; Olivier, P. β-Cyclodextrin: 52-Week toxicity studies in the rat and dog. *Food Chem. Toxicol.* **1995**, *33*, 367–376. [CrossRef]
7. Loftsson, T.; Hreinsdóttir, D.; Másson, M. Evaluation of cyclodextrin solubilization of drugs. *Int. J. Pharm.* **2005**, *302*, 18–28. [CrossRef]
8. Gidwani, B.; Vyas, A. A Comprehensive Review on Cyclodextrin-Based Carriers for Delivery of Chemotherapeutic Cytotoxic Anticancer Drugs. *BioMed Res. Int.* **2015**, *2015*, 198268. [CrossRef]
9. Saokham, P.; Muankaew, C.; Jansook, P.; Loftsson, T. Solubility of Cyclodextrins and Drug/Cyclodextrin Complexes. *Molecules* **2018**, *23*, 1161. [CrossRef]

10. Schönbeck, C.; Gaardahl, K.; Houston, B. Drug Solubilization by Mixtures of Cyclodextrins: Additive and Synergistic Effects. *Mol. Pharm.* **2019**, *16*, 648–654. [CrossRef]

11. Jambhekar, S.S.; Breen, P. Cyclodextrins in pharmaceutical formulations II: Solubilization, binding constant, and complexation efficiency. *Drug Discov. Today* **2016**, *21*, 363–368. [CrossRef] [PubMed]

12. Loftsson, T. Drug solubilization by complexation. *Int. J. Pharm.* **2017**, *531*, 276–280. [CrossRef] [PubMed]

13. Adeoye, O.; Cabral-Marques, H. Cyclodextrin nanosystems in oral drug delivery: A mini review. *Int. J. Pharm.* **2017**, *531*, 521–531. [CrossRef] [PubMed]

14. Muankaew, C.; Loftsson, T. Cyclodextrin-Based Formulations: A Non-Invasive Platform for Targeted Drug Delivery. *Basic Clin. Pharmacol. Toxicol.* **2018**, *122*, 46–55. [CrossRef] [PubMed]

15. Patel, M.R.; Lamprou, D.A.; Vavia, P. Synthesis, Characterization, and Drug Delivery Application of Self-assembling Amphiphilic Cyclodextrin. *AAPS PharmSciTech* **2020**, *21*, 11. [CrossRef] [PubMed]

16. Doan, V.T.H.; Lee, J.H.; Takahashi, R.; Nguyen, P.T.M.; Nguyen, V.A.T.; Pham, H.T.T.; Fujii, S.; Sakurai, K. Cyclodextrin-based nanoparticles encapsulating α-mangostin and their drug release behavior: Potential carriers of α-mangostin for cancer therapy. *Polym. J.* **2020**, *52*, 457–466. [CrossRef]

17. Crini, G.; Fourmentin, S.; Fenyvesi, É.; Torri, G.; Fourmentin, M.; Morin-Crini, N. Cyclodextrins, from molecules to applications. *Environ. Chem. Lett.* **2018**, *16*, 1361–1375. [CrossRef]

18. Lai, W.-F. Design of cyclodextrin-based systems for intervention execution. In *Delivery of Therapeutics for Biogerontological Interventions*; Elsevier BV: Amsterdam, The Netherlands, 2019; pp. 49–59.

19. Davis, M.E. The First Targeted Delivery of siRNA in Humans via a Self-Assembling, Cyclodextrin Polymer-Based Nanoparticle: From Concept to Clinic. *Mol. Pharm.* **2009**, *6*, 659–668. [CrossRef]

20. Malhotra, M.; Gooding, M.; Evans, J.C.; O'Driscoll, D.; Darcy, R.; O'Driscoll, C.M. Cyclodextrin-siRNA conjugates as versatile gene silencing agents. *Eur. J. Pharm. Sci.* **2018**, *114*, 30–37. [CrossRef]

21. Wankar, J.; Kotla, N.G.; Gera, S.; Rasala, S.; Pandit, A.; Rochev, Y.A. Recent Advances in Host–Guest Self-Assembled Cyclodextrin Carriers: Implications for Responsive Drug Delivery and Biomedical Engineering. *Adv. Funct. Mater.* **2020**, *30*, 1909049. [CrossRef]

22. Van De Manakker, F.; Vermonden, T.; Van Nostrum, C.F.; Hennink, W.E. Cyclodextrin-Based Polymeric Materials: Synthesis, Properties, and Pharmaceutical/Biomedical Applications. *Biomacromolecules* **2009**, *10*, 3157–3175. [CrossRef] [PubMed]

23. Feng, Y.; Xiao, Q.; Zhang, Y.; Li, F.; Li, Y.; Li, C.; Wang, Q.; Shi, L.; Lin, H. Neodymium-doped NaHoF4 nanoparticles as near-infrared luminescent/T2-weighted MR dual-modal imaging agents in vivo. *J. Mater. Chem. B* **2017**, *5*, 504–510. [CrossRef] [PubMed]

24. Ye, M.; Qian, Y.; Shen, Y.; Hu, H.; Sui, M.; Tang, J. Facile synthesis and in vivo evaluation of biodegradable dendritic MRI contrast agents. *J. Mater. Chem.* **2012**, *22*, 14369–14377. [CrossRef]

25. Bartlett, D.W.; Su, H.; Hildebrandt, I.J.; Weber, W.A.; Davis, M.E. Impact of tumor-specific targeting on the biodistribution and efficacy of siRNA nanoparticles measured by multimodality in vivo imaging. *Proc. Natl. Acad. Sci. USA* **2007**, *104*, 15549–15554. [CrossRef] [PubMed]

26. Wang, X.; Parvathaneni, V.; Shukla, S.K.; Kanabar, D.D.; Muth, A.; Gupta, V. Cyclodextrin Complexation for Enhanced Stability and Non-invasive Pulmonary Delivery of Resveratrol—Applications in Non-small Cell Lung Cancer Treatment. *AAPS PharmSciTech* **2020**, *21*, 1–14. [CrossRef] [PubMed]

27. Rekharsky, M.V.; Inoue, Y. Complexation Thermodynamics of Cyclodextrins. *Chem. Rev.* **1998**, *98*, 1875–1917. [CrossRef] [PubMed]

28. Hădărugă, N.G.; Bandur, G.N.; David, I.; Hădărugă, D.I. A review on thermal analyses of cyclodextrins and cyclodextrin complexes. *Environ. Chem. Lett.* **2018**, *17*, 349–373. [CrossRef]

29. Liu, L.; Guo, Q.-X. The Driving Forces in the Inclusion Complexation of Cyclodextrins. *J. Incl. Phenom. Macrocycl. Chem.* **2002**, *42*, 1–14. [CrossRef]

30. Hashidzume, A.; Yamaguchi, H.; Harada, A. Cyclodextrin-Based Rotaxanes: From Rotaxanes to Polyrotaxanes and Further to Functional Materials. *Eur. J. Org. Chem.* **2019**, *2019*, 3344–3357. [CrossRef]

31. Hashidzume, A.; Yamaguchi, H.; Harada, A. Cyclodextrin-based molecular machines. In *Topics in Current Chemistry*; Springer: Berlin, Germany, 2014; Volume 354, pp. 71–110.

32. Lai, W.-F.; Rogach, A.L.; Wong, W.-Y. Chemistry and engineering of cyclodextrins for molecular imaging. *Chem. Soc. Rev.* **2017**, *46*, 6379–6419. [CrossRef]

33. Aime, S.; Botta, M.; Panero, M.; Grandi, M.; Uggeri, F. Inclusion complexes between β-cyclodextrin and β-benzyloxy-α-propionic derivatives of paramagnetic DOTA- and DPTA-Gd(III) complexes. *Magn. Reson. Chem.* **1991**, *29*, 923–927. [CrossRef]

34. Aime, S.; Benetollo, F.; Bombieri, G.; Colla, S.; Fasano, M.; Paoletti, S. Non-ionic Ln(III) chelates as MRI contrast agents: Synthesis, characterisation and 1H NMR relaxometric investigations of bis(benzylamide)diethylenetriaminepentaacetic acid Lu(III) and Gd(III) complexes. *Inorg. Chim. Acta* **1997**, *254*, 63–70. [CrossRef]

35. Bryson, J.M.; Chu, W.-J.; Lee, J.-H.; Reineke, T.M. A β-Cyclodextrin "Click Cluster" Decorated with Seven Paramagnetic Chelates Containing Two Water Exchange Sites. *Bioconj. Chem.* **2008**, *19*, 1505–1509. [CrossRef]

36. Barge, A.; Cravotto, G.; Robaldo, B.; Gianolio, E.; Aime, S. New CD derivatives as self-assembling contrast agents for magnetic resonance imaging (MRI). *J. Incl. Phenom. Macrocycl. Chem.* **2007**, *57*, 489–495. [CrossRef]

37. Carrera, C.; Digilio, G.; Baroni, S.; Burgio, D.; Consol, S.; Fedeli, F.; Longo, D.L.; Mortillaro, A.; Aime, S. Synthesis and characterization of a Gd(iii) based contrast agent responsive to thiol containing compounds. *Dalton Trans.* **2007**, 4980–4987. [CrossRef] [PubMed]

38. Gomes, P.M.O.; Silva, A.M.; Silva, V.L. Pyrazoles as Key Scaffolds for the Development of Fluorine-18-Labeled Radiotracers for Positron Emission Tomography (PET). *Molecules* **2020**, *25*, 1722. [CrossRef]

39. Liu, Q.; Chen, M.; Sun, Y.; Chen, G.; Yang, T.; Gao, Y.; Zhang, X.; Li, F. Multifunctional rare-earth self-assembled nanosystem for tri-modal upconversion luminescence /fluorescence /positron emission tomography imaging. *Biomaterials* **2011**, *32*, 8243–8253. [CrossRef]

40. Hajdu, I.; Angyal, J.; Szikra, D.; Kertész, I.; Malanga, M.; Fenyvesi, É.; Szente, L.; Vecsernyés, M.; Bácskay, I.; Váradi, J.; et al. Radiochemical synthesis and preclinical evaluation of 68Ga-labeled NODAGA-hydroxypropyl-beta-cyclodextrin (68Ga-NODAGA-HPBCD). *Eur. J. Pharm. Sci.* **2019**, *128*, 202–208. [CrossRef]

41. Trencsényi, G.; Kis, A.; Szabó, J.P.; Ráti, Á.; Csige, K.; Fenyvesi, É.; Szente, L.; Malanga, M.; Méhes, G.; Emri, M.; et al. In vivo preclinical evaluation of the new 68Ga-labeled beta-cyclodextrin in prostaglandin E2 (PGE2) positive tumor model using positron emission tomography. *Int. J. Pharm.* **2020**, *576*, 118954. [CrossRef]

42. Areses, P.; Agüeros, M.T.; Quincoces, G.; Collantes, M.; Richter, J.Á.; López-Sánchez, L.M.; Sánchez-Martínez, M.; Irache, J.M.; Peñuelas, I. Molecular Imaging Techniques to Study the Biodistribution of Orally Administered 99mTc-Labelled Naive and Ligand-Tagged Nanoparticles. *Mol. Imaging Biol.* **2011**, *13*, 1215–1223. [CrossRef]

43. Perret, P.; Bacot, S.; Gèze, A.; Maurin, A.G.D.; Debiossat, M.; Soubies, A.; Blanc-Marquis, V.; Choisnard, L.; Boutonnat, J.; Ghezzi, C.; et al. Biodistribution and preliminary toxicity studies of nanoparticles made of Biotransesterified β–cyclodextrins and PEGylated phospholipids. *Mater. Sci. Eng. C* **2018**, *85*, 7–17. [CrossRef] [PubMed]

44. Yao, Y.; Liu, X.; Liu, T.; Zhou, J.; Zhu, J.; Sun, G.; He, D. Preparation of inclusion complex of perfluorocarbon compound with β-cyclodextrin for ultrasound contrast agent. *RSC Adv.* **2015**, *5*, 6305–6310. [CrossRef]

45. Weber, J.; Beard, P.C.; Bohndiek, S.E. Contrast agents for molecular photoacoustic imaging. *Nat. Methods* **2016**, *13*, 639–650. [CrossRef] [PubMed]

46. Laramie, M.D.; Smith, M.K.; Marmarchi, F.; McNally, L.R.; Henary, M. Small Molecule Optoacoustic Contrast Agents: An Unexplored Avenue for Enhancing In Vivo Imaging. *Molecules* **2018**, *23*, 2766. [CrossRef]

47. Yu, Z.; Wang, M.; Pan, W.; Wang, H.; Li, N.; Tang, B. Tumor microenvironment-triggered fabrication of gold nanomachines for tumor-specific photoacoustic imaging and photothermal therapy. *Chem. Sci.* **2017**, *8*, 4896–4903. [CrossRef]

48. Wang, Y.; Thompson, J.M.; Ashbaugh, A.G.; Khodakivskyi, P.; Budin, G.; Sinisi, R.; Heinmiller, A.; Van Oosten, M.; Van Dijl, J.M.; Van Dam, G.M.; et al. Preclinical Evaluation of Photoacoustic Imaging as a Novel Noninvasive Approach to Detect an Orthopaedic Implant Infection. *J. Am. Acad. Orthop. Surg.* **2017**, *25*, S7–S12. [CrossRef]

49. Maji, S.K.; Sreejith, S.; Joseph, J.; Lin, M.; He, T.; Tong, Y.; Sun, H.; Yu, S.W.-K.; Zhao, Y. Upconversion Nanoparticles as a Contrast Agent for Photoacoustic Imaging in Live Mice. *Adv. Mater.* **2014**, *26*, 5633–5638. [CrossRef] [PubMed]

50. Yin, W.; Tian, G.; Ren, W.; Yan, L.; Jin, S.; Gu, Z.; Zhou, L.; Li, J.; Zhao, Y. Design of multifunctional alkali ion doped CaF2 upconversion nanoparticles for simultaneous bioimaging and therapy. *Dalton Trans.* **2014**, *43*, 3861–3870. [CrossRef]

51. Lin, W.; Zhang, X.; Qian, L.; Yao, N.; Pan, Y.; Zhang, L. Doxorubicin-Loaded Unimolecular Micelle-Stabilized Gold Nanoparticles as a Theranostic Nanoplatform for Tumor-Targeted Chemotherapy and Computed Tomography Imaging. *Biomacromolecules* **2017**, *18*, 3869–3880. [CrossRef]

52. Zhang, C.; Wang, S.-B.; Chen, Z.-X.; Fan, J.-X.; Zhong, Z.; Zhang, X.-Z. A tungsten nitride-based degradable nanoplatform for dual-modal image-guided combinatorial chemo-photothermal therapy of tumors. *Nanoscale* **2019**, *11*, 2027–2036. [CrossRef]

53. Zhou, J.; Lu, Z.; Shan, G.; Wang, S.; Liao, Y. Gadolinium complex and phosphorescent probe-modified NaDyF4 nanorods for T1- and T2-weighted MRI/CT/phosphorescence multimodality imaging. *Biomaterials* **2014**, *35*, 368–377. [CrossRef] [PubMed]

54. Champagne, P.-L.; Barbot, C.; Zhang, P.; Han, X.; Gaamoussi, I.; Hubert-Roux, M.; Bertolesi, G.E.; Gouhier, G.; Ling, C.-C. Synthesis and Unprecedented Complexation Properties of β-Cyclodextrin-Based Ligand for Lanthanide Ions. *Inorg. Chem.* **2018**, *57*, 8964–8977. [CrossRef] [PubMed]

55. Kotková, Z.; Kotek, J.; Jirák, D.; Jendelová, P.; Herynek, V.; Berková, Z.; Hermann, P.; Lukeš, I. Cyclodextrin-Based Bimodal Fluorescence/MRI Contrast Agents: An Efficient Approach to Cellular Imaging. *Chem. A Eur. J.* **2010**, *16*, 10094–10102. [CrossRef] [PubMed]

56. Kotková, Z.; Helm, L.; Kotek, J.; Hermann, P.; Lukeš, I. Gadolinium complexes of monophosphinic acid DOTA derivatives conjugated to cyclodextrin scaffolds: Efficient MRI contrast agents for higher magnetic fields. *Dalton Trans.* **2012**, *41*, 13509–13519. [CrossRef]

57. Freed, J.H. Dynamic effects of pair correlation functions on spin relaxation by translational diffusion in liquids. II. Finite jumps and independent T1 processes. *J. Chem. Phys.* **1978**, *68*, 4034–4037. [CrossRef]

58. Aime, S.; Botta, M.; Frullano, L.; Crich, S.G.; Giovenzana, G.B.; Pagliarin, R.; Palmisano, G.; Sisti, M. Contrast Agents for Magnetic Resonance Imaging: A Novel Route to Enhanced Relaxivities Based on the Interaction of a Gd III Chelate with Poly-b-cyclodextrins. *Chem. Eur. J.* **1999**, *5*, 1254–1260. [CrossRef]

59. Aime, S.; Botta, M.; Fedeli, F.; Gianolio, E.; Terreno, E.; Anelli, P. High-Relaxivity Contrast Agents for Magnetic Resonance Imaging Based on Multisite Interactions between a β-Cyclodextrin Oligomer and Suitably Functionalized Gd III Chelates. *Chem. Eur. J.* **2001**, *7*, 5262–5269. [CrossRef]

60. Battistini, E.; Gianolio, E.; Gref, R.; Couvreur, P.; Füzerová, S.; Othman, M.; Aime, S.; Badet, B.; Durand, P.; Patrick, C. High-Relaxivity Magnetic Resonance Imaging (MRI) Contrast Agent Based on Supramolecular Assembly between a Gadolinium Chelate, a Modified Dextran, and Poly-β-Cyclodextrin. *Chem. A Eur. J.* **2008**, *14*, 4551–4561. [CrossRef]

61. Martinelli, J.; Fekete, M.; Tei, L.; Botta, M. Cleavable β-cyclodextrin nanocapsules incorporating GdIII-chelates as bioresponsive MRI probes. *Chem. Commun.* **2011**, *47*, 3144–3146. [CrossRef]

62. Lahrech, H.; Perles-Barbacaru, A.-T.; Aous, S.; Le Bas, J.-F.; Debouzy, J.-C.; Gadelle, A.; Fries, P.H. Cerebral Blood Volume Quantification in a C6 Tumor Model Using Gadolinium per (3,6-Anhydro) α-Cyclodextrin as a New Magnetic Resonance Imaging Preclinical Contrast Agent. *J. Cereb. Blood Flow Metab.* **2008**, *28*, 1017–1029. [CrossRef]

63. Sun, M.; Zhang, H.-Y.; Liu, B.-W.; Liu, Y. Construction of a Supramolecular Polymer by Bridged Bis(permethyl-β-cyclodextrin)s with Porphyrins and Its Highly Efficient Magnetic Resonance Imaging. *Macromolecules* **2013**, *46*, 4268–4275. [CrossRef]

64. Sun, M.; Zhang, H.; Hu, X.; Liu, B.; Liu, Y. Hyperbranched Supramolecular Polymer of Tris(permethyl-β-cyclodextrin)s with Porphyrins: Characterization and Magnetic Resonance Imaging. *Chin. J. Chem.* **2014**, *32*, 771–776. [CrossRef]

65. Gomori, J.M.; I Grossman, R.; Yu-Ip, C.; Asakura, T. NMR relaxation times of blood: Dependence on field strength, oxidation state, and cell integrity. *J. Comput. Assist. Tomogr.* **1987**, *11*, 684–690. [CrossRef] [PubMed]

66. Peters, A.M.; Brookes, M.J.; Hoogenraad, F.G.; Gowland, P.A.; Francis, S.T.; Morris, P.G.; Bowtell, R. T2* measurements in human brain at 1.5, 3 and 7 T. *Magn. Reson. Imaging* **2007**, *25*, 748–753. [CrossRef]

67. Zhou, Z.; Han, Z.; Lu, Z.-R. A targeted nanoglobular contrast agent from host–guest self-assembly for MR cancer molecular imaging. *Biomaterials* **2016**, *85*, 168–179. [CrossRef]

68. Liu, T.; Li, X.; Qian, Y.; Hu, X.; Liu, S. Multifunctional pH-Disintegrable micellar nanoparticles of asymmetrically functionalized β-cyclodextrin-Based star copolymer covalently conjugated with doxorubicin and DOTA-Gd moieties. *Biomaterials* **2012**, *33*, 2521–2531. [CrossRef]
69. Skinner, P.J.; Beeby, A.; Dickins, R.S.; Parker, D.; Aime, S.; Botta, M. Conjugates of cyclodextrins with charged and neutral macrocyclic europium, terbium and gadolinium complexes: Sensitised luminescence and relaxometric investigations and an example of supramolecular relaxivity enhancement. *J. Chem. Soc. Perkin Trans. 2* **2000**, 1329–1338. [CrossRef]
70. Fredy, J.W.; Scelle, J.; Guenet, A.; Morel, E.; De Beaumais, S.A.; Ménand, M.; Marvaud, V.; Bonnet, C.S.; Tóth, É.; Sollogoub, M.; et al. Cyclodextrin Polyrotaxanes as a Highly Modular Platform for the Development of Imaging Agents. *Chem. A Eur. J.* **2014**, *20*, 10915–10920. [CrossRef]
71. Su, H.; Liu, Y.; Wang, D.; Wu, C.; Xia, C.; Gong, Q.; Song, B.; Ai, H. Amphiphilic starlike dextran wrapped superparamagnetic iron oxide nanoparticle clsuters as effective magnetic resonance imaging probes. *Biomaterials* **2013**, *34*, 1193–1203. [CrossRef]
72. Orcujeni, M.; Kaboudin, B.; Xia, W.; Jönsson, P.; Ossipov, D.A. Conjugation of cyclodextrin to magnetic Fe3O4 nanoparticles via polydopamine coating for drug delivery. *Prog. Org. Coat.* **2018**, *114*, 154–161. [CrossRef]
73. Zhou, Z.; Mondjinou, Y.; Hyun, S.-H.; Kulkarni, A.; Lu, Z.-R.; Thompson, D.H. Gd3+-1,4,7,10-Tetraazacyclododecane-1,4,7-triacetic-2-hydroxypropyl-β-cyclodextrin/Pluronic Polyrotaxane as a Long Circulating High Relaxivity MRI Contrast Agent. *ACS Appl. Mater. Interfaces* **2015**, *7*, 22272–22276. [CrossRef] [PubMed]
74. Zhou, X.; Ye, M.; Han, Y.; Tang, J.; Qian, Y.; Hu, H.; Shen, Y. Enhancing MRI of liver metastases with a zwitterionized biodegradable dendritic contrast agent. *Biomater. Sci.* **2017**, *5*, 1588–1595. [CrossRef] [PubMed]
75. Zhang, Q.; Lu, Y.; Xu, X.; Li, S.; Du, Y.; Yu, R.-S. MR molecular imaging of HCC employing a regulated ferritin gene carried by a modified polycation vector. *Int. J. Nanomed.* **2019**, *14*, 3189–3201. [CrossRef] [PubMed]
76. Mortezazadeh, T.; Gholibegloo, E.; Alam, N.R.; Dehghani, S.; Haghgoo, S.; Ghanaati, H.; Khoobi, M. Gadolinium (III) oxide nanoparticles coated with folic acid-functionalized poly(β-cyclodextrin-co-pentetic acid) as a biocompatible targeted nano-contrast agent for cancer diagnostic: In vitro and in vivo studies. *MAGMA Magn. Reson. Mater. Phys. Biol. Med.* **2019**, *32*, 487–500. [CrossRef]
77. Han, Y.; Zhou, X.; Qian, Y.; Hu, H.; Zhou, Z.; Liu, X.; Tang, J.; Shen, Y. Hypoxia-targeting dendritic MRI contrast agent based on internally hydroxy dendrimer for tumor imaging. *Biomaterials* **2019**, *213*, 119195. [CrossRef]
78. Hane, F.T.; Robinson, M.; Lee, B.Y.; Bai, O.; Leonenko, Z.; Albert, M.S. Recent Progress in Alzheimer's Disease Research, Part 3: Diagnosis and Treatment. *J. Alzheimers Dis.* **2017**, *57*, 645–665. [CrossRef]
79. Schröder, L.; Lowery, T.J.; Hilty, C.; Wemmer, D.E.; Pines, A. Molecular Imaging Using a Targeted Magnetic Resonance Hyperpolarized Biosensor. *Science* **2006**, *314*, 446–449. [CrossRef]
80. Hane, F.T.; Li, T.; Smylie, P.; Pellizzari, R.M.; Plata, J.A.; DeBoef, B.; Albert, M.S. In vivo detection of cucurbit[6]uril, a hyperpolarized xenon contrast agent for a xenon magnetic resonance imaging biosensor. *Sci. Rep.* **2017**, *7*, 41027. [CrossRef]
81. Albert, M.S.; Cates, G.D.; Driehuys, B.; Happer, W.; Saam, B.; Springer, C.S.; Wishnia, A. Biological magnetic resonance imaging using laser-polarized 129Xe. *Nat. Cell Biol.* **1994**, *370*, 199–201. [CrossRef]
82. Norquay, G.; Collier, G.J.; Rao, M.; Stewart, N.J.; Wild, J.M. Xe129 -Rb Spin-Exchange Optical Pumping with High Photon Efficiency. *Phys. Rev. Lett.* **2018**, *121*, 153201. [CrossRef]
83. Nikolaou, P.; Coffey, A.M.; Walkup, L.L.; Gust, B.M.; Whiting, N.; Newton, H.; Barcus, S.; Muradyan, I.; Dabaghyan, M.; Moroz, G.D.; et al. Near-unity nuclear polarization with an open-source 129Xe hyperpolarizer for NMR and MRI. *Proc. Natl. Acad. Sci. USA* **2013**, *110*, 14150–14155. [CrossRef] [PubMed]
84. Comment, A.; Jannin, S.; Hyacinthe, J.-N.; Miéville, P.; Sarkar, R.; Ahuja, P.; Vasos, P.R.; Montet, X.; Lazeyras, F.; Vallee, J.-P.; et al. Hyperpolarizing Gases via Dynamic Nuclear Polarization and Sublimation. *Phys. Rev. Lett.* **2010**, *105*, 1–4. [CrossRef] [PubMed]
85. Kurhanewicz, J.; Vigneron, D.B.; Ardenkjaer-Larsen, J.H.; Bankson, J.A.; Brindle, K.; Cunningham, C.H.; Gallagher, F.A.; Keshari, K.R.; Kjaer, A.; Laustsen, C.; et al. Hyperpolarized 13C MRI: Path to Clinical Translation in Oncology. *Neoplasia* **2019**, *21*, 1–16. [CrossRef] [PubMed]
86. Keshari, K.R.; Kurhanewicz, J.; Macdonald, J.M.; Wilson, D.M. Generating contrast in hyperpolarized 13C MRI using ligand–receptor interactions. *Analyst* **2012**, *137*, 3427–3429. [CrossRef] [PubMed]

87. Caracciolo, F.; Carretta, P.; Filibian, M.; Melone, L. Dynamic Nuclear Polarization of β-Cyclodextrin Macromolecules. *J. Phys. Chem. B* **2017**, *121*, 2584–2593. [CrossRef] [PubMed]

88. Caracciolo, F.; Paioni, A.L.; Filibian, M.; Melone, L.; Carretta, P. Proton and Carbon-13 Dynamic Nuclear Polarization of Methylated β-Cyclodextrins. *J. Phys. Chem. B* **2018**, *122*, 1836–1845. [CrossRef]

89. Wang, Y.; Dmochowski, I.J. An Expanded Palette of Xenon-129 NMR Biosensors. *Acc. Chem. Res.* **2016**, *49*, 2179–2187. [CrossRef]

90. Shapiro, M.G.; Ramirez, R.M.; Sperling, L.J.; Sun, G.; Sun, J.; Pines, A.; Schaffer, D.V.; Bajaj, V.S. Genetically encoded reporters for hyperpolarized xenon magnetic resonance imaging. *Nat. Chem.* **2014**, *6*, 629–634. [CrossRef]

91. Song, Y.-Q.; Goodson, A.P.B.M.; Taylor, R.E.; Laws, D.D.; Navon, G.; Pines, A. Selective Enhancement of NMR Signals forα-Cyclodextrin with Laser-Polarized Xenon. *Angew. Chem. Int. Ed.* **1997**, *36*, 2368–2370. [CrossRef]

92. Karas, S. The Synthesis of Rotaxane Probes for Magnetic Resonance Imaging (MRI). Master's Thesis, University of Rhode Island, Kingston, RI, USA, 2016.

93. Hane, F.T.; Smylie, P.S.; Julia, R.; Ruberto, J.; Dowhos, K.; Ball, I.; Tomanek, B.; DeBoef, B.; Albert, M.S. HyperCEST detection of cucurbit[6]uril in whole blood using an ultrashort saturation Pre-pulse train. *Contrast Media Mol. Imaging* **2016**, *11*, 285–290. [CrossRef]

94. Finbloom, J.A.; Slack, C.C.; Bruns, C.J.; Jeong, K.; Wemmer, D.E.; Pines, A.; Francis, M.B. Rotaxane-mediated suppression and activation of cucurbit[6]uril for molecular detection by 129Xe hyperCEST NMR. *Chem. Commun.* **2016**, *52*, 3119–3122. [CrossRef] [PubMed]

95. Hane, F.T.; Fernando, A.; Prete, B.R.J.; Peloquin, B.; Karas, S.; Chaudhuri, S.; Chahal, S.; Shepelytskyi, Y.; Wade, A.; Li, T.; et al. Cyclodextrin-Based Pseudorotaxanes: Easily Conjugatable Scaffolds for Synthesizing Hyperpolarized Xenon-129 Magnetic Resonance Imaging Agents. *ACS Omega* **2018**, *3*, 677–681. [CrossRef] [PubMed]

96. Cavalieri, F.; El Hamassi, A.; Chiessi, E.; Paradossi, G.; Villa, R.; Zaffaroni, N. Tethering Functional Ligands onto Shell of Ultrasound Active Polymeric Microbubbles. *Biomacromolecules* **2006**, *7*, 604–611. [CrossRef] [PubMed]

97. Yeo, S.K.; Shepelytskyi, Y.; Grynko, V.; Albert, M.S. Molecular Imaging of Fluorinated Probes for Tau Protein and Amyloid-β Detection. *Molecules* **2020**, *25*, 3413. [CrossRef] [PubMed]

98. Schluep, T.; Hwang, J.; Hildebrandt, I.J.; Czernin, J.; Hang, C.; Choi, J.; Alabi, C.A.; Mack, B.C.; Davis, M.E. Pharmacokinetics and tumor dynamics of the nanoparticle IT-101 from PET imaging and tumor histological measurements. *Proc. Natl. Acad. Sci. USA* **2009**, *106*, 11394–11399. [CrossRef]

99. Hou, S.; Choi, J.-S.; Garcia, M.A.; Xing, Y.; Chen, K.-J.; Chen, Y.-M.; Jiang, Z.K.; Ro, T.; Wu, L.; Stout, D.B.; et al. Pretargeted Positron Emission Tomography Imaging That Employs Supramolecular Nanoparticles with in Vivo Bioorthogonal Chemistry. *ACS Nano* **2016**, *10*, 1417–1424. [CrossRef]

100. Yaméogo, J.B.G.; Géze, A.; Choisnard, L.; Putaux, J.-L.; Semdé, R.; Wouessidjewe, D. Progress in developing amphiphilic cyclodextrin-based nanodevices for drug delivery. *Curr. Top. Med. Chem.* **2014**, *14*, 526–541. [CrossRef] .

101. Yaméogo, J.B.; Gèze, A.; Choisnard, L.; Putaux, J.-L.; Mazet, R.; Passirani, C.; Keramidas, M.; Coll, J.-L.; Lautram, N.; Bejaud, J.; et al. Self-assembled biotransesterified cyclodextrins as potential Artemisinin nanocarriers. II: In vitro behavior toward the immune system and in vivo biodistribution assessment of unloaded nanoparticles. *Eur. J. Pharm. Biopharm.* **2014**, *88*, 683–694. [CrossRef]

102. Yu, G.; Yang, Z.; Fu, X.; Yung, B.C.; Yang, J.; Mao, Z.; Shao, L.; Hua, B.; Liu, Y.; Zhang, F.; et al. Polyrotaxane-based supramolecular theranostics. *Nat. Commun.* **2018**, *9*, 1–13. [CrossRef]

103. Sauer, R.-S.; Rittner, H.L.; Roewer, N.; Sohajda, T.; Shityakov, S.; Brack, A.; Broscheit, J.-A. A Novel Approach for the Control of Inflammatory Pain: Prostaglandin E2 Complexation by Randomly Methylated β-Cyclodextrins. *Anesth. Analg.* **2017**, *124*, 675–685. [CrossRef]

104. Zhu, C.; Ma, X.; Ma, D.; Zhang, T.; Gu, N. Crosslinked Dextran Gel Microspheres with Computed Tomography Angiography and Drug Release Function. *J. Nanosci. Nanotechnol.* **2018**, *18*, 2931–2937. [CrossRef]

105. Lin, W.; Yao, N.; Qian, L.; Zhang, X.; Chen, Q.; Wang, J.; Zhang, L.-J. pH-responsive unimolecular micelle-gold nanoparticles-drug nanohybrid system for cancer theranostics. *Acta Biomater.* **2017**, *58*, 455–465. [CrossRef] [PubMed]

106. Lin, W.; Yang, C.; Xue, Z.; Huang, Y.; Luo, H.; Zu, X.; Zhang, L.; Yi, G. Controlled construction of gold nanoparticles in situ from β-cyclodextrin based unimolecular micelles for in vitro computed tomography imaging. *J. Colloid Interface Sci.* **2018**, *528*, 135–144. [CrossRef] [PubMed]
107. Lee, N.; Choi, S.H.; Hyeon, T. Nano-Sized CT Contrast Agents. *Adv. Mater.* **2013**, *25*, 2641–2660. [CrossRef]
108. Xi, D.; Dong, S.; Meng, X.; Lu, Q.; Meng, L.; Ye, J. Gold nanoparticles as computerized tomography (CT) contrast agents. *RSC Adv.* **2012**, *2*, 12515–12524. [CrossRef]
109. Shilo, M.; Reuveni, T.; Motiei, M.; Popovtzer, R. Nanoparticles as computed tomography contrast agents: Current status and future perspectives. *Nanomedicine* **2012**, *7*, 257–269. [CrossRef] [PubMed]
110. Gu, J.; Bednarz, B.; Caracappa, P.F.; Xu, X.G. The development, validation and application of a multi-detector CT (MDCT) scanner model for assessing organ doses to the pregnant patient and the fetus using Monte Carlo simulations. *Phys. Med. Biol.* **2009**, *54*, 2699–2717. [CrossRef] [PubMed]
111. Semeniuk, O.; Grynko, O.; DeCrescenzo, G.; Juska, G.; Wang, K.; Reznik, A. Characterization of polycrystalline lead oxide for application in direct conversion X-ray detectors. *Sci. Rep.* **2017**, *7*, 1–10. [CrossRef]
112. Wolterink, J.M.; Leiner, T.; Viergever, M.A.; Isgum, I. Generative Adversarial Networks for Noise Reduction in Low-Dose CT. *IEEE Trans. Med. Imaging* **2017**, *36*, 2536–2545. [CrossRef]
113. Grynko, O.; Juska, G.; Reznik, A. An Engineering of multilayered lead oxide photoconductor for lag-free X-ray digital detector. In Proceedings of the IEEE Nuclear Science Symposium and Medical Imaging Conference (NSS/MIC), Manchester, UK, 26 October–2 November 2019. [CrossRef]
114. Patino, M.; Prochowski, A.; Agrawal, M.D.; Simeone, F.J.; Gupta, R.; Hahn, P.F.; Sahani, D.V. Material Separation Using Dual-Energy CT: Current and Emerging Applications. *Radiographics* **2016**, *36*, 1087–1105. [CrossRef] [PubMed]
115. Albrecht, M.H.; Vogl, T.J.; Martin, S.S.; Nance, J.W.; Duguay, T.M.; Wichmann, J.L.; De Cecco, C.N.; Varga-Szemes, A.; Van Assen, M.; Tesche, C.; et al. Review of Clinical Applications for Virtual Monoenergetic Dual-Energy CT. *Radiology* **2019**, *293*, 260–271. [CrossRef] [PubMed]

Publisher's Note: MDPI stays neutral with regard to jurisdictional claims in published maps and institutional affiliations.

Review

The Role of β-Cyclodextrin in the Textile Industry—Review

Fabricio Maestá Bezerra [1,*], **Manuel José Lis** [2,*], **Helen Beraldo Firmino** [3],
Joyce Gabriella Dias da Silva [4], **Rita de Cassia Siqueira Curto Valle** [5],
José Alexandre Borges Valle [5], **Fabio Alexandre Pereira Scacchetti** [1] and **André Luiz Tessaro** [6]

[1] Textile Engineering (COENT), Universidade Tecnológica Federal do Paraná (UTFPR),
    Apucarana 86812-460, Paraná, Brazil; fabiosachetti@utfpr.edu.br
[2] INTEXTER-UPC, Terrassa, 0822 Barcelona, Spain
[3] Postgraduate Program in Materials Science & Engineering (PPGCEM), Universidade Tecnológica Federal do
    Paraná (UTFPR), Apucarana 86812-460, Paraná, Brazil; helenfirmino@yahoo.com
[4] Postgraduate Program in Environmental Engineering (PPGEA), Universidade Tecnológica Federal do
    Paraná (UTFPR), Apucarana 86812-460, Paraná, Brazil; g.joycedias@gmail.com
[5] Department of Textile Engineering, Universidade Federal de Santa Catarina (UFSC),
    Blumenau 89036-002, Santa Catarina, Brazil; rita.valle@ufsc.br (R.d.C.S.C.V.);
    alexandre.valle@ufsc.br (J.A.B.V.)
[6] Chemistry graduation (COLIQ), Universidade Tecnológica Federal do Paraná (UTFPR),
    Apucarana 86812-460, Paraná, Brazil; andretessaro@utfpr.edu.br
*   Correspondence: fabriciom@utfpr.edu.br (F.M.B.); manuel-jose.lis@upc.edu (M.J.L.)

Academic Editors: Marina Isidori, Margherita Lavorgna and Rosa Iacovino
Received: 30 June 2020; Accepted: 8 August 2020; Published: 9 August 2020

Abstract: β-Cyclodextrin (β-CD) is an oligosaccharide composed of seven units of D-(+)-glucopyranose joined by α-1,4 bonds, which is obtained from starch. Its singular trunk conical shape organization, with a well-defined cavity, provides an adequate environment for several types of molecules to be included. Complexation changes the properties of the guest molecules and can increase their stability and bioavailability, protecting against degradation, and reducing their volatility. Thanks to its versatility, biocompatibility, and biodegradability, β-CD is widespread in many research and industrial applications. In this review, we summarize the role of β-CD and its derivatives in the textile industry. First, we present some general physicochemical characteristics, followed by its application in the areas of dyeing, finishing, and wastewater treatment. The review covers the role of β-CD as an auxiliary agent in dyeing, and as a matrix for dye adsorption until chemical modifications are applied as a finishing agent. Finally, new perspectives about its use in textiles, such as in smart materials for microbial control, are presented.

Keywords: cyclodextrin; dyeing; textile finishing; textile wastewater

---

1. Introduction

Since the first publication on cyclodextrins (CDs) in 1891, and the first patent in 1953, many technological advances have occurred, and the application of CDs has expanded [1]. According to Szejtli [2], over the years, CDs have been used in many diverse areas, and are identified, among all the receptor molecules, as the most important.

This scenario is no different in the textile sector, which constantly seeks technological innovation, especially in the dyeing, finishing, and water treatment sectors. With the market and consumers increasingly demanding environmental improvements, the development of new features combined with green processes has become a constant challenge [3]. Among the various materials that can be used for this purpose CDs stand out; they are oligosaccharides made up of glucose units that are organized in a conical trunk shape, providing a well-defined cavity for the formation of host–guest complexes with a series of molecules [4]. This versatility allows complexation with drugs, dyes, insecticides, essential oils, cosmetics, and other compounds [5–9], allowing this class of molecules to assume a leading role in the textile industry.

For the period from 1948 until today, since the term cyclodextrin started to be used as a research topic, 46,989 research papers have been reported by SCOPUS, and this number is continually increasing, as shown in Figure 1. This growth became significant in 1996, when the terms cyclodextrin and textile were combined and used as a research topic. These data were downloaded on 6 June 2020.

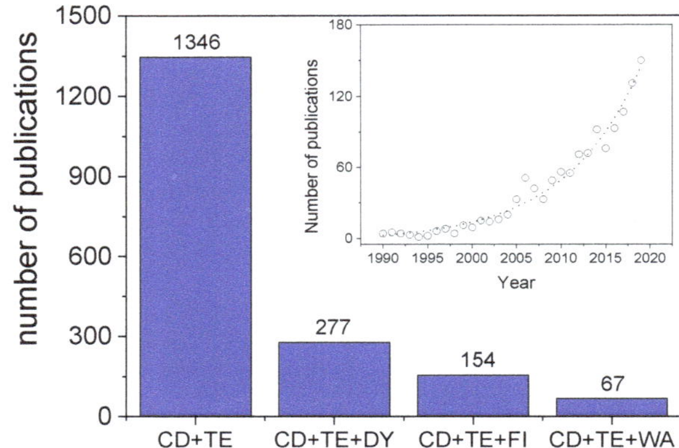

Figure 1. Number of publications available from SCOPUS when cyclodextrin (CD); textile (TE); dyeing (DY); textile finishing (FI); and textile wastewater (WA) are selected as keywords.

Furthermore, due to the presence of numerous hydroxyl groups either in the interior or exterior, CDs are susceptible to the addition of new functional groups, which may yield new properties and functionalities. Additionally, CDs have a set of outstanding characteristics, such as high biodegradability, high biocompatibility, and approval by Food and Drug and Administration (FDA), which makes them human and environmental-friendly [10]. Therefore, this review presents an overview of the use of cyclodextrins, especially beta CD and its derivatives, in the textile field. Although some general physicochemical characteristics are presented, the scope of the work is focused on the application of CDs in the areas of dyeing, finishing, and wastewater treatment.

2. General Characteristics of Cyclodextrins

Initially known as Schardinger dextrins [11], the widespread use of CDs as hosts in supramolecular chemistry is relatively recent. Because they are natural products, CDs are biocompatible and accepted in biological applications; therefore, there is a growing interest in them both scientifically and industrially [12]. The optimization of methods for obtaining and applying CDs is, as a result, constantly evolving [7].

CDs are obtained through the enzymatic degradation of potatoes, corn and rice starch, which gives a mixture of linear, branched, or cyclic dextrins [13]. Initially, the cyclization reaction of the starch glucopyranose linear chains occurs by the enzyme cyclomaltodextrin-glucanotransferase (CGTase) [14], produced for example by *Bacilus firmus*. This step results in a mixture of α-CD, β-CD and γ-CD, composed of six, seven and eight units of D-(+)-glucopyranose, respectively, joined by α-1,4 bonds [15].

Subsequently, the separation and purification of these three CDs are required [5,16]. Among the methods used for this purpose, the most simple and widely used to isolate α-, β- and γ-CD is selective precipitation, forming inclusion complexes with an appropriate guest molecule—for example, α, β and γ-CD crystallize with 1-decanol, toluene, and cyclohexadec-8-en-1-one, respectively [7]. However, separation has a relatively high cost, making the entire synthesis process expensive. Fortunately, over time, research intertwined with the production of CGTase has evolved and allowed the isolation of α, β and γ-CGTase, increasing yield and consequently decreasing the production costs of the CDs [7].

As a structural consequence of the glucose units connecting through α-1,4 bonds, CDs occur as conical trunk shaped structures which are capable of solubilizing and encapsulating hydrophobic molecules in an aqueous environment [4,17,18].

The structure of CDs consists of primary hydroxyl groups (C6) located at the end of the rings, and secondary hydroxyls (C2 and C3) located at the outer edge of the rings. Ether type oxygen and polar hydrogen groups (C3 and C5) are present inside the trunk of the CDs. While the external hydroxyls are responsible for the relative solubility of CDs in water and micro-heterogeneous environments, the glycosidic oxygen bridges and, consequently, their pairs of non-binding electrons facing the interior of the cavity give this region, in addition to its Lewis basic character, hydrophobicity [18,19], making it capable of complexing nonpolar molecules [20]. The chirality caused by the five chiral carbons of the D-glucose unit associated with the rigidity of the macrocycle due to the intramolecular hydrogen interactions between the 2- and 3-hydroxyl groups are fundamental characteristics of the chemistry of CDs. Table 1 lists the physical properties of natural CDs.

Table 1. Some physicochemical properties of cyclodextrins [2,21].

Properties	α-CD	β-CD	γ-CD
Empirical formula	$C_{36}H_{60}O_{30}$	$C_{42}H_{70}O_{35}$	$C_{48}H_{80}O_{40}$
Molecular weight (g/mol)	972	1135	1297
Glucopyranose units	6	7	8
Cavity diameter (nm)	0.47–0.57	0.60–0.78	0.83–0.95
Internal cavity volume (nm^3)	1740	2620	4720
Number of water molecules in the cavity	6	11	17
Aqueous solubility (g/L)	129.5	18.4	249.2
Temperature of degradation (°C)	278	298	267

The data presented in Table 1 indicate an apparent regularity in some properties, however, irregularities have been observed regarding the degradation temperature and solubility. Szejtli [2] has suggested that the lower solubility of β-CD is associated with intramolecular hydrogen bonds occurring at the edge. Although it has the lowest solubility, β-CD and its derivatives are the most used due to factors such as simplicity in obtaining it, lower price, reduced sensitivity and irritability to skin, and the absence of mutagenic effects [22].

The limitations imposed by the reduced solubility combined with the expressive attractiveness cause CD derivatives to be synthesized industrially. The CD derivatives that are most industrially produced include methylated β-CD, heptakis (2,6-dimethyl)-β-CD, heptakis (2,3,6-trimethyl)-β-CD, hydroxypropyl-β-CD, peracetylated β-CD, sulfobutyl ether-CD, and sulfated CD [10,23,24]. All have greater solubility in water compared to natural CDs, expanding the spectrum of applications in the controlled release of drugs, increasing blood solubility and bioavailability of medicines and textile deodorants, and assisting in polymerization [5,21,25]. Studies on their toxicology, mutagenicity, teratogenicity, and carcinogenicity have been carried out, and have shown negative results [7,26,27]. CDs also have hemolytic activity in vitro; β-CD has the highest and γ-CD the lowest activity [27,28].

The industrial applications of CDs are very diverse; they have been used in the pharmaceutical industry, in agriculture, in the textile area, in food technology, in chemical and biological analysis, in tenvironmental protection, and in cosmetics [6,9,29,30].

CDs play a significant role in the textile industry, as they can be used to remove surfactants from washed textile materials [31], as leveling agents in dyeing [32–34], in textile finishing [35–39], and in wastewater treatment [40–43].

Host-Guest Complex Formation

A striking feature of CDs is that they can form inclusion complexes with a variety of organic or inorganic compounds, allowing for the subsequent controlled release of these active compounds [44,45]. The fundamental factor for the guest molecule to be able to form a complex with the CD (host molecule) is its suitability within the cavity, which can be integral or partial [38,46,47]. Note that in the complexation, no covalent chemical bonds occur between the guest-molecule, nor is the compound closed within the macromolecular structure, which makes this type of complexation unique in terms of behavior as a modeler for the release of compounds [21,37].

Thus, the appropriate choice of the CD to be used for the possible formation of a complex is of great value. For small molecules, it is easier to form stable complexes with α-CD and β-CD, due to the compatibility of the volume of the guest molecule and the size of the CD cavity (Table 1). In the case of γ-CD, if the guest molecule is too small, the fit becomes unfavorable due to the much larger size of the cavity [19,48].

The general trend of CD complexation thermodynamics can be understood based on the concept of size; that is, by the analysis of the size and shape of the included molecule, and critical factors in the van der Waals interactions. Therefore, due to the fact that the cavity diameter of α-CD is much smaller than that of β-CD, and because the van der Waals forces are dependent on the distance between the molecules, it is expected that the forces induced by the complexation of extended chain molecules will be greater for α-CD than for β-CD [49].

As long as the fit-size requirements are satisfied, a number of other factors contribute to the complexation thermodynamics of the guest molecule in CDs. Considering only the aqueous environment, the following can be mentioned: (i) the entry of the hydrophobic portion of the guest-molecule into the CD cavity, (ii) the dehydration of the guest molecule and the exclusion of water molecules from the interior of the cavity, (iii) interactions of the hydrogen bonds between specific groups of the guest molecule and the OH of the receptor, and (iv) changes in conformation and/or stress reduction [49]. Although the preference for inclusion is of the hydrophobic portion (i), since charged species and hydrophilic groups are located in the bulk, certain groups with a hydrophilic character, such as phenolic OH, penetrate the cavity [50] and interact (iii) with the receptor.

According to Venturi et al. [48], after complexation in an aqueous environment, the new chemical environment experienced by the guest molecule causes changes in its chemical reactivity. In numerous cases, an increase in stability, reduction in volatility, stabilization against light, heat and oxidation [47,51], solubility of the guest molecule in the solution, increase in the speed of dissolution [52,53], and bioavailability [54,55] were observed. However, depending on the experimental condition and type of CD, the inclusion can be deleterious for the guest, for example, enhancing the chlorpromazine photodegradation as observed by Wang et al. [56].

In terms of the stoichiometry of the inclusion complex, the four most common types of complexes are considered in CDs: guest molecules with a 1:1, 1:2, 2:1 and 2:2 ratio [57]. However, Pinho et al. [10] point out that the most common cases of complexation are 1:1 and 1:2. These configurations are dependent on the size and structural aspect of the guest-molecule in relation to the cavity of the CDs, allowing the formation of stable inclusion complexes [58].

However, Rama et al. [59] highlight that the chemical composition of the guest molecule, as well as its solubility, ionization state, and molecular mass, in addition to the conditions of the medium, such as the pH, temperature, solvent used, and other parameters, influence the process. The choice of the appropriate medium, working temperature, pH, and other factors will determine the best conditions for the interaction between the CD and the guest molecule [60,61]. Voncina et al. [62] highlight that an increase in temperature in the dyeing of polyacrylonitrile with cationic dyes using β-CD improves complexation, which reaffirms the importance of these parameters in the process. Other determining factors are related to the type of cyclodextrin used and the method of preparation: physical mixing [63], kneading [64], atomization [65], lyophilization [66], or coprecipitation [67].

The mechanism of the formation of inclusion complexes can be divided into several steps; an illustration is shown in Figure 2. In the complexation of a substance in aqueous solution, the ends of the isolated CD cavity are opened in such a way that the guest molecule can enter the CD ring from both sides. There is, in principle, the absence of the guest molecule, and the slightly non-polar cavity, which acts as a host, is occupied by water molecules that are energetically unfavorable, as seen in Figure 2a. Given the nature of the polar–non-polar interaction, they can be easily replaced by a guest molecule that is less polar than water [14,68] (Figure 2b).

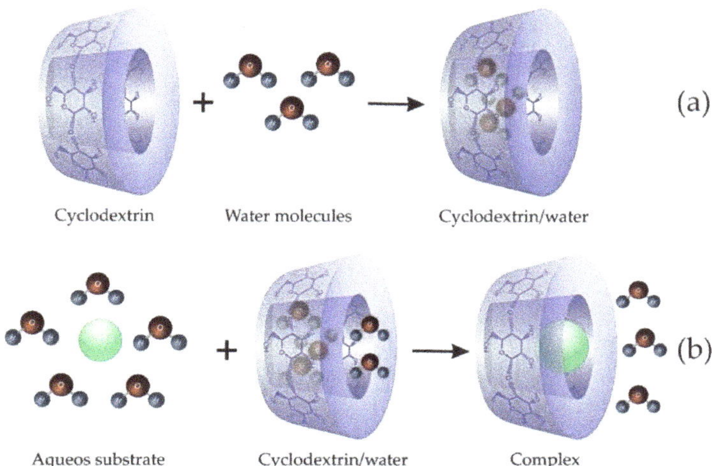

Figure 2. Complexation system. (a) Inclusion of water molecules in the cyclodextrin cavity; (b) complexation mechanism of the guest molecule in aqueous medium.

The molecules interact with each other as they are influenced by forces arising from the characteristics that are specific to each substance. Then, a complex phenomenon of molecular interaction occurs, since each interaction corresponds to a set of distinct forces [48]. Complexation is characterized by the absence of formation and the breaking of covalent bonds [69]. The driving force of the process is the increase in the entropy caused by the exit of water molecules present in the cavity and their consequent freedom [21]. Other forces also contribute to the maintenance of the complex, such as the release of the ring tension (especially for α-cyclodextrin), van der Waals interactions, hydrogen bonds, and changes in the surface tension of the solvent used as a medium for complexation [37,70].

3. Application of Cyclodextrins in the Textile Area

The wide range of applications of CDs has attracted the attention of many industries; however, according to Venturini et al. [48], initiatives for the industrial application of CDs have not been widely considered for three reasons: their scarce quantity and high prices, incomplete toxicological studies, and the fact that the knowledge obtained about CDs was not yet broad enough to envision their use in industry. The 1970s and 1980s were of fundamental importance for the diffusion of CDs in industry. Several studies have been successful in the production of CDs and their derivatives, and reliable tests have reduced doubts about their toxicity [2]. Their introduction into textile-related studies took on increasing relevance from the 1990s, according to SCOPUS data.

Bhaskara-Amrit et al. [31] emphasize that CDs have a very important role in textile processing and innovation; their use provides immediate opportunities for the development of environmentally friendly products and eco-textiles, in addition to having great potential in various applications. Cyclodextrins can be applied in the areas of spinning [71], pretreatment [72], dyeing [62,68,73], finishing [44,74–78], and dye removal [40,79–81], with dyeing, finishing, and water treatment being the most applicable in the textile area registered so far.

3.1. Dyeing Process

The use of cyclodextrins in the dyeing process can include their use as a dyeing aid, forming a complex with the dye [33,82], or as a chemical modification of the surface [83,84]. Figure 3 shows these two processes.

CDs can form a variety of inclusion complexes with textile dyes (Table 2), and this inclusion changes their properties, directly influencing the quality of the dyeing. Therefore, cyclodextrins can be used as auxiliaries in the dyeing process.

Table 2. Examples of dyeing CD applications, as a textile aid and as a chemical modifier.

Application of Cyclodextrins	Fiber	Dye	Reference
Auxiliary agent	Polyester	Disperse	[73]
		Synthetic	[85]
		Disperse Orange 30, Disperse Red 167, Disperse Blue 79	[86]
	Polyamide 6	Methylene Blue	[84]
		Disperse Red 60	[87]
		Disperse	[88]
		Synthetic	[32,89,90]
	Nylon, polyester and cotton	Synthetic, reactive and disperse dye	[91]
	Cellulose Acetate	Azo disperse	[92]
	Polyacrylic	Basic Blue 4	[62]
	Cotton	Direct	[82]
	Wool	Natural (*Bixa orellana*)	[93]
Chemical modification	Polyester	Pigment inks carbon black, magenta, yellow and cyan	[94]
		Disperse Red 60, Disperse Yellow, Disperse Blue 56, Disperse Red 343	[83]
	Cellulose Acetate	Disperse Red 60 and 82	[75]
	Vinylon fibre	Reactive Red 2	[95]
	Cotton	Acid	[96]
	Cotton and cotton/polyester	Basic Red 14, Basic Blue 3, Basic Yellow 24 and 13	[97]
	Polyester/Wool	Disperse Red 54 and 167, Disperse blue 183	[98]
	Polypropylene	Disperse, acid and reactive	[99]

3.1.1. Cyclodextrin as an Auxiliary Agent in Dyeing

The introduction of new auxiliaries in the textile industry is feasible when they are used in low concentrations, are biodegradable, and do not affect the quality of the effluent. In addition to being biodegradable, CDs do not cause problems in textile effluents, and they improve the biodegradability of many toxic organic substances [17,68]. Their use is due to their formation of complexes with dyes, and they can be used to improve the uniformity of dyes and washing processes [31,82,86]. However, for the cyclodextrin to act as an auxiliary, the formation of the CD:dye complex is essential; if it is not formed, there will be no improvement in the quality of the dyeing [75,95].

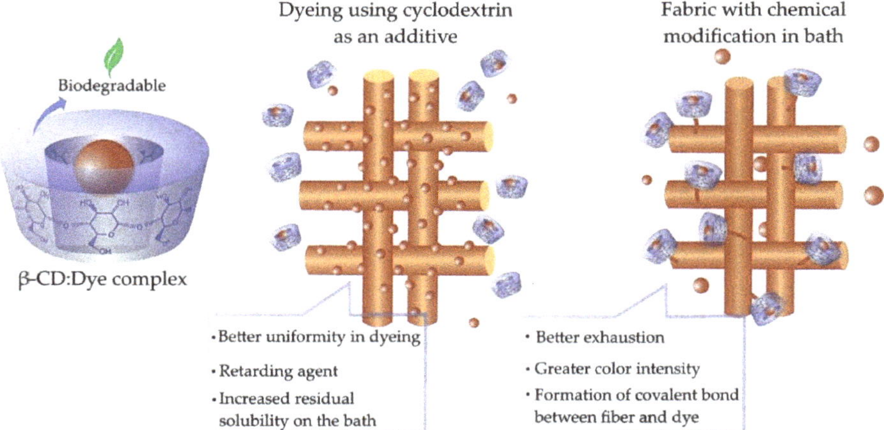

Figure 3. Use of cyclodextrins in dyeing.

In the dyeing of polyester fibers, dyes of the dispersed type are used for the coloring of the substrate [100]. These dyes are poorly soluble in water and generally require the use of surfactants [73], which are products from non-renewable sources that cause environmental problems and must be replaced to reduce damage [85]. Therefore, cyclodextrins are an alternative to these products that can maintain the quality of the coloring of the textile article [68].

Carpignano et al. [73] conducted studies on the application of β-CD with dispersed type dyes and polyester, and stated that the presence of β-CD positively affects color uniformity, intensity, and bath exhaustion when compared to dyeing using commercial surfactants. The insertion of cyclodextrins into the dyeing process decreases the amount of free dye molecules [33], causing the dyeing rate to decrease and favoring leveling [91]. This is due to the fact that the complex (CD:dye) has a molar mass greater than that of free dye, hindering its diffusion into the fiber, thus favoring the dye delay mechanism, which causes better leveling [101].

Another important synthetic fiber in the textile area is polyamide. This fiber presents, at the ends of its chains, carboxylic and amine groups, which gives it a substantivity for several classes of anionic dyes [102]. Commercially, acid dye is the most used due to the dye–fiber interaction in the acid medium, the leveling results, and the achieved colors [103]. Dispersed dye is seldom used due to its low adsorption and the possibility of a barre effect; therefore, in order to be able to use dispersed dyes for the dyeing of polyamide, it is necessary to improve the leveling and adsorption of this dye by the fiber. This can be achieved when using cyclodextrins as an auxiliary agent in dyeing [32].

Ferreira et al. [87] studied the dyeing of a polyamide 6 microfiber using dispersed dye complexed with cyclodextrins, and found that the complex changes the dyeing kinetics, improving its distribution in the fiber. There are also changes related to the thermodynamics of dyeing, since the dyeing also proved to be more intense, with greater adsorption of the dye by the fiber related to the increase in the dispersibility of the dye in the aqueous phase [88]. Similar results were found by Savarino et al. [89] when they dyed polyamide 6 with dispersed dyes, showing changes in the kinetic and thermodynamic phases of the dyeing. This indicates that cyclodextrins can replace additives from non-renewable sources and improve the dyeing and the effluent generated.

With regards to natural fibers, cotton is one of the most important textile fibers [104] and, in its dyeing, the use of reactive and direct dyes stands out. Reactive dye has structure groups that covalently bond with the fiber, improving the solidity; however, they have low affinity, requiring high amounts of electrolytes for good dyeing to occur [105]. Direct dye, on the other hand, has a high affinity for the cotton fiber [106] and it is often necessary to use retarding agents, such as alkaline salts, to prevent stains on the fabric and thus achieve better leveling [107].

The use of cyclodextrins can help to solve these dyeing problems. Parlat et al. [91] dyed cotton with reactive dye using cyclodextrins as an auxiliary. In this case, as a result of the complexation of the reactive dye, there was a good diffusion of the dye into the fiber, increasing its uniformity and color intensity.

In the works of Cireli et al. [82], the insertion of CDs occurred in the process of dyeing cotton with direct dye. The CDs acted as a retarding agent, forming complexes with the dye molecules, causing the dyeing speed to decrease, which improved the leveling.

Other works performed dyeing using β-CD, such as those of Voncina et al. [62], who dyed polyacrylonitrile with cationic dye and observed an improvement in color intensity and exhaustion when compared to the use of quaternary ammonium. Shibusawa et al. [92], who dyed cellulose acetate with dispersed dye, found that the complex formed between CD: dye changed the speed at which the chemical balance of the process was achieved, making it slower.

In general, cyclodextrins inserted as an auxiliary affect both the properties of the dyes and the dyeing kinetics, allowing improvements in exhaustion, uniformity, and in the quality of the effluent water. However, it is worth mentioning that this is only achieved when inclusion of the dye in the cavity of the CD is achieved.

3.1.2. Dyeing Chemical Modification

Some textile fibers present difficulty in dyeing due to the terminal groups present in their chains, causing some dyes to fail to create interactions, as is the case with polypropylene fibers [99] and vinylon fibers [95]. Other fibers present selectivity for dyes, such as cotton, which is not dyed by acidic and dispersed dyes [108]. However, promoting the modification of the surface of these fibers can cause new possibilities for the interactions between the dye and the fibers [109].

Cyclodextrins are polymers that can cause this chemical modification through incorporation into the fiber [99]. This incorporation can be seen as a pre-treatment for the dyeing or as a finishing, depending on the actions taken after modification. In this section, only the modifications for dyeing will be addressed and, in the next, finishing will be explored.

With cyclodextrins incorporated into the fabric, new groups and pores through which the dyes can fix become available. One fiber that presents difficulty in dyeing is cellulose acetate fiber, due to its compact structure, low content of polar groups, and hydrophobicity [110]. These factors make it difficult for dyes to diffuse in the fiber. To obtain better results in the dyeing process, Raslan et al. [75] treated the cellulose acetate fabric (38.5% acetyl) with monochlorotriazinyl-β-cyclodextrin (MCT-β-CD) using the padding technique to improve its dyeability. As a result, they were able to perform dyeing at a low temperature, improving the color intensity, and they also increased the diffusion of the dye within the fiber by about 70%.

In the case of polyester fibers, some authors have performed the process of acetylation [83] or coating [97] to modify the surface with CDs. This results in an improvement in the solidity of the dyeing [98], in addition to the possibility of dyeing with other classes of dyes. Zhang et al. [97], after performing the modification of polyester fiber, dyed this fabric with cationic dye. The fabric showed a gain in hydrophilicity, a reduction in the dyeing temperature to 70 °C, and interaction between the crosslinking carboxylate groups and the cationic dye, in addition to its complexation by the CDs.

Another work that used the modification of the polyester surface with cyclodextrins was carried out by Chen et al. [94]. In this work, the modification enabled a 47% increase in the color intensity in the stamping process, a fact associated with the greater sharpness and depth achieved by the dyes. In addition, the CDs, when chemically bonded to the fabric, can act as an anti-migration agent, because during the drying or curing of polyester fabrics dispersed dyes tend to migrate to the fabric surface and the CDs act as a dye sequestrant, consequently preventing this dyeing defect [83].

In the case of the modification of cotton fiber with cyclodextrins, several routes are possible, but the most used is esterification using citric acid or 1,2,3,4-butane tetra-carboxylic acid (BTCA) [111] as crosslinking. These changes will be covered in more detail in the next section. Rehan et al. [96] carried out the modification of cotton fiber with CDs and citric acid to perform dyeing with acid dye. These dyes present low affinity for the dyeing of cellulosic fiber [108]. After the modification, the authors realized that the dye was adsorbed by the cyclodextrins, which allowed the dyeing to achieve satisfactory solidity.

In general, the modification of the fiber surface through the insertion of cyclodextrins increases the adsorption of dye and allows a greater variability of dye classes in fibers that have no affinity, often achieving better color standards in multi-fiber items [97,98] and improving the efficiency of the dyeing process for fibers that require greater use of auxiliaries to achieve the proper color standard.

3.2. Textile Finishing

In the area of textile finishing, cyclodextrins can have many applications; they are able to absorb unpleasant odors, and act as an encapsulation agent for essential oils [38,76,78,112,113], vitamins [114], hormones [77] and biocides [6,115] in order to preserve compounds and/or control their release, as shown in Figure 4. The loading of active ingredients allows the incorporation of specific and desired functions into textile materials, which may act differently under particular uses, such as in medicine [116], cosmetics [117], and engineering [118].

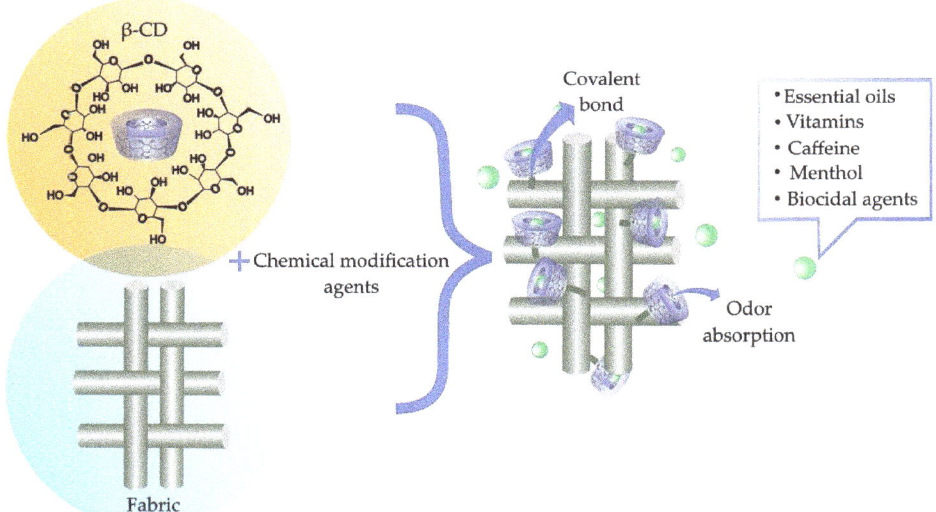

Figure 4. Use of cyclodextrins in textile finishing.

In numerous cases, the complexation of active ingredients by CDs improves their physicochemical properties, controls their release, maintains bioavailability, increases shelf life, provides storage conditions, reduces environmental toxicity, increases chemical stability, protects against oxidation, and favors resistance to repeated washing [6,7,114,119,120].

In order to make it possible to incorporate these active molecules into the textile substrate, there is a need to fix the CDs in the fiber. Several methods have been proposed for the permanent fixation of CDs into textile fibers, and in some cases, there is a need for a first step—the modification of the cyclodextrins—so that they can be incorporated into the fabric. The selection of the best method for fixing CDs into a textile substrate depends on different factors, the main ones being reactivity of the cyclodextrins to the final application, and the type of fiber [23,121].

3.2.1. Preparation of Cyclodextrins

Cyclodextrins are capable of forming complexes with a wide range of molecules, but they cannot form a direct covalent bond with textile materials; therefore, some cyclodextrin derivatives have been synthesized with reactive groups to allow them to chemically bond to various substrates [122], as shown in Figure 5.

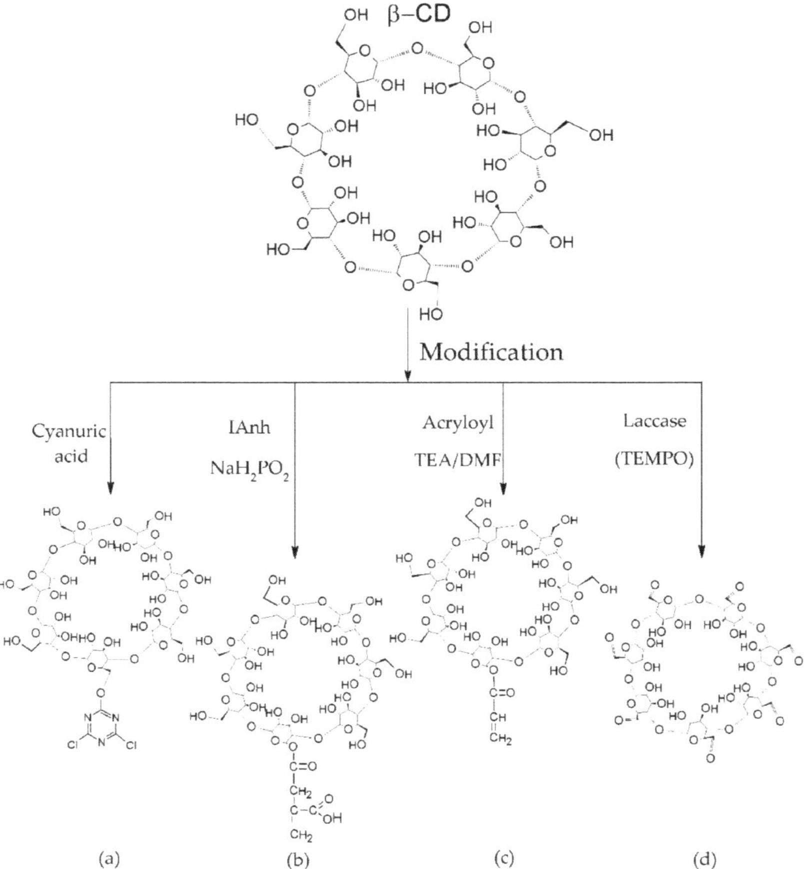

Figure 5. Modification of β-CD by: (**a**) cyanuric acid; (**b**) itaconic anhydride (IAnh); (**c**) acryloyl chloride and (**d**) laccase/2,2,6,6-tetramethylpiperidine-1-oxyl (TEMPO) enzyme.

One of the most common reactive derivatives of cyclodextrins is MCT-β-CD, as seen in Figure 5a, synthesized through the reaction between cyanuric chloride and β-cyclodextrin [123]. MCT-β-CD is the most interesting derivative used on cellulosic substrates due to the simple bonding process in relatively mild conditions. The monochlorotriazine groups incorporated into the CDs react by a nucleophilic substitution mechanism, and form covalent bonds with the hydroxyl groups of the cellulose [124]. Another product that can be synthesized from MCT-β-CD is the cyclodextrin polymer (6^{A}-O-triazine-β-cyclodextrin), produced by polycondensation using β-CD and cyanuric chloride [125].

Formation occurs due to nucleophilic substitution, in which the hydroxyl groups of the CDs react with the chlorine contained in the cyanuric chloride, and thus form the β-CD copolymer [125]. From the formation of this compound it is possible to create interactions with the hydroxyl groups present in the textile fibers; this occurs by substitution.

The modification of CDs can also be performed using itaconic acid (Figure 5b) containing carboxyl and vinyl groups. This bifunctional compound can be linked to the CDs via an esterification reaction, and its vinyl group can perform polymerization by free radicals [5,122]. Itaconic anhydride is obtained from itaconic acid at 110 °C in the presence of sodium hypophosphite [122]. From the modification of the CDs, the end containing the itaconic anhydride is able to bond with the textile fibers through covalent reactions.

Another CD modification for incorporation in textiles can be carried out via a reaction with acryloyl derivative (Figure 5c). The CDs are dissolved in dimethylformamide (DMF), mixed by stirring with triethanolamine (TEA), and reacted with acryloyl chloride dissolved in DMF, forming an acryloyl ester derivative [126]. The compound has a vinyl group on the side chain that is able to react with hydroxyl groups, and can be incorporated into the fibers [23,124].

In addition to the reaction through the incorporation of new chemical groups into the CDs, to make them more reactive hydroxyl groups can be oxidized, as can be seen in Figure 5d. The hydroxyl groups in the polysaccharides can be oxidized by a laccase/2,2,6,6-tetramethylpiperidine-1-oxyl enzyme catalyzed to convert the hydroxyl groups of the CDs into aldehyde groups that are capable of reacting with the amino groups of polyamide, silk, and wool [127].

3.2.2. Grafting of Cyclodextrins onto Textile Substrates

The most common procedure in the application of cyclodextrin into textiles is esterification, which can be done using modified cyclodextrins (Figure 5), or through a reaction using dimethylol urea [128], citric acid [111], BTCA [78,129], or other products.

Esterification can be defined as a nucleophilic substitution reaction of the acyl group catalyzed by a mineral acid, involving a carboxylic acid and an alcohol [130]. From there, a proton transitions from one oxygen to another, resulting in a second tetrahedral intermediate, and converts the -OH group into a leaving group, culminating in the loss of a proton that regenerates the acid catalyst, originating the ester [131].

Figure 6 shows the procedure for incorporating MCT-β-CD into cellulosic fiber. The interaction occurs due to the availability of the chlorine group present in MCT-β-CD and the hydroxyl group of cellulose, thus representing a second order nucleophilic substitution reaction [132].

MCT-β-CD is fixed on cellulosic fibers in alkaline conditions and, due to the covalent bond between the cellulosic chain and MCT-β-CD, the durability of β-CD in textile products is excellent [23,133].

Ibrahim et al. [134] also used MCT-β-CD for the functionalization of wool by a method of fixation in foularding. Due to the presence of -OH groups in the protein, it is also possible to perform nucleophilic substitution. As with polyamide fabrics and polyester/cotton blends, this β-CD derivative has also been grafted, making the fabric antibacterial and a receptor for drugs and essential oils, in addition to improving thermal stability and dyeability [128,130].

The MCT compound was also used to make polyester a functional fabric, made from alkaline hydrolysis, which created reactive hydroxyl groups on the surface of the polyester fibers able to react with MCT-β-CD covalently [39]. From the interaction with the cyclodextrins, the modified polyester can adsorb bioactive molecules [112].

Cyclodextrin compounds treated with itaconic anhydride can bind to cellulosic and polyamide fibers. In the case of cellulosic fibers, the fabric must be treated with a mixture of nitric acid (1%) and cholic ammonium nitrate to generate free radicals and, after drying, the cotton is treated with a derivative of CD itaconate, which is able to covalently bond to cellulosic fibers [5,122].

In addition to the processes using modified cyclodextrins, esterification between cyclodextrins and textile fibers can be achieved. In this case, the esterification reaction requires a crosslinking agent such as citric acid, BTCA, or other polycarboxylic acids [135]. The disadvantage of using citric acid is the yellowing of the cellulosic fabric in the curing phase [136]. This process includes two steps; in the first, a cyclic anhydride is formed between two groups of adjacent carboxylic acids and, in the second, the esterification reaction occurs between the acid anhydrides previously formed and the hydroxyl groups of the macromolecules of the fiber and of the cyclodextrins, to form ester bonds [23].

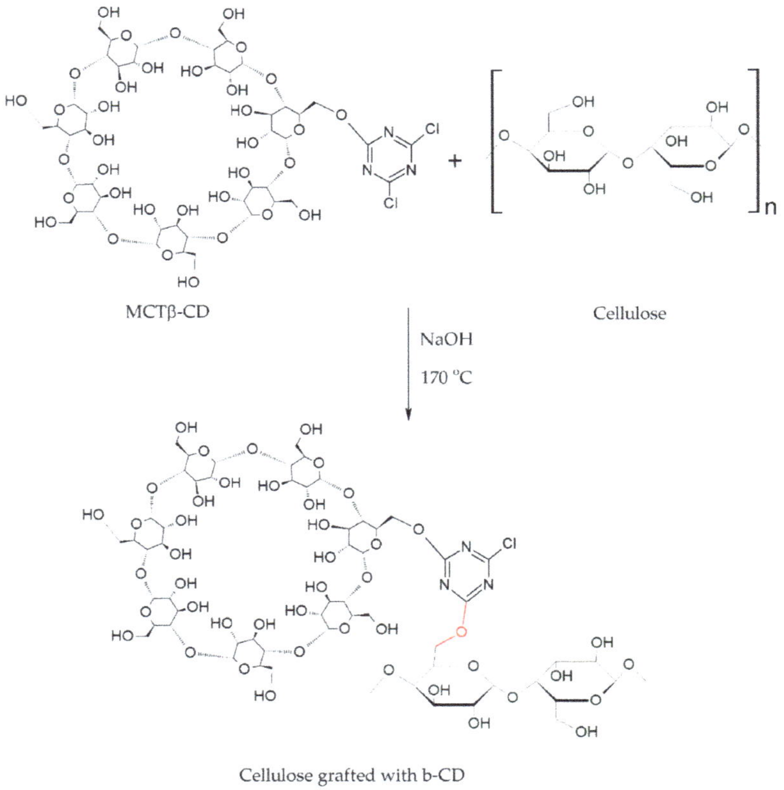

Figure 6. Nucleophilic substitution reaction of MCT-β-CD with cellulose.

Figure 7 illustrates the bonding between CDs, through BTCA as a crosslinking agent, and -OH groups of fibers.

For the esterification reaction to occur, both sodium hypophosphite and the cure are used as catalysts [38]. The same process can be performed on other fibers that have -OH groups, such as cellulose, silk, polyamide, and wool [78].

Regarding the insertion of cyclodextrins into polyester fibers, they can be functionalized by forming a network of CDs that cover the fiber, forming a reticulated coating between β-CD and BTCA through a polyesterification reaction [38,137].

As shown in Figure 5d, the hydroxyl groups of the CD can be oxidized by enzymes, converting them into aldehyde groups, which are able to react with the amine groups of the wool fibers through a Schiff-based reaction [127]. Figure 8 shows this reaction.

In this way, the application of CDs in fibrous polymers occurs. The substrate undergoes a change at the surface that can transform it, in the future, into functionalized fabrics after the complexation of the bioactive molecules by the CDs present on the surface of the materials.

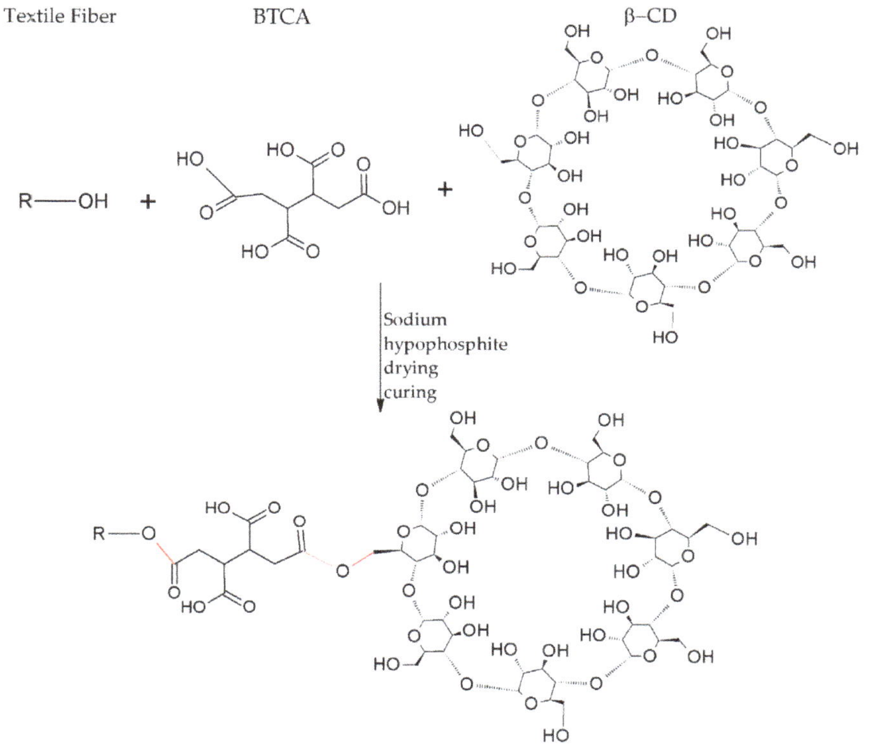

Figure 7. Direct connection of the β-CD to the textile fiber via crosslinking.

Table 3 shows some studies that used cyclodextrin for the functionalization of finished textiles.

Table 3. Studies that used cyclodextrin to graft finishes in textiles.

Fiber	Effect	Active Molecule	Reference
Cotton	Antimicrobial	Octenidine dihydrochloride	[138]
		Silver	[139,140]
		Phenolic compounds	[76,141]
		Ketoconazole	[115]
		ZnO, TiO2 and Ag nanoparticles	[142]
		Miconazole nitrate	[143]
		Triclosan	[144]
	Fragrance, antimicrobial	Essential Oils	[74,145]
	Insect repellent	Cypermethrin and Prallethrin	[146]
	Nocturnal regulation of sleep and antioxidant properties	Melatonin	[77]
	For coetaneous affections	Hydrocortisone acetate	[147]
Polyamide	Perfume, moisturize and UV-protect.	2-ethoxynaphtalene (neroline)	[119]
	Antibiotics	Ciprofloxacin	[148]

Table 3. *Cont.*

Fiber	Effect	Active Molecule	Reference
Tencel	Sunscreen	Octyl methoxycinnamate	[149]
	Fragrance, antimicrobial and insect repellent	Vanillin, benzoic acid andIodine, *N,N*-diethyl-m-toluamide and dimethyl-phthalate	[44]
Polyester	Antibiotics	Ciprofloxacin	[150]
	Antimicrobial	Curcumin	[151]
		4-tert-butylbenzoic acid	[152]
Wool	Insect repellent	Citronella essential oil	[78]
Cotton and Polyester	Insect repellent	Citronella essential oil	[38]
Cotton, wool and polyester	Fragrance	β-citronellol, camphor, menthol, cis-jasmone and benzyl acetate	[153]

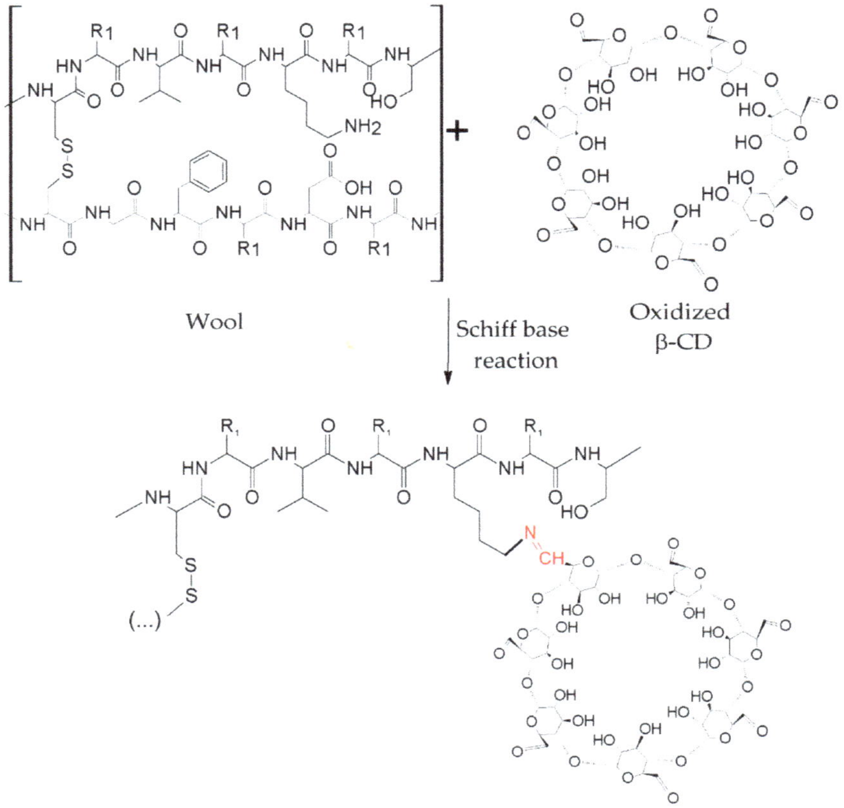

Figure 8. Functionalization of wool fibers with β-CD after oxidation.

3.3. New Trends in Textile Finishes Using Cyclodextrins

The use of citric acid as a reticulating agent was also a strategy adopted by Castriciano et al. [154] to design polypropylene fabric finished with hydroxypropyl β-CD. After complexation with tetra-anionic 5,10,15,20-tetrakis(4-sulfonatophenyl)-21H,23H-porphine (TPPS), the textile device was evaluated as a biocidal agent via antimicrobial Photodynamic Therapy (aPDT)—an alternative treatment to overcome the drug resistance associated with the indiscriminate use of antibiotics. The base of aPDT is the irradiation of a photosensitizer (PS) in the presence of oxygen, to generate reactive oxygen species (ROS) which attack the microorganisms at the target site (Figure 9). The PP-CD/TPPS fabric, containing 0.022 ± 0.0019 mg cm^{-2} of the TPPS, was capable of photokilling 99.98% of Gram-positive S. aureus, with low adhesion of bacteria to the textile. The aPDT approach was also used by Yao et al. [155] to develop biocidal materials based on beta cyclodextrins modified with hyaluronic acid (HA) for coating purposes. After the inclusion of PS methylene blue (MB), HA-CD/MB was tested against S. aureus, eradicating 99% of the bacteria at 0.53 ± 0.06 µg cm^{-2}. The use of aPDT in textile finishing may represent a new class of smart textiles with high anti-microorganism potential.

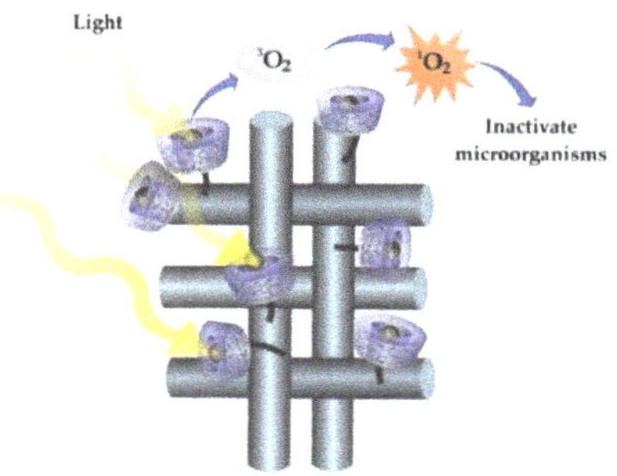

Figure 9. Finishing textiles with photodynamic potential.

3.4. Cyclodextrins in Textile Effluent Treatment

Cyclodextrins, in addition to being used as additives for the dyeing process when seeking improvements in washing, color intensity, and leveling, and as a functionalization agent, can be used to remove dyes and auxiliaries present in industrial effluents [40,156]. In the wastewater from the dyeing process, the presence of several types of dyes, surfactants, and salts can be an issue [157]. The dyes used in dyeing are compounds that are stable to oxidizing agents and light, have a complex structure, are non-biodegradable, and are highly soluble in water. Therefore, they are difficult to remove and can easily enter the ecosystem, affecting flora and fauna [79,158–162].

Various technologies for the treatment of water from the textile industry are used, such as photocatalytic oxidation [163], electrochemical oxidation [164], membrane separation [165], coagulation/flocculation [166], ozonation [167], and biological treatment [168], among others; however, there are restrictions regarding these processes, due to the high energy consumption and sludge generation. Thus, the search for processes that can eliminate residues from the dyeing and finishing processes is essential to alleviate major environmental problems. Lin et al. [169] and Crini et al. [170] point out that, among the different treatment systems, adsorption should be highlighted. It has been increasingly used, mainly due to its adaptability, easy operation, and low cost.

Among the adsorbents used, cyclodextrins are seen as a promising product [40] due to the high reactivity of the hydroxyl groups present in CDs for the adsorption process [171]. In addition, other advantages are related to its biodegradability, non-toxicity, availability [172], and the possibility of them interacting with the hydrophobic chain of surfactants, keeping them within its cavity [173]. In this way, they can form an insoluble CD:dye:surfactant system that can be removed from the water [79]. In general, Crini et al. [170] showed that the use of cyclodextrins as a dye adsorbent can be carried out by two methods, shown in Figure 10.

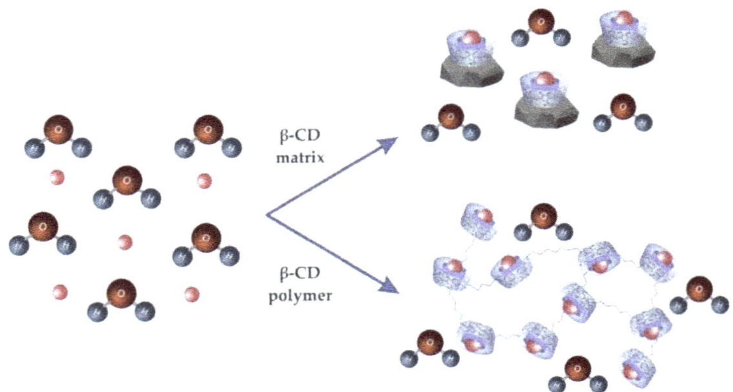

Figure 10. The role of β-CD as a dye adsorbent.

In the first method, cyclodextrins are incorporated into an insoluble matrix (nanoparticles, composites, nanotubes, and others), while in the second, CDs form an insoluble polymer capable of adsorbing the dyes. Table 4 shows some studies that have used cyclodextrin for the removal of textile dyes.

Table 4. Use of cyclodextrins as a removal agent in the textile process.

Method	Dye	Reference
Cyclodextrin incorporated into a matrix	Crystal Violet	[162]
	Reactive Black 5	[80]
	Methylene Blue	[174]
	Methyl Orange	[175]
	Safranin O, Brilliant Green and Methylene Blue	[161]
	Methylene Blue and Safranine T	[176]
	Methylene Blue, Acid Blue 113, Methyl Orange and Disperse Red 1	[177]
	Remazol Red 3BS, Remazol Blue RN, Remazol Yellow gelb 3RS 133	[178]
	Methyl Blue	[169]
β-cyclodextrin polymer	Acid Blue 25, Reactive Blue 19, Disperse Blue 3, Basic Blue 3 and Direct Red 81	[81]
	Basic Blue 3, Basic Violet 3 and Basic Violet 10	[170]
	Direct Violent 51, Methyl Orange, And Tropaeolin 000	[179]
	Congo Red and Methylene Blue	[41]
	Evans Blue, Chicago Sky Blue, Benzidine, P-hloroaniline	[180]
	Methylene Blue And Methyl Orange	[42]
	Congo Red, Methylene Blue, Methylene Orange	[156]
	Basic Orange 2, Rhodamine B, Methylene blue trihydrate, and Bisphenol A	[40]
	Methyl orange, Congo Red, Rhodamine B	[181]
	Direct Red 83:1	[171]

3.4.1. Cyclodextrin Matrix

A material used for the adsorption of dyes present in effluents must have a high adsorption capacity, ease of regeneration, mechanical resistance, and ability to adsorb a variety of dyes [182]. This last characteristic is often neglected, with experiments being carried out on solutions that contain only one type of dye; however, Debnath et al. [161] showed that most wastewater contains a mixture of different dyes, which affect the behavior of the adsorption system differently to a single dye system.

Therefore, cyclodextrins, due to their well-defined structure, can guarantee high reactivity for the adsorption of various dyes [171], and this can be improved if they are inserted into other adsorbent materials. These hybrid materials have a high adsorption capacity due to large specific surfaces and pore volume [43].

One of the techniques used for the production of promising adsorbent materials is electrospinning [183]. Abd-Elhamid et al. [162] produced nanocomposites using polyacrylic as an incorporation matrix and graphene oxide and cyclodextrin as adsorbent materials. According to the author, the nanocomposite is easy to prepare and has a high sorption capacity and is easy to remove from water. The combination of cyclodextrins and graphene to obtain a hybrid adsorbent material was also used by Liu et al. [176], who, in this case, also used poly(acrylic acid). This nanocomposite showed efficiency in pollutant adsorption, water dispersibility due to the hydrophilicity of the polymer, ease of regeneration, and a small loss of adsorption capacity.

Cyclodextrins can also be used in the production of biosorbents, together with chitosan. Chitosan is a compound rich in hydroxyl and amino groups, which allows interactions with organic and inorganic compounds [184,185]. However, chitosan, if not modified, can dissolve in acidic solutions because of the protonation of amino acids, hindering the adsorption of dyes [186]. To avoid such a problem, chitosan can be crosslinked with carboxylic acids and, to improve this biosorbent, Zhao et al. [187] chemically incorporated cyclodextrins into chitosan by means of esterification using citric acid, obtaining a biosorbent with a high capacity for adsorption of reactive dyes from textile effluents.

Chen et al. [157] showed that some researchers have grafted β-CD into insoluble solids, such as zeolite, activated carbon, silica gel and magnetic materials, obtaining good adsorption results. These characteristics show that adsorbent materials with cyclodextrin incorporation have great potential for applications in wastewater treatment, due to their large amount of hydroxyl groups, hydrophobic cavity, and interactions with organic and inorganic compounds.

3.4.2. Cyclodextrin Polymers

The synthesis of cyclodextrin polymers, especially those that are insoluble in water, has aroused growing interest given their applications in water treatment. Among the various methods of obtaining them, deprotonation stands out, in which the hydroxyl anion can be used in SN2 type polymerization reactions, direct dehydration in the presence of appropriate diodes and diacids, and condensation in the presence of a series of linkers [188]. In addition to polymerization, some studies have used β-CD for the development of organic-inorganic hybrid systems for the removal of dyes, such as magnetic CD polymers [40,189] and Halloysite−Cyclodextrin Nanosponges [190].

Crini et al. [81], using epichlorohydrin as a crosslinking agent for obtaining β-CD polymers, evaluated their efficiency in removing various dyes (acid blue (AB25), basic blue (BB3), reactive blue (RB19), dispersive blue (DB3) and direct red (DR81)). The capability to remove dyes by these polymers followed the order AB25 > RB19 > DB3 > DR81 >> BB3, with AB25 being close to 100% removed. The same author also prepared β-CD/carboxy methylcellulose polymers using the same crosslinking agent for the removal of Basic Blue 3, Basic Violet 3 and Basic Violet 10. Kinetic and equilibrium studies suggested that the process occurs by chemisorption, with an adsorptive capacity of 53.2, 42.4 and 35.8 mg of dye per gram of polymer for BV 10, BB 3 and BV 3, respectively [160].

Pellicer et al. [171] also used epichlorohydrin as a crosslinking agent to synthesize polymers of β-CD and HP-β-CD, which were used to remove the azo dye Direct Red 83:1. The adsorption capacity of the polymer synthesized from β-CD was approximately six times greater than that obtained using HP-β-CD.

Ozmen and Yilmaz [191] used β-CD polymer, prepared using 4-4-methylene-bis-phenyldiisocyanate (MDI), to remove Congo red dye. The authors observed 80% removal after one hour of contact in solution at pH 5.8. The same authors, using MDI and hexamethylene diisocyanate (HMDI) with crosslinking agents, synthesized β-CD polymers and evaluated their adsorptive capacities against the azo dyes Evans Blue and Chicago Sky Blue. At pH 2, the polymers showed around 50% removal.

Jiang et al. [192] synthesized a new polymer of β-CD for the removal of methylene blue. The strategy used by the authors was the use of tetrafluoroterephtalonitrile (TFPN) as a crosslinking agent, which, after being hydrolyzed, generates sites of carboxylic acids that interact electrostatically with the MB at the appropriate pH. A maximum adsorption capacity of 672 mg/g of the polymer was observed and, even after four cycles of adsorption/desorption, the capacity of the material remained high. The same group of researchers used a similar strategy for the synthesis of β-CD polymers, however, the nitrile groups of TFPN were modified with ethanolamine. This strategy enabled the selective removal of MO in a mixture of MO and MB. The polymer also showed a high adsorptive capacity for MO (602 mg/g) and Congo red (1085 mg/g). Recently, the selective removal of the anionic dye Orange G in a mixture with methylene blue has also been carried out by modifying the TFPN nitriles to form amide groups [41]. An innovative strategy using molecularly imprinted polymers (MPI) from chitosan and β-CD was used for the selective separation of Remazol Red 3BS in a trichromatic mixture. This new polymer also showed a high adsorption capacity after four cycles of use [168].

Some multifunctional CD polymers have also been developed for the simultaneous removal of dyes and other contaminants (bisphenol and heavy metals). Zhou et al. [42] synthesized a polymer of β-CD using citric acid as a crosslinking agent, which, after esterification, was grafted with 2-dimethylamino ethyl methacrylate monomer (DMAEMA) for the polymerization reaction. This elegant strategy allows modulating of the zeta potential of the adsorbent with the pH, enabling its electrostatic interaction with anionic (MO) or cationic (MB) dye. Simultaneously, the material can adsorb Bisphenol A inside the CD, and its interaction with the CD is unchanged between pH 2 to 10. The adsorption capacity at equilibrium for Bisphenol A was 79.0 mg/g, while the adsorption capacity of MO and MB was, respectively, 165.8 and 335.5 mg/g.

Zhao et al. [193] presented an elegant strategy for the treatment of industrial wastewater by means of a bifunctional adsorbent, consisting of a polymer of ethylene diamine tetra-acetic acid and β-CD (EDTA-β-CD). This bifunctional agent can simultaneously remove metals and dyes from wastewater, since β-CD has the ability to include dyes while EDTA becomes a site for metals. In experiments with binary systems containing Cu^{2+} and dyes (methylene blue, safranin O or crystal violet), the authors observed an increase in the adsorption capacity of the metal, but no significant change in the adsorption of the dyes, compared to experiments in systems with the isolated metal. The increase in the adsorption of the metal in binary Cu^{2+}-dye systems was attributed to the presence of the complexed dye in the CD, which provides extra groups containing nitrogen that become new sites for the adsorption of metals.

Despite the efficiency of the CD-based polymer in removing dyes and other agents in the textile process, some important points should be highlighted. Most of the studies presented above still need to be applied at a high scale level (in a real industrial system). Another important issue that should be emphasized is the regenerability of the CDs, making the process more ecofriendly and viable, with a lower cost.

4. Final Considerations and Future Perspectives

The increasing use of CDs in the textile industry is the result, among other factors, of the versatility of these cyclic molecules and the benefits of their use across the productive chain of this sector. Their unique ability to form an inclusion complex with a wide variety of molecules allows their use in several sectors. CDs are able to include dyes, repellents, insecticides, essential oils, caffeine, vitamins, drugs and surfactants, among other substances. Although they are used in the spinning and pretreatment areas, it is in the dyeing, finishing, and water treatment processes that β-CD and its derivatives have the greatest applicability.

The advantages of using CDs in dyeing include changes in bath exhaustion, color uniformity, less effluent treatment, dye savings, and the fact that they are biodegradable. They can be used as a dyeing aid, or as a surface modifying agent that absorbs more dye.

With regard to finishing, different types can be made with CDs, expanding the range of applications for these textiles and giving rise to a new class of materials called functional or intelligent textiles.

It is foreseeable that the use of CDs will continue to expand to keep up with the demands for differentiated products, and fill the gap that still exists in the literature around their application in the textile area, aiming at the optimization of the processes and viable results for industrial use.

This functionalization of CDs in substrates opens the door for the development of new products, such as medical textiles. With the new reality caused by the SARS-CoV-2 pandemic, the development of antiviral textiles is on the rise, and many of these new materials could be generated from technologies that use CDs. Furthermore, the transposition of new medical treatment technologies into textile materials from CDs is already a reality. An example is the use of β-CD for the development of textiles aiming at the photodynamic inactivation of microorganisms.

Finally, the capacity of CDs to adsorb and separate pollutants (dyes, metals, surfactants, etc.) from industrial waste is important with regards to environmentally sustainable industrial processes. In addition to adaptability and ease of operation, their biodegradability and lack of toxicity make CDs stand out in different areas.

Without a doubt, the use of CDs in basic and applied research around the development of new materials is fundamental, and should be the focus of many future studies seeking sustainable alternatives in the textile area.

Author Contributions: Conceptualization, F.M.B.; methodology, H.B.F., R.d.C.S.C.V. and J.A.B.V.; investigation, H.B.F., J.G.D.d.S. and F.A.P.S.; data curation, R.d.C.S.C.V. and J.A.B.V.; writing—original draft preparation, F.M.B., H.B.F., J.G.D.d.S. and A.L.T.; writing—review and editing, M.J.L., R.d.C.S.C.V., J.A.B.V., F.A.P.S. and A.L.T.; visualization, H.B.F., J.G.D.d.S., F.A.P.S. and A.L.T.; supervision, F.M.B.; project administration, F.M.B. and M.J.L.; funding acquisition, M.J.L. All authors have read and agreed to the published version of the manuscript.

Funding: This research received no external funding.

Acknowledgments: INTEXTER-UPC, UTFPR-AP and National Council for Scientific and Technological Development (*CNPq*). The author José Alexandre Borges Valle are grateful to CAPES-PRINT, project number 88887.310560/2018-00.

References

1. Villiers, A. Sur la fermentation de la fécule par l'action du ferment butyrique. *C. R. Acad. Sci.* **1891**, *112*, 536–537.

2. Szejtli, J. Introduction and General Overview of Cyclodextrin Chemistry. *Chem. Rev.* **1998**, *98*, 1743–1754. [CrossRef] [PubMed]

3. Alonso, C.; Martí, M.; Barba, C.; Lis, M.; Rubio, L.; Coderch, L. Skin penetration and antioxidant effect of cosmeto-textiles with gallic acid. *J. Photochem. Photobiol. B Boil.* **2016**, *156*, 50–55. [CrossRef]

4. Rasheed, A. Cyclodextrins as Drug Carrier Molecule: A Review. *Sci. Pharm.* **2008**, *76*, 567–598. [CrossRef]

5. Radu, C.-D.; Parteni, O.; Ochiuz, L. Applications of cyclodextrins in medical textiles—Review. *J. Control. Release* **2016**, *224*, 146–157. [CrossRef]

6. Nardello-Rataj, V.; Leclercq, L. Encapsulation of biocides by cyclodextrins: Toward synergistic effects against pathogens. *Beilstein J. Org. Chem.* **2014**, *10*, 2603–2622. [CrossRef] [PubMed]

7. Leclercq, L. 17 - Smart medical textiles based on cyclodextrins for curative or preventive patient care. In *Active Coatings for Smart Textiles*; Hu, J., Ed.; Woodhead Publishing Series in Textiles; Woodhead Publishing: Cambridge, UK, 2016; pp. 391–427. ISBN 978-0-08-100263-6.

8. Crini, G. Review: A History of Cyclodextrins. *Chem. Rev.* **2014**, *114*, 10940–10975. [CrossRef]

9. Buschmann, H.-J.; Schollmeyer, E. Applications of cyclodextrins in cosmetic products: A review. *J. Cosmet. Sci.* **2002**, *53*, 185–191.

10. Pinho, E.; Grootveld, M.; Soares, G.; Henriques, M. Cyclodextrins as encapsulation agents for plant bioactive compounds. *Carbohydr. Polym.* **2014**, *101*, 121–135. [CrossRef]

11. French, D. The Schardinger Dextrins. In *Advances in Carbohydrate Chemistry*; Wolfrom, M.L., Tipson, R.S., Eds.; Academic Press: New York, NY, USA, 1957; Volume 12, pp. 189–260.

12. Fernández, M.A.; Silva, O.F.; Vico, R.V.; de Rossi, R.H. Complex systems that incorporate cyclodextrins to get materials for some specific applications. *Carbohydr. Res.* **2019**, *480*, 12–34. [CrossRef]

13. Loftsson, T.; Duchêne, D. Cyclodextrins and their pharmaceutical applications. *Int. J. Pharm.* **2007**, *329*, 1–11. [CrossRef] [PubMed]

14. Matioli, G. *CICLODEXTRINAS E SUAS APLICAÇÕES EM: Alimentos, Fármacos, Cosméticos, Agricultura, Biotecnologia, Química Analítica e Produtos Gerais|Eduem—Editora da UEM*, 1st ed.; Eduem: Maringá, Brazil, 2000; ISBN 85-85545-46-1.

15. Challa, R.; Ahuja, A.; Ali, J.; Khar, R.K. Cyclodextrins in drug delivery: An updated review. *AAPS PharmSciTech* **2005**, *6*, E329–E357. [CrossRef] [PubMed]

16. Schöffer, J.D.N.; Klein, M.P.; Rodrigues, R.C.; Hertz, P.F. Continuous production of β-cyclodextrin from starch by highly stable cyclodextrin glycosyltransferase immobilized on chitosan. *Carbohydr. Polym.* **2013**, *98*, 1311–1316. [CrossRef] [PubMed]

17. Andreaus, J.; Dalmolin, M.C.; Junior, I.B.D.O.; Barcellos, I.O. Application of cyclodextrins in textile processes. *Química Nova* **2010**, *33*, 929–937. [CrossRef]

18. Amiri, S.; Amiri, S. *Cyclodextrins: Properties and Industrial Applications*; John Wiley&Sons: Hoboken, NJ, USA, 2017; ISBN 978-1-119-24760-9.

19. Harata, K. Structural Aspects of Stereodifferentiation in the Solid State. *Chem. Rev.* **1998**, *98*, 1803–1828. [CrossRef]

20. Faisal, Z.; Kunsági-Máté, S.; Lemli, B.; Szente, L.; Bergmann, D.; Humpf, H.-U.; Poór, M. Interaction of Dihydrocitrinone with Native and Chemically Modified Cyclodextrins. *Molecules* **2019**, *24*, 1328. [CrossRef]

21. Del Valle, E.M.M. Cyclodextrins and their uses: A review. *Process. Biochem.* **2004**, *39*, 1033–1046. [CrossRef]

22. Singh, N.; Yadav, M.; Khanna, S.; Sahu, O. Sustainable fragrance cum antimicrobial finishing on cotton: Indigenous essential oil. *Sustain. Chem. Pharm.* **2017**, *5*, 22–29. [CrossRef]

23. Shabbir, M.; Ahmed, S.; Sheikh, J.N. *Frontiers of Textile Materials: Polymers, Nanomaterials, Enzymes, and Advanced Modification Techniques*; Scrivener Publishing: Beverly, MA, USA, 2020; ISBN 978-1-119-62036-5.

24. Semeraro, P.; Rizzi, V.; Fini, P.; Matera, S.; Cosma, P.; Franco, E.; García, R.; Ferrándiz, M.; Núñez, E.; Gabaldón, J.A.; et al. Interaction between industrial textile dyes and cyclodextrins. *Dye. Pigment.* **2015**, *119*, 84–94. [CrossRef]

25. Gao, S.; Liu, Y.; Jiang, J.; Ji, Q.; Fu, Y.; Zhao, L.; Li, C.; Ye, F. Physicochemical properties and fungicidal activity of inclusion complexes of fungicide chlorothalonil with β-cyclodextrin and hydroxypropyl-β-cyclodextrin. *J. Mol. Liq.* **2019**, *293*, 111513. [CrossRef]

26. Saenger, W. Cyclodextrin Inclusion Compounds in Research and Industry. *Angew. Chem. Int. Ed.* **1980**, *19*, 344–362. [CrossRef]

27. Stella, V.J.; He, Q. Cyclodextrins. *Toxicol. Pathol.* **2008**, *36*, 30–42. [CrossRef] [PubMed]

28. Irie, T.; Uekama, K. Pharmaceutical Applications of Cyclodextrins. III. Toxicological Issues and Safety Evaluation. *J. Pharm. Sci.* **1997**, *86*, 147–162. [CrossRef] [PubMed]

29. Cunha-Filho, M.S.S.D.; SÁ-BARRETO, L.C.L. Utilização de ciclodextrinas na formação de complexos de inclusão de interesse farmacêutico. *Revista de Ciências Farmacêuticas Básica e Aplicada* **2007**, *28*, 1–9.

30. Bilensoy, E. *Cyclodextrins in Pharmaceutics, Cosmetics, and Biomedicine: Current and Future Industrial Applications*; John Wiley & Sons, Inc.: Hoboken, NJ, USA, 2011.

31. Bhaskara-Amrit, U.R.; Agrawal, P.B.; Warmoeskerken, M.M.C.G. Applications of β-cyclodextrins in textiles. *Autex Res. J.* **2011**, *11*, 94–101.

32. Savarino, P.; Viscardi, G.; Quagliotto, P.; Montoneri, E.; Barni, E. Reactivity and effects of cyclodextrins in textile dyeing. *Dye. Pigment.* **1999**, *42*, 143–147. [CrossRef]

33. Bezerra, F.M.; Moraes, F.F.D.; Santos, W.L.F.; Santos, J.C. Emprego de β-ciclodextrina como auxiliar no tingimento de fibras têxteis. *Química Têxt.* **2013**, *27*, 12–21.

34. Grigoriu, A.; Luca, C.; Grigoriu, A. Cyclodextrins Applications in the Textile Industry. *Cellul. Chem. Technol.* **2007**, *1*, 103–112.

35. Crupi, V.; Ficarra, R.; Guardo, M.; Majolino, D.; Stancanelli, R.; Venuti, V. UV–vis and FTIR–ATR spectroscopic techniques to study the inclusion complexes of genistein with β-cyclodextrins. *J. Pharm. Biomed. Anal.* **2007**, *44*, 110–117. [CrossRef]

36. Abdel-Halim, E.S.; Abdel-Mohdy, F.A.; Fouda, M.M.G.; El-Sawy, S.M.; Hamdy, I.A.; Al-Deyab, S.S. Antimicrobial activity of monochlorotriazinyl-β-cyclodextrin/chlorohexidin diacetate finished cotton fabrics. *Carbohydr. Polym.* **2011**, *86*, 1389–1394. [CrossRef]

37. Marques, H.M.C. A review on cyclodextrin encapsulation of essential oils and volatiles. *Flavour Fragr. J.* **2010**, *25*, 313–326. [CrossRef]

38. Lis, M.J.; Carmona, O.G.; Carmona, C.G.; Bezerra, F.M. Inclusion Complexes of Citronella Oil with β-Cyclodextrin for Controlled Release in Biofunctional Textiles. *Polymers* **2018**, *10*, 1324. [CrossRef] [PubMed]

39. Popescu, V.; Vasluianu, E.; Popescu, G. Quantitative analysis of the multifunctional finishing of cotton fabric with non-formaldehyde agents. *Carbohydr. Polym.* **2014**, *111*, 870–882. [CrossRef] [PubMed]

40. Hu, X.; Hu, Y.; Xu, G.; Li, M.; Zhu, Y.; Jiang, L.; Tu, Y.; Zhu, X.; Xie, X.; Li, A. Green synthesis of a magnetic β-cyclodextrin polymer for rapid removal of organic micro-pollutants and heavy metals from dyeing wastewater. *Environ. Res.* **2020**, *180*, 108796. [CrossRef] [PubMed]

41. Xu, M.-Y.; Jiang, H.-L.; Xie, Z.-W.; Li, Z.-T.; Xu, D.; He, F.-A. Highly efficient selective adsorption of anionic dyes by modified β-cyclodextrin polymers. *J. Taiwan Inst. Chem. Eng.* **2020**, *108*, 114–128. [CrossRef]

42. Zhou, Y.; Hu, Y.; Huang, W.; Cheng, G.; Cui, C.; Lu, J. A novel amphoteric β-cyclodextrin-based adsorbent for simultaneous removal of cationic/anionic dyes and bisphenol A. *Chem. Eng. J.* **2018**, *341*, 47–57. [CrossRef]

43. Teng, M.; Li, F.; Zhang, B.; Taha, A.A. Electrospun cyclodextrin-functionalized mesoporous polyvinyl alcohol/SiO2 nanofiber membranes as a highly efficient adsorbent for indigo carmine dye. *Colloids Surf. A: Physicochem Eng. Asp.* **2011**, *385*, 229–234. [CrossRef]

44. Nostro, P.L.; Fratoni, L.; Ridi, F.; Baglioni, P. Surface treatments on Tencel fabric: Grafting with β-cyclodextrin. *J. Appl. Polym. Sci.* **2003**, *88*, 706–715. [CrossRef]

45. Perchyonok, V.T.; Oberholzer, T. Cyclodextrins as Oral Drug Carrier Molecular Devices: Origins, Reasons and In-vitro Model Applications. *Curr. Org. Chem.* **2012**, *16*, 2365–2378. [CrossRef]

46. Acartürk, F.; Çelebi, N. Cyclodextrins as Bioavailability Enhancers. In *Cyclodextrins in Pharmaceutics, Cosmetics, and Biomedicine*; John Wiley & Sons, Ltd.: Hoboken, NJ, USA, 2011; pp. 45–64.

47. Andersen, F.M.; Bundgaard, H. Inclusion complexation of metronidazole benzoate with β-cyclodextrin and its depression of anhydrate-hydrate transition in aqueous suspensions. *Int. J. Pharm.* **1984**, *19*, 189–197. [CrossRef]

48. Venturini, C.D.G.; Nicolini, J.; Machado, C.; Machado, V.G. Propriedades e aplicações recentes das ciclodextrinas. *Química Nova* **2008**, *31*, 360–368. [CrossRef]

49. Rekharsky, M.V.; Inoue, Y. Complexation Thermodynamics of Cyclodextrins. *Chem. Rev.* **1998**, *98*, 1875–1918. [CrossRef] [PubMed]

50. Ross, P.D.; Rekharsky, M.V. Thermodynamics of hydrogen bond and hydrophobic interactions in cyclodextrin complexes. *Biophys. J.* **1996**, *71*, 2144–2154. [CrossRef]

51. Loftsson, T.; Björnsdóttir, S.; Pálsdóttir, G.; Bodor, N. The effects of 2-hydroxypropyl-β-cyclodextrin on the solubility and stability of chlorambucil and melphalan in aqueous solution. *Int. J. Pharm.* **1989**, *57*, 63–72. [CrossRef]

52. Pitha, J.; Milecki, J.; Fales, H.; Pannell, L.; Uekama, K. Hydroxypropyl-β-cyclodextrin: Preparation and characterization; effects on solubility of drugs. *Int. J. Pharm.* **1986**, *29*, 73–82. [CrossRef]

53. Loftsson, T. Cyclodextrins and the Biopharmaceutics Classification System of Drugs. *J. Incl. Phenom.* **2002**, *44*, 63–67. [CrossRef]

54. Chow, D.D.; Karara, A.H. Characterization, dissolution and bioavailability in rats of ibuprofen-β-cyclodextrin complex system. *Int. J. Pharm.* **1986**, *28*, 95–101. [CrossRef]

55. Vila-Jato, J.L.; Blanco, J.; Torres, J.J. Biopharmaceutical aspects of the tolbutamide-beta-cyclodextrin inclusion compound. *Farm. Prat.* **1988**, *43*, 37–45.

56. Wang, Z.; Landy, D.; Sizun, C.; Cézard, C.; Solgadi, A.; Przybylski, C.; de Chaisemartin, L.; Herfindal, L.; Barratt, G.; Legrand, F.-X. Cyclodextrin complexation studies as the first step for repurposing of chlorpromazine. *Int. J. Pharm.* **2020**, *584*, 119391. [CrossRef]

57. Kfoury, M.; Landy, D.; Fourmentin, S. Characterization of Cyclodextrin/Volatile Inclusion Complexes: A Review. *Molecules* **2018**, *23*, 1204. [CrossRef]
58. Takahashi, K. Organic Reactions Mediated by Cyclodextrins. *Chem. Rev.* **1998**, *98*, 2013–2034. [CrossRef] [PubMed]
59. Rama, A.C.R.; Veiga, F.; Figueiredo, I.V.; Sousa, A.; Caramona, M. Aspectos biofarmacêuticos da formulação de medicamentos para neonatos: Fundamentos da complexação de indometacina com hidroxipropil-beta-ciclodextrina para tratamento oral do fechamento do canal arterial. *Revista Brasileira de Ciências Farmacêuticas* **2005**, *41*, 281–299. [CrossRef]
60. Charumanee, S.; Titwan, A.; Sirithunyalug, J.; Weiss-Greiler, P.; Wolschann, P.; Viernstein, H.; Okonogi, S. Thermodynamics of the encapsulation by cyclodextrins. *J. Chem. Technol. Biotechnol.* **2006**, *81*, 523–529. [CrossRef]
61. Ameen, H.M.; Kunsági-Máté, S.; Bognár, B.; Szente, L.; Poór, M.; Lemli, B. Thermodynamic Characterization of the Interaction between the Antimicrobial Drug Sulfamethazine and Two Selected Cyclodextrins. *Molecules* **2019**, *24*, 4565. [CrossRef] [PubMed]
62. Vončina, B.; Vivod, V.; Jaušovec, D. β-Cyclodextrin as retarding reagent in polyacrylonitrile dyeing. *Dye. Pigment.* **2007**, *74*, 642–646. [CrossRef]
63. Hedges, A.R. Industrial Applications of Cyclodextrins. *Chem. Rev.* **1998**, *98*, 2035–2044. [CrossRef]
64. Cirri, M.; Maestrelli, F.; Orlandini, S.; Furlanetto, S.; Pinzauti, S.; Mura, P. Determination of stability constant values of flurbiprofen-cyclodextrin complexes using different techniques. *J. Pharm. Biomed. Anal.* **2005**, *37*, 995–1002. [CrossRef]
65. Vozone, C.M.; Marques, H.M.C. Complexation of Budesonide in Cyclodextrins and Particle Aerodynamic Characterization of the Complex Solid Form for Dry Powder Inhalation. *J. Incl. Phenom.* **2002**, *44*, 111–116. [CrossRef]
66. Cao, F.; Guo, J.; Ping, Q. The physicochemical characteristics of freeze-dried scutellarin-cyclodextrin tetracomponent complexes. *Drug Dev. Ind. Pharm.* **2005**, *31*, 747–756. [CrossRef]
67. Miro, A.; Ungaro, F.; Quaglia, F. Cyclodextrins as Smart Excipients in Polymeric Drug Delivery Systems. In *Cyclodextrins in Pharmaceutics, Cosmetics, and Biomedicine*; John Wiley & Sons, Ltd.: New Jersey, NY, USA, 2011; pp. 65–89.
68. Buschmann, H.-J.; Denter, U.; Knittel, D.; Schollmeyer, E. The Use of Cyclodextrins in Textile Processes—An Overview. *J. Text. Inst.* **1998**, *89*, 554–561. [CrossRef]
69. Loftsson, T.; Brewster, M.E. Pharmaceutical Applications of Cyclodextrins. 1. Drug Solubilization and Stabilization. *J. Pharm. Sci.* **1996**, *85*, 1017–1025. [CrossRef] [PubMed]
70. Alzate-Sánchez, D.M.; Smith, B.J.; Alsbaiee, A.; Hinestroza, J.P.; Dichtel, W.R. Cotton Fabric Functionalized with a β-Cyclodextrin Polymer Captures Organic Pollutants from Contaminated Air and Water. *Chem. Mater.* **2016**, *28*, 8340–8346. [CrossRef]
71. Tonelli, A.E. Cyclodextrins as a means to nanostructure and functionalize polymers. *J. Incl. Phenom. Macrocycl. Chem.* **2008**, *60*, 197–202. [CrossRef]
72. Hodul, P.; Duris, M.; Králik, M. Inclusion complexes of B-cyclodextrin with non-ionic surfactants in textile preparation process. *Vlakna a Text.* **1996**, *3*, 16–19.
73. Carpignano, R.; Parlati, S.; Piccinini, P.; Savarino, P.; Giorgi, M.R.D.; Fochi, R. Use of β-cyclodextrin in the dyeing of polyester with low environmental impact. *Color. Technol.* **2010**, *126*, 201–208. [CrossRef]
74. Sricharussin, W.; Sopajaree, C.; Maneerung, T.; Sangsuriya, N. Modification of cotton fabrics with β-cyclodextrin derivative for aroma finishing. *J. Text. Inst.* **2009**, *100*, 682–687. [CrossRef]
75. Raslan, W.M.; El-Aref, A.T.; Bendak, A. Modification of cellulose acetate fabric with cyclodextrin to improve its dyeability. *J. Appl. Polym. Sci.* **2009**, *112*, 3192–3198. [CrossRef]
76. Scacchetti, F.A.P.; Pinto, E.; Soares, G.M.B. Functionalization and characterization of cotton with phase change materials and thyme oil encapsulated in beta-cyclodextrins. *Prog. Org. Coatings* **2017**, *107*, 64–74. [CrossRef]
77. Mihailiasa, M.; Caldera, F.; Li, J.; Peila, R.; Ferri, A.; Trotta, F. Preparation of functionalized cotton fabrics by means of melatonin loaded β-cyclodextrin nanosponges. *Carbohydr. Polym.* **2016**, *142*, 24–30. [CrossRef]
78. Bezerra, F.M.; Carmona, O.G.; Carmona, C.G.; Plath, A.S.; Lis, M. Biofunctional wool using β-cyclodextrins as vehiculizer of citronella oil. *Process. Biochem.* **2019**, *77*, 151–158. [CrossRef]

79. Chang, Y.; Dou, N.; Liu, M.; Jiang, M.; Men, J.; Cui, Y.; Li, R.; Zhu, Y. Efficient removal of anionic dyes from aqueous solution using CTAB and β-cyclodextrin-induced dye aggregation. *J. Mol. Liq.* **2020**, *309*, 113021. [CrossRef]

80. Keskin, N.O.S.; Celebioglu, A.; Sarioglu, O.F.; Uyar, T.; Tekinay, T. Encapsulation of living bacteria in electrospun cyclodextrin ultrathin fibers for bioremediation of heavy metals and reactive dye from wastewater. *Colloids Surf. B Biointerfaces* **2018**, *161*, 169–176. [CrossRef] [PubMed]

81. Crini, G. Studies on adsorption of dyes on beta-cyclodextrin polymer. *Bioresour. Technol.* **2003**, *90*, 193–198. [CrossRef]

82. Cireli, A.; Yurdakul, B. Application of cyclodextrin to the textile dyeing and washing processes. *J. Appl. Polym. Sci.* **2006**, *100*, 208–218. [CrossRef]

83. Park, J.S.; Kim, I.-S. Use of β-cyclodextrin in an antimigration coating for polyester fabric. *Color. Technol.* **2013**, *129*, 347–351. [CrossRef]

84. Kacem, I.; Laurent, T.; Blanchemain, N.; Neut, C.; Chai, F.; Haulon, S.; Hildebrand, H.F.; Martel, B. Dyeing and antibacterial activation with methylene blue of a cyclodextrin modified polyester vascular graft. *J. Biomed. Mater. Res. Part A* **2014**, *102*, 2942–2951. [CrossRef]

85. Dardeer, H.M.; El-sisi, A.A.; Emam, A.A.; Hilal, N.M. Synthesis, Application of a Novel Azo Dye and Its Inclusion Complex with Beta-cyclodextrin onto Polyester Fabric. *Int. J. Text. Sci.* **2017**, *6*, 79–87.

86. Dutra, F.V.; Santos, K.R.M.D.; Bezerra, F.M. Complexação de corantes dispersos utilizando ß-ciclodextrina em tricromia para poliéster. *Química Têxt.* **2019**, *43*, 40–49.

87. Ferreira, B.T.M.; Espinoza-Quiñones, F.R.; Borba, C.E.; Módenes, A.N.; Santos, W.L.F.; Bezerra, F.M. Use of the β-Cyclodextrin Additive as a Good Alternative for the Substitution of Environmentally Harmful Additives in Industrial Dyeing Processes. *Fibers Polym.* **2020**, *21*, 1266–1274. [CrossRef]

88. Savarino, P.; Piccinini, P.; Montoneri, E.; Viscardi, G.; Quagliotto, P.; Barni, E. Effects of additives on the dyeing of nylon-6 with dyes containing hydrophobic and hydrophilic moieties. *Dye. Pigment.* **2000**, *47*, 177–188. [CrossRef]

89. Savarino, P.; Parlati, S.; Buscaino, R.; Piccinini, P.; Degani, I.; Barni, E. Effects of additives on the dyeing of polyamide fibres. Part I: β-cyclodextrin. *Dye. Pigment.* **2004**, *60*, 223–232. [CrossRef]

90. Savarino, P.; Parlati, S.; Buscaino, R.; Piccinini, P.; Barolo, C.; Montoneri, E. Effects of additives on the dyeing of polyamide fibres. Part II: Methyl-β-cyclodextrin. *Dye. Pigment.* **2006**, *69*, 7–12. [CrossRef]

91. Parlati, S.; Gobetto, R.; Barolo, C.; Arrais, A.; Buscaino, R.; Medana, C.; Savarino, P. Preparation and application of a β-cyclodextrin-disperse/reactive dye complex. *J. Incl. Phenom. Macrocycl. Chem.* **2007**, *57*, 463–470. [CrossRef]

92. Shibusawa, T.; Okamoto, J.; Abe, K.; Sakata, K.; Ito, Y. Inclusion of azo disperse dyes by cyclodextrins at dyeing temperature. *Dye. Pigment.* **1998**, *36*, 79–91. [CrossRef]

93. Dutra, F.V.; Caruzi, B.B.; Kawasaki, I.; Silva, T.L.; Bezerra, F.M. Utilização de β-ciclodextrina no tingimento de lã com Extratode Urucum (BiXa orellana). *Química Têxt.* **2014**, *37*, 58–64.

94. Chen, L.; Wang, C.; Tian, A.; Wu, M. An attempt of improving polyester inkjet printing performance by surface modification using β-cyclodextrin. *Surf. Interface Anal.* **2012**, *44*, 1324–1330. [CrossRef]

95. Lu, M.; Liu, Y.P. Dyeing Kinetics of Vinylon Modified with β-Cyclodextrin. *Fibres Text. East. Eur.* **2011**, *5*, 88.

96. Rehan, M.; Mahmoud, S.A.; Mashaly, H.M.; Youssef, B.M. β-Cyclodextrin assisted simultaneous preparation and dyeing acid dyes onto cotton fabric. *React. Funct. Polym.* **2020**, *151*, 104573. [CrossRef]

97. Zhang, W.; Ji, X.; Wang, C.; Yin, Y. One-bath one-step low-temperature dyeing of polyester/cotton blended fabric with cationic dyes via β-cyclodextrin modification. *Text. Res. J.* **2019**, *89*, 1699–1711. [CrossRef]

98. Ibrahim, N.A.; El-Zairy, E.M.R. Union disperse printing and UV-protecting of wool/polyester blend using a reactive β-cyclodextrin. *Carbohydr. Polym.* **2009**, *76*, 244–249. [CrossRef]

99. Ghoul, Y.E.; Martel, B.; Achari, A.E.; Campagne, C.; Razafimahefa, L.; Vroman, I. Improved dyeability of polypropylene fabrics finished with β-cyclodextrin–citric acid polymer. *Polym. J.* **2010**, *42*, 804–811. [CrossRef]

100. Burkinshaw, S.M.; Liu, K.; Salihu, G. The wash-off of dyeings using interstitial water Part 5: Residual dyebath and wash-off liquor generated during the application of disperse dyes and reactive dyes to polyester/cotton fabric. *Dye. Pigment.* **2019**, *171*, 106367. [CrossRef]

101. Hou, A.; Chen, B.; Dai, J.; Zhang, K. Using supercritical carbon dioxide as solvent to replace water in polyethylene terephthalate (PET) fabric dyeing procedures. *J. Clean. Prod.* **2010**, *18*, 1009–1014. [CrossRef]

102. Burkinshaw, S.M.; Lagonika, K. Sulphur dyes on nylon 6,6. Part 3. Preliminary studies of the nature of dye–fibre interaction. *Dye. Pigment.* **2006**, *69*, 185–191. [CrossRef]

103. Burkinshaw, S.M.; Son, Y.-A. A comparison of the colour strength and fastness to repeated washing of acid dyes on standard and deep dyeable nylon 6,6. *Dye. Pigment.* **2006**, *70*, 156–163. [CrossRef]

104. Xie, K.; Liu, H.; Wang, X. Surface modification of cellulose with triazine derivative to improve printability with reactive dyes. *Carbohydr. Polym.* **2009**, *78*, 538–542. [CrossRef]

105. Burkinshaw, S.M.; Mignanelli, M.; Froehling, P.E.; Bide, M.J. The use of dendrimers to modify the dyeing behaviour of reactive dyes on cotton. *Dye. Pigment.* **2000**, *47*, 259–267. [CrossRef]

106. Burkinshaw, S.M.; Salihu, G. The role of auxiliaries in the immersion dyeing of textile fibres part 2: Analysis of conventional models that describe the manner by which inorganic electrolytes promote direct dye uptake on cellulosic fibres. *Dye. Pigment.* **2019**, *161*, 531–545. [CrossRef]

107. Burkinshaw, S.M.; Salihu, G. The role of auxiliaries in the immersion dyeing of textile fibres: Part 4 theoretical model to describe the role of liquor ratio in dyeing cellulosic fibres with direct dyes in the absence and presence of inorganic electrolyte. *Dye. Pigment.* **2019**, *161*, 565–580. [CrossRef]

108. Hunger, K. *Industrial Dyes: Chemistry, Properties, Applications*; Wiley-VCH: Weinheim, Germany, 2003; ISBN 978-3-527-30426-4.

109. Wang, H.; Lewis, D.M. Chemical modification of cotton to improve fibre dyeability. *Color. Technol.* **2002**, *118*, 159–168. [CrossRef]

110. Udrescu, C.; Ferrero, F.; Periolatto, M. Ultrasound-assisted dyeing of cellulose acetate. *Ultrason. Sonochem.* **2014**, *21*, 1477–1481. [CrossRef] [PubMed]

111. Martel, B.; Ruffin, D.; Weltrowski, M.; Lekchiri, Y.; Morcellet, M. Water-soluble polymers and gels from the polycondensation between cyclodextrins and poly(carboxylic acid)s: A study of the preparation parameters. *J. Appl. Polym. Sci.* **2005**, *97*, 433–442. [CrossRef]

112. Arias, M.J.L.; Coderch, L.; Martí, M.; Alonso, C.; Carmona, O.G.; Carmona, C.G.; Maesta, F. Vehiculation of Active Principles as a Way to Create Smart and Biofunctional Textiles. *Materials* **2018**, *11*, 2152. [CrossRef] [PubMed]

113. Ciobanu, A.; Mallard, I.; Landy, D.; Brabie, G.; Nistor, D.; Fourmentin, S. Retention of aroma compounds from Mentha piperita essential oil by cyclodextrins and crosslinked cyclodextrin polymers. *Food Chem.* **2013**, *138*, 291–297. [CrossRef]

114. Peila, R.; Migliavacca, G.; Aimone, F.; Ferri, A.; Sicardi, S. A comparison of analytical methods for the quantification of a reactive β-cyclodextrin fixed onto cotton yarns. *Cellulose* **2012**, *19*, 1097–1105. [CrossRef]

115. Hedayati, N.; Montazer, M.; Mahmoudirad, M.; Toliyat, T. Ketoconazole and Ketoconazole/β-cyclodextrin performance on cotton wound dressing as fungal skin treatment. *Carbohydr. Polym.* **2020**, *240*, 116267. [CrossRef]

116. El-Ghoul, Y. Biological and microbiological performance of new polymer-based chitosan and synthesized amino-cyclodextrin finished polypropylene abdominal wall prosthesis biomaterial. *Text. Res. J.* **2020**, 0040517520926624. [CrossRef]

117. McQueen, R.H.; Vaezafshar, S. Odor in textiles: A review of evaluation methods, fabric characteristics, and odor control technologies. *Text. Res. J.* **2019**, *90*, 1–17. [CrossRef]

118. Kadam, V.; Kyratzis, I.L.; Truong, Y.B.; Wang, L.; Padhye, R. Air filter media functionalized with β-Cyclodextrin for efficient adsorption of volatile organic compounds. *J. Appl. Polym. Sci.* **2020**, *137*, 49228. [CrossRef]

119. Azizi, N.; Ben Abdelkader, M.; Chevalier, Y.; Majdoub, M. New β-Cyclodextrin-Based Microcapsules for Textiles Uses. *Fibers Polym.* **2019**, *20*, 683–689. [CrossRef]

120. Wang, C.X.; Chen, S.L. Aromachology and its application in the textile field. *Fibres Text. East. Eur.* **2005**, *13*, 41–44.

121. Agrawal, P.B.; Warmoeskerken, M.M.C.G. Permanent fixation of β-cyclodextrin on cotton surface—An assessment between innovative and established approaches. *J. Appl. Polym. Sci.* **2012**, *124*, 4090–4097. [CrossRef]

122. Nazi, M.; Malek, R.M.A.; Kotek, R. Modification of β-cyclodextrin with itaconic acid and application of the new derivative to cotton fabrics. *Carbohydr. Polym.* **2012**, *88*, 950–958. [CrossRef]

123. Zhang, F.; Islam, M.S.; Berry, R.M.; Tam, K.C. β-Cyclodextrin-Functionalized Cellulose Nanocrystals and Their Interactions with Surfactants. *ACS Omega* **2019**, *4*, 2102–2110. [CrossRef] [PubMed]

124. Reuscher, H.; Hirsenkorn, R. BETA W7 MCT—New ways in surface modification. *J. Incl. Phenom. Macrocycl. Chem.* **1996**, *25*, 191–196. [CrossRef]
125. Shown, I.; Murthy, C.N. Grafting of cotton fiber by water-soluble cyclodextrin-based polymer. *J. Appl. Polym. Sci.* **2009**, *111*, 2056–2061. [CrossRef]
126. Nichifor, M.; Constantin, M.; Mocanu, G.; Fundueanu, G.; Branisteanu, D.; Costuleanu, M.; Radu, C.D. New multifunctional textile biomaterials for the treatment of leg venous insufficiency. *J. Mater. Sci. Mater. Med.* **2009**, *20*, 975–982. [CrossRef]
127. Yu, Y.; Wang, Q.; Yuan, J.; Fan, X.; Wang, P. A novel approach for grafting of β-cyclodextrin onto wool via laccase/TEMPO oxidation. *Carbohydr. Polym.* **2016**, *153*, 463–470. [CrossRef]
128. Haji, A.; Mehrizi, M.K.; Akbarpour, R. Optimization of β-cyclodextrin grafting on wool fibers improved by plasma treatment and assessment of antibacterial activity of berberine finished fabric. *J. Incl. Phenom. Macrocycl. Chem.* **2015**, *81*, 121–133. [CrossRef]
129. Montazer, M.; Jolaei, M.M. β-Cyclodextrin stabilized on three-dimensional polyester fabric with different crosslinking agents. *J. Appl. Polym. Sci.* **2010**, *116*, 210–217. [CrossRef]
130. Abdel-Halim, E.S.; Abdel-Mohdy, F.A.; Al-Deyab, S.S.; El-Newehy, M.H. Chitosan and monochlorotriazinyl-β-cyclodextrin finishes improve antistatic properties of cotton/polyester blend and polyester fabrics. *Carbohydr. Polym.* **2010**, *82*, 202–208. [CrossRef]
131. Shi. H.; Wang, Y.; Hua, R. Acid-catalyzed carboxylic acid esterification and ester hydrolysis mechanism: Acylium ion as a sharing active intermediate via a spontaneous trimolecular reaction based on density functional theory calculation and supported by electrospray ionization-mass spectrometry. *Phys. Chem. Chem. Phys.* **2015**, *17*, 30279–30291. [CrossRef]
132. Radu, C.-D.; Salariu, M.; Avadanei, M.; Ghiciuc, C.; Foia, L.; Lupusoru, E.C.; Ferri, A.; Ulea, E.; Lipsa, F. Cotton-made cellulose support for anti-allergic pajamas. *Carbohydr. Polym.* **2013**, *95*, 479–486. [CrossRef] [PubMed]
133. Khanna, S.; Chakraborty, J.N. Optimization of monochlorotriazine β-cyclodextrin grafting on cotton and assessment of release behavior of essential oils from functionalized fabric. *Fash. Text.* **2017**, *4*, 6. [CrossRef]
134. Ibrahim, N.A.; Abdalla, W.A.; El-Zairy, E.M.R.; Khalil, H.M. Utilization of monochloro-triazine β-cyclodextrin for enhancing printability and functionality of wool. *Carbohydr. Polym.* **2013**, *92*, 1520–1529. [CrossRef] [PubMed]
135. Martel, B.; Weltrowski, M.; Ruffin, D.; Morcellet, M. Polycarboxylic acids as crosslinking agents for grafting cyclodextrins onto cotton and wool fabrics: Study of the process parameters. *J. Appl. Polym. Sci.* **2002**, *83*, 1449–1456. [CrossRef]
136. Liu, S. Bio-Functional Textiles. In *Handbook of Medical Textiles*; Textiles; Woodhead Publishing: Cambridge, UK, 2011; pp. 336–359. ISBN 978-0-85709-369-1.
137. Blanchemain, N.; Karrout, Y.; Tabary, N.; Neut, C.; Bria, M.; Siepmann, J.; Hildebrand, H.F.; Martel, B. Methyl-β-cyclodextrin modified vascular prosthesis: Influence of the modification level on the drug delivery properties in different media. *Acta Biomater.* **2011**, *7*, 304–314. [CrossRef]
138. Abdel-Halim, E.S.; Al-Deyab, S.S.; Alfaifi, A.Y.A. Cotton fabric finished with β-cyclodextrin: Inclusion ability toward antimicrobial agent. *Carbohydr. Polym.* **2014**, *102*, 550–556. [CrossRef]
139. Bajpai, M.; Gupta, P.; Bajpai, S.K. Silver (I) ions loaded cyclodextrin-grafted-cotton fabric with excellent antimicrobial property. *Fibers Polym.* **2010**, *11*, 8–13. [CrossRef]
140. Hebeish, A.; El-Shafei, A.; Sharaf, S.; Zaghloul, S. In situ formation of silver nanoparticles for multifunctional cotton containing cyclodextrin. *Carbohydr. Polym.* **2014**, *103*, 442–447. [CrossRef]
141. Rukmani, A.; Sundrarajan, M. Inclusion of antibacterial agent thymol on β-cyclodextrin-grafted organic cotton. *J. Ind. Text.* **2011**, *42*, 132–144. [CrossRef]
142. Selvam, S.; Gandhi, R.R.; Suresh, J.; Gowri, S.; Ravikumar, S.; Sundrarajan, M. Antibacterial effect of novel synthesized sulfated β-cyclodextrin crosslinked cotton fabric and its improved antibacterial activities with ZnO, TiO2 and Ag nanoparticles coating. *Int. J. Pharm.* **2012**, *434*, 366–374. [CrossRef] [PubMed]
143. Wang, J.; Cai, Z. Incorporation of the antibacterial agent, miconazole nitrate into a cellulosic fabric grafted with β-cyclodextrin. *Carbohydr. Polym.* **2008**, *72*, 695–700. [CrossRef]
144. Cabrales, L.; Abidi, N.; Hammond, A.; Hamood, A. Cotton Fabric Functionalization with Cyclodextrins. *Surfaces* **2012**, *6*, 14.

145. Khanna, S.; Sharma, S.; Chakraborty, J.N. Performance assessment of fragrance finished cotton with cyclodextrin assisted anchoring hosts. *Fash. Text.* **2015**, *2*, 19. [CrossRef]
146. Abdel-Mohdy, F.A.; Fouda, M.M.G.; Rehan, M.F.; Aly, A.S. Repellency of controlled-release treated cotton fabrics based on cypermethrin and prallethrin. *Carbohydr. Polym.* **2008**, *73*, 92–97. [CrossRef]
147. Radu, C.D.; Parteni, O.; Popa, M.; Muresan, I.E.; Ochiuz, L.; Bulgariu, L.; Munteanu, C.; Istrate, B.; Ulea, E. Comparative Study of a Drug Release from a Textile to Skin. *J. Pharm. Drug Deliv. Res.* **2016**, *2015*, 1–8. [CrossRef]
148. El-Ghoul, Y.; Blanchemain, N.; Laurent, T.; Campagne, C.; El Achari, A.; Roudesli, S.; Morcellet, M.; Martel, B.; Hildebrand, H.F. Chemical, biological and microbiological evaluation of cyclodextrin finished polyamide inguinal meshes. *Acta Biomater.* **2008**, *4*, 1392–1400. [CrossRef]
149. Scalia, S.; Tursilli, R.; Bianchi, A.; Nostro, P.L.; Bocci, E.; Ridi, F.; Baglioni, P. Incorporation of the sunscreen agent, octyl methoxycinnamate in a cellulosic fabric grafted with β-cyclodextrin. *Int. J. Pharm.* **2006**, *308*, 155–159. [CrossRef]
150. Blanchemain, N.; Karrout, Y.; Tabary, N.; Bria, M.; Neut, C.; Hildebrand, H.F.; Siepmann, J.; Martel, B. Comparative study of vascular prostheses coated with polycyclodextrins for controlled ciprofloxacin release. *Carbohydr. Polym.* **2012**, *90*, 1695–1703. [CrossRef]
151. Shlar, I.; Droby, S.; Rodov, V. Antimicrobial coatings on polyethylene terephthalate based on curcumin/cyclodextrin complex embedded in a multilayer polyelectrolyte architecture. *Colloids Surf. B Biointerfaces* **2018**, *164*, 379–387. [CrossRef] [PubMed]
152. Martin, A.; Tabary, N.; Leclercq, L.; Junthip, J.; Degoutin, S.; Aubert-Viard, F.; Cazaux, F.; Lyskawa, J.; Janus, L.; Bria, M.; et al. Multilayered textile coating based on a β-cyclodextrin polyelectrolyte for the controlled release of drugs. *Carbohydr. Polym.* **2013**, *93*, 718–730. [CrossRef] [PubMed]
153. Martel, B.; Morcellet, M.; Ruffin, D.; Vinet, F.; Weltrowski, L. Capture and Controlled Release of Fragrances by CD Finished Textiles. *J. Incl. Phenom.* **2002**, *44*, 439–442. [CrossRef]
154. Castriciano, M.A.; Zagami, R.; Casaletto, M.P.; Martel, B.; Trapani, M.; Romeo, A.; Villari, V.; Sciortino, M.T.; Grasso, L.; Guglielmino, S.; et al. Poly(carboxylic acid)-Cyclodextrin/Anionic Porphyrin Finished Fabrics as Photosensitizer Releasers for Antimicrobial Photodynamic Therapy. *Biomacromolecules* **2017**, *18*, 1134–1144. [CrossRef]
155. Yao, T.; Wang, J.; Xue, Y.; Yu, W.; Gao, Q.; Ferreira, L.; Ren, K.-F.; Ji, J. A photodynamic antibacterial spray-coating based on the host–guest immobilization of the photosensitizer methylene blue. *J. Mater. Chem. B* **2019**, *7*, 5089–5095. [CrossRef]
156. Jiang, H.-L.; Xu, M.-Y.; Xie, Z.-W.; Hai, W.; Xie, X.-L.; He, F.-A. Selective adsorption of anionic dyes from aqueous solution by a novel β-cyclodextrin-based polymer. *J. Mol. Struct.* **2020**, *1203*, 127373. [CrossRef]
157. Chen, J.; Liu, M.; Pu, Y.; Wang, C.; Han, J.; Jiang, M.; Liu, K. The preparation of thin-walled multi-cavities β-cyclodextrin polymer and its static and dynamic properties for dyes removal. *J. Environ. Manag.* **2019**, *245*, 105–113. [CrossRef]
158. Afkhami, A.; Saber-Tehrani, M.; Bagheri, H. Modified maghemite nanoparticles as an efficient adsorbent for removing some cationic dyes from aqueous solution. *Desalination* **2010**, *263*, 240–248. [CrossRef]
159. Zare, K.; Gupta, V.K.; Moradi, O.; Makhlouf, A.S.H.; Sillanpää, M.; Nadagouda, M.N.; Sadegh, H.; Shahryari-ghoshekandi, R.; Pal, A.; Wang, Z.; et al. A comparative study on the basis of adsorption capacity between CNTs and activated carbon as adsorbents for removal of noxious synthetic dyes: A review. *J. Nanostruct. Chem.* **2015**, *5*, 227–236. [CrossRef]
160. Sansuk, S.; Srijaranai, S.; Srijaranai, S. A New Approach for Removing Anionic Organic Dyes from Wastewater Based on Electrostatically Driven Assembly. *Environ. Sci. Technol.* **2016**, *50*, 6477–6484. [CrossRef]
161. Debnath, S.; Ballav, N.; Maity, A.; Pillay, K. Competitive adsorption of ternary dye mixture using pine cone powder modified with β-cyclodextrin. *J. Mol. Liq.* **2017**, *225*, 679–688. [CrossRef]
162. Abd-Elhamid, A.I.; El-Aassar, M.R.; El Fawal, G.F.; Soliman, H.M.A. Fabrication of polyacrylonitrile/β-cyclodextrin/graphene oxide nanofibers composite as an efficient adsorbent for cationic dye. *Environ. NanoTechnol. Monit. Manag.* **2019**, *11*, 100207. [CrossRef]
163. Batista, L.M.B.; dos Santos, A.J.; da Silva, D.R.; de Melo Alves, A.P.; Garcia-Segura, S.; Martínez-Huitle, C.A. Solar photocatalytic application of NbO2OH as alternative photocatalyst for water treatment. *Sci. Total Environ.* **2017**, *596–597*, 79–86. [CrossRef] [PubMed]

164. Basha, C.A.; Selvakumar, K.V.; Prabhu, H.J.; Sivashanmugam, P.; Lee, C.W. Degradation studies for textile reactive dye by combined electrochemical, microbial and photocatalytic methods. *Sep. Purif. Technol.* **2011**, *79*, 303–309. [CrossRef]

165. Chollom, M.N.; Rathilal, S.; Pillay, V.L.; Alfa, D. The applicability of nanofiltration for the treatment and reuse of textile reactive dye effluent. *Water SA* **2015**, *41*, 398–405. [CrossRef]

166. Yeap, K.L.; Teng, T.T.; Poh, B.T.; Morad, N.; Lee, K.E. Preparation and characterization of coagulation/flocculation behavior of a novel inorganic–organic hybrid polymer for reactive and disperse dyes removal. *Chem. Eng. J.* **2014**, *243*, 305–314. [CrossRef]

167. Gosavi, V.D.; Sharma, S. A General Review on Various Treatment Methods for Textile Wastewater. Available online: /paper/A-General-Review-on-Various-Treatment-Methods-for-Gosavi-Sharma/e8d649881280265b4bf146b25f6e5eb2ddb99afb (accessed on 14 June 2020).

168. Ahmed, M.B.; Zhou, J.L.; Ngo, H.H.; Guo, W.; Thomaidis, N.S.; Xu, J. Progress in the biological and chemical treatment technologies for emerging contaminant removal from wastewater: A critical review. *J. Hazard. Mater.* **2017**, *323*, 274–298. [CrossRef]

169. Lin, Q.; Gao, M.; Chang, J.; Ma, H. Adsorption properties of crosslinking carboxymethyl cellulose grafting dimethyldiallylammonium chloride for cationic and anionic dyes. *Carbohydr. Polym.* **2016**, *151*, 283–294. [CrossRef]

170. Crini, G. Kinetic and equilibrium studies on the removal of cationic dyes from aqueous solution by adsorption onto a cyclodextrin polymer. *Dye. Pigment.* **2008**, *77*, 415–426. [CrossRef]

171. Pellicer, J.A.; Rodríguez-López, M.I.; Fortea, M.I.; Lucas-Abellán, C.; Mercader-Ros, M.T.; López-Miranda, S.; Gómez-López, V.M.; Semeraro, P.; Cosma, P.; Fini, P.; et al. Adsorption Properties of β- and Hydroxypropyl-β-Cyclodextrins Cross-Linked with Epichlorohydrin in Aqueous Solution. A Sustainable Recycling Strategy in Textile Dyeing Process. *Polymers* **2019**, *11*, 252. [CrossRef]

172. Crini, G. Recent developments in polysaccharide-based materials used as adsorbents in wastewater treatment. *Prog. Polym. Sci.* **2005**, *30*, 38–70. [CrossRef]

173. Eli, W.; Chen, W.; Xue, Q. The association of anionic surfactants with β-cyclodextrin. An isothermal titration calorimeter study. *J. Chem. Thermodyn.* **1999**, *31*, 1283–1296. [CrossRef]

174. Zhao, R.; Wang, Y.; Li, X.; Sun, B.; Jiang, Z.; Wang, C. Water-insoluble sericin/β-cyclodextrin/PVA composite electrospun nanofibers as effective adsorbents towards methylene blue. *Colloids Surf. B Biointerfaces* **2015**, *136*, 375–382. [CrossRef] [PubMed]

175. Yilmaz, A.; Yilmaz, E.; Yilmaz, M.; Bartsch, R.A. Removal of azo dyes from aqueous solutions using calix[4]arene and β-cyclodextrin. *Dye. Pigment.* **2007**, *74*, 54–59. [CrossRef]

176. Liu, J.; Liu, G.; Liu, W. Preparation of water-soluble β-cyclodextrin/poly(acrylic acid)/graphene oxide nanocomposites as new adsorbents to remove cationic dyes from aqueous solutions. *Chem. Eng. J.* **2014**, *257*, 299–308. [CrossRef]

177. Mohammadi, A.; Veisi, P. High adsorption performance of β-cyclodextrin-functionalized multi-walled carbon nanotubes for the removal of organic dyes from water and industrial wastewater. *J. Environ. Chem. Eng.* **2018**, *6*, 4634–4643. [CrossRef]

178. Kyzas, G.Z.; Lazaridis, N.K.; Bikiaris, D.N. Optimization of chitosan and β-cyclodextrin molecularly imprinted polymer synthesis for dye adsorption. *Carbohydr. Polym.* **2013**, *91*, 198–208. [CrossRef]

179. Ozmen, E.Y.; Sezgin, M.; Yilmaz, A.; Yilmaz, M. Synthesis of β-cyclodextrin and starch based polymers for sorption of azo dyes from aqueous solutions. *Bioresour. Technol.* **2008**, *99*, 526–531. [CrossRef]

180. Yilmaz, E.; Memon, S.; Yilmaz, M. Removal of direct azo dyes and aromatic amines from aqueous solutions using two β-cyclodextrin-based polymers. *J. Hazard. Mater.* **2010**, *174*, 592–597. [CrossRef]

181. Li, X.; Zhou, M.; Jia, J.; Jia, Q. A water-insoluble viologen-based β-cyclodextrin polymer for selective adsorption toward anionic dyes. *React. Funct. Polym.* **2018**, *126*, 20–26. [CrossRef]

182. Kekes, T.; Tzia, C. Adsorption of indigo carmine on functional chitosan and β-cyclodextrin/chitosan beads: Equilibrium, kinetics and mechanism studies. *J. Environ. Manag.* **2020**, *262*, 110372. [CrossRef] [PubMed]

183. Chen, P.-Y.; Tung, S.-H. One-Step Electrospinning to Produce Nonsolvent-Induced Macroporous Fibers with Ultrahigh Oil Adsorption Capability. *Macromolecules* **2017**, *50*, 2528–2534. [CrossRef]

184. Akinyeye, O.J.; Ibigbami, T.B.; Odeja, O. Effect of Chitosan Powder Prepared from Snail Shells to Remove Lead (II) Ion and Nickel (II) Ion from Aqueous Solution and Its Adsorption Isotherm Model. *Am. J. Appl. Chem.* **2016**, *4*, 146. [CrossRef]

185. Chatterjee, S.; Lee, D.S.; Lee, M.W.; Woo, S.H. Nitrate removal from aqueous solutions by cross-linked chitosan beads conditioned with sodium bisulfate. *J. Hazard. Mater.* **2009**, *166*, 508–513. [CrossRef] [PubMed]
186. Cestari, A.R.; Vieira, E.F.S.; Mota, J.A. The removal of an anionic red dye from aqueous solutions using chitosan beads—The role of experimental factors on adsorption using a full factorial design. *J. Hazard. Mater.* **2008**, *160*, 337–343. [CrossRef] [PubMed]
187. Zhao, J.; Zou, Z.; Ren, R.; Sui, X.; Mao, Z.; Xu, H.; Zhong, Y.; Zhang, L.; Wang, B. Chitosan adsorbent reinforced with citric acid modified β-cyclodextrin for highly efficient removal of dyes from reactive dyeing effluents. *Eur. Polym. J.* **2018**, *108*, 212–218. [CrossRef]
188. Krause, R.W.M.; Mamba, B.B.; Bambo, F.M.; Malefetse, T.J. Cyclodextrin polymers: Synthesis and Application in Water Treatment. In *Cyclodextrins: Chemistry and Physics*; Transworld Research Network: Kerala, India, 2010; ISBN 978-81-7895-430-1.
189. Vahedi, S.; Tavakoli, O.; Khoobi, M.; Ansari, A.; Faramarzi, M.A. Application of novel magnetic β-cyclodextrin-anhydride polymer nano-adsorbent in cationic dye removal from aqueous solution. *J. Taiwan Inst. Chem. Eng.* **2017**, *80*, 452–463. [CrossRef]
190. Massaro, M.; Colletti, C.G.; Lazzara, G.; Guernelli, S.; Noto, R.; Riela, S. Synthesis and Characterization of Halloysite–Cyclodextrin Nanosponges for Enhanced Dyes Adsorption. *ACS Sustain. Chem. Eng.* **2017**, *5*, 3346–3352. [CrossRef]
191. Ozmen, E.Y.; Yilmaz, M. Use of β-cyclodextrin and starch based polymers for sorption of Congo red from aqueous solutions. *J. Hazard. Mater.* **2007**, *148*, 303–310. [CrossRef]
192. Jiang, H.-L.; Lin, J.-C.; Hai, W.; Tan, H.-W.; Luo, Y.-W.; Xie, X.-L.; Cao, Y.; He, F.-A. A novel crosslinked β-cyclodextrin-based polymer for removing methylene blue from water with high efficiency. *Colloids Surf. A Physicochem. Eng. Asp.* **2019**, *560*, 59–68. [CrossRef]
193. Zhao, F.; Repo, E.; Yin, D.; Meng, Y.; Jafari, S.; Sillanpää, M. EDTA-Cross-Linked β-Cyclodextrin: An Environmentally Friendly Bifunctional Adsorbent for Simultaneous Adsorption of Metals and Cationic Dyes. *Environ. Sci. Technol.* **2015**, *49*, 10570–10580. [CrossRef] [PubMed]

MDPI

St. Alban-Anlage 66

4052 Basel

Switzerland

Tel. +41 61 683 77 34

Fax +41 61 302 89 18

www.mdpi.com

Molecules Editorial Office

E-mail: molecules@mdpi.com

www.mdpi.com/journal/molecules